DISASTERS IN THE PHILIPPINES

Before and After Haiyan

Edited by
Glenda Tibe Bonifacio
and Roxanna Balbido Epe

BRISTOL
UNIVERSITY
PRESS

First published in Great Britain in 2023 by

Bristol University Press
University of Bristol
1–9 Old Park Hill
Bristol
BS2 8BB
UK
t: +44 (0)117 374 6645
e: bup-info@bristol.ac.uk

Details of international sales and distribution partners are available at bristoluniversitypress.co.uk

British Library Cataloguing in Publication Data
A catalogue record for this book is available from the British Library

ISBN 978-1-5292-2290-6 hardcover
ISBN 978-1-5292-2291-3 ePub
ISBN 978-1-5292-2292-0 ePdf

Cover design: Hayes Design and Advertising
Front cover image: Malapascua island après super typhon yolanda, Philippines
© 123RF.com
Bristol University Press use environmentally responsible print partners.
Printed in Great Britain by CPI Group (UK) Ltd, Croydon, CR0 4YY

FSC
www.fsc.org
MIX
Paper | Supporting responsible forestry
FSC® C013604

For

Dulce

Ian

Vangie

Lives lost from super typhoon Yolanda

Lives lost from all disasters in the Philippines

and

Those who live with memories of disasters

Contents

List of Figures and Tables

Figures

Tables

Notes on Contributors

Kaira Zoe Alburo-Cañete specializes in Critical Development Studies focusing on the intersections of gender, environment, and disasters. For the past ten years, she has been involved in policy and development research on various contemporary issues and their interplays, such as understanding local perspectives of resilience, promoting sexual and reproductive health in humanitarian settings, examining human trafficking following disasters, and supporting gender and disability inclusion in disaster risk reduction and recovery. She is currently a postdoctoral research fellow for the Humanitarian Governance Project at the International Institute of Social Studies Erasmus University Rotterdam (2022–2024).

Xavier Venn A. Asuncion is a university researcher at the Center for Local and Regional Governance (CLRG), a constituent unit of the University of the Philippines-National College of Public Administration and Governance (UP-NCPAG). He has served as a lecturer in Local Governance and Regional Administration at UP NCPAG since 2020. He has mostly worked on research; capacity building; and consultancy work/projects in the area of local development planning; local fiscal administration; and Disaster Risk Reduction-Climate Change Adaptation (DRR-CCA). Xavier Asuncion completed a BA in Public Administration (University of the Philippines) and MSc in Urban Management and Development (Institute for Housing and Urban Development Studies, Erasmus University Rotterdam, EUR). He is currently pursuing a second MA in Urban and Regional Planning at the University of the Philippines-School of Urban and Regional Planning (UP-SURP).

Glenda Tibe Bonifacio is a full professor in women and gender studies at the University of Lethbridge, Canada. She is the author of *Pinay on the Prairies: Filipino Women and Transnational Identities* (UBC Press). In collaboration with international scholars, she has edited and co-edited eight books related to gender and migration, feminism, religion, youth migration, and immigration in Canada. Her current research project centers on gender, foreign aid, and post-disaster communities in the Philippines funded by the

Social Sciences and Humanities Research Council (SSHRC) in Canada. She is an advisory board member of the Small Island Cultures Research Initiative (SICRI).

Corazon Clarin is a woman with a disability who serves as the convener of the Cebu Disability-Inclusive Disaster Risk Reduction Network. Since the network's establishment in 2015, she has continued to promote advocacy among government and non-government organizations in the Philippines and expand the network's membership. She is also the Chair of the local chapter of Women with Disabilities LEAP to social and economic progress (WOWLEAP) in the province of Cebu.

Marivic Codiñera is an environmental planner currently affiliated with the Philippine Augustinian Development Office in Cebu, Philippines. In 2015, Marivic was instrumental in organizing and establishing the Cebu Disability-inclusive Disaster Risk Reduction Network in her capacity as a training officer of A2D Project-Research Group for Alternatives to Development Inc. in a project funded by Arbeiter-Samariter-Bund under their Philippine Community Resilience Program.

Anita G. Cular is an associate professor in economics at the University of the Philippines Visayas (UPV). Her teaching and research interests include money and banking, ecotourism, women and gender, and evaluation or impact assessment.

Roxanna Balbido Epe is a PhD candidate in Cultural, Social, and Political Thought at the University of Lethbridge. Her research focuses on gender, power, displacement, and social protection in Mindanao, Philippines. She has previously worked with national government departments, such as the National Irrigation Administration and the Department of Agriculture-Bohol Agricultural Promotion Center, as well as different national and international organizations including Save the Children International, United Nations Population Fund, Population Council (Manila), Save the Children Philippines, Research Institute for Mindanao Culture (RIMCU) select projects, and the Manga Development Cooperation Network (MDCN) in the Philippines.

Ervina A. Espina is an assistant professor in psychology at the Division of Social Sciences, University of the Philippines Visayas Tacloban College in Tacloban City, Philippines. She completed her doctoral degree program in Social-Organizational Psychology at the Ateneo de Manila University, Philippines. Her research interests cover a range of topics on disaster, space, and women. She is part of local and international research publications on disaster preparedness, community drug rehabilitation program, space, disaster, and women. She is also a member of a volunteer core group that provides mental health and psychosocial support for disaster survivors and those affected by the COVID-19 pandemic.

Dhino B. Geges is an assistant professor at the Department of Social Development Services, College of Human Ecology, University of the Philippines Los Baños (UPLB). He completed a BSc in Human Ecology and an MA in Development Management and Governance from UPLB, and a PhD in Asia Pacific Studies at Ritsumeikan Asia Pacific University, Japan. His research and service endeavors are mostly centered on disaster risk reduction and management (DRRM), social entrepreneurship, organizational development, and community organizing. He has published several articles and presented papers both in local and international conferences on the impacts, sustainability, and resiliency mechanisms of livelihood and enterprise development programs as part of the development efforts in the Philippines and the Asia-Pacific region. His previous work involved the Philippine Business for Social Progress (PBSP), a corporate-led social development foundation that implemented various CSR programs in the country. He handled various development projects for almost five years, particularly on the Strategic Private Public Sector Partnership for Urban Poverty Reduction (STEP-UP).

Rowena S. Guiang is an assistant professor in psychology at the Division of Social Sciences, University of the Philippines Visayas Tacloban College in Tacloban City, Leyte, Philippines. She is a clinical psychologist and is engaged in providing mental health and psychosocial support (MHPSS) services and training in MHPSS skills. She was involved in designing modules for an intervention program to facilitate recovery and promote resilience among survivors of Haiyan. Her research interests are in the areas of mental health, experiences of women living with HIV, women, and disaster.

Dennis D. Gupa is a theater director and an assistant professor in the Theatre and Film Department at the University of Winnipeg. He received his PhD in Applied Theatre at the University of Victoria and was a SSHRC Vanier Scholar. His research interest is on the intersection of climate change, Indigenous ecological knowledge, and applied theater. He obtained an MFA Theatre (Directing) degree from the University of British Columbia and an MA in Theatre Arts at the University of the Philippines. His performance project, "Gossip with Whales," was presented at the COP26/The 2021 United Nations Climate Change Conference's Virtual Ocean Pavilion. He won the Performance Studies international's Dwight Conquergood Award (PSi Laureate), the Ada Slaight Drama in Education Award, and the National Commission for Culture and the Arts' Doctoral Award. Dennis is one of the Asia Society's Philippines 21 Fellows for the Arts and Culture. He is included in The Cultural Centre of the Philippines' *Encyclopedia of Philippine Arts (Theatre Volume)* for his contribution to the contemporary theatrical heritage of the Philippines.

Shiely Anne O. Julian earned her degree in Economics (cum laude) at the University of the Philippines (UP) School of Economics in 2019. Aside from her academic pursuits, she kept herself busy in various student organizations such as the Organization of Business Economics Majors and COPE UP, a mental health organization in UP Diliman. At present, she is working as a credit officer at the Bank of the Philippine Islands.

Erlidia F. Llamas-Clark is a Professor V and Attending Consultant at the University of the Philippines (UP)—Philippine General Hospital. She is an obstetrician-gynecologist specializing in ultrasound, with a Master of Public Health and a PhD in Epidemiology and Population Health. Dr. Clark is involved with clinical practice, faculty work, and multidisciplinary research with studies related to women. Her PhD research focused on the sociocultural determinants and health impacts of climate change, extreme weather events on food and nutrition security of women and children in the Philippines during Tropical Storm Ketsana and Typhoon Parma in 2009. She is an Affiliate Research Professor at the National Institutes of Health – Institute of Child Health and Human Development and a Research and Extension Fellow of the UP-Resilience Institute, and is the immediate past Chair of the Philippine Obstetrical and Gynecological Society Subcommittee on Climate Change and Disaster Risk Preparedness.

Francine Blaise M. Loja finished her undergraduate studies in June 2019 with a degree in Economics (cum laude) from the University of the Philippines (UP) School of Economics. In conjunction with her academic work, she actively participated in student organizations such as the UP Economics Society and the School of Economics Student Council Core. She is currently a third-year Juris Doctor student at the University of Santo Tomas, Faculty of Civil Law in the Philippines.

Ladylyn Mangada is a professor of political science at the University of the Philippines, Visayas Tacloban College. She has 20 years' experience in local governance, community-driven development, and local politics. Her research has focused on issues related to elections, law enforcement, and disaster risk reduction.

Maria Cecilia T. Medina holds a BA in Sociology (cum laude) and a MA in Asian Studies from the University of the Philippines (UP), and a PhD in

Sociology from Xavier University, Philippines. She is an associate professor of Philippine and Asian Studies at the Asian Center, UP Diliman, and serves as an associate editor of *Social Sciences Diliman: A Philippine Journal of Society and Change* (SSD). She is a regular member of the Division of Social Sciences of the National Research Council of the Philippines (NRCP), the Philippine Sociological Society, and the International Sociological Association (ISA) and its Research Committees on the Sociology of Religion (RC 22) and Sociology of Disasters (RC 39). She is also a member of the International Honor Society of Phi Kappa Phi and the International Honor Society of Pi Gamma Mu. Her publications and research interests are on Asian religions and cultures, Indigenous peoples, peace and development, and Southeast Asian Studies.

Ara Joy Pacoma is a PhD student at Chulalongkorn University and a research associate at Stockholm Environment Institute Asia. Her research interests include adaptive governance, post-disaster social capital, and resilience.

Yvonne Su is an assistant professor in the Department of Equity Studies at York University. She is an expert on post-disaster recovery, forced migration and poverty, and inequality. Yvonne has published on climate change-induced displacement, mainstreaming climate change resilience into urban policy, applying the capabilities approach to the global governance of migration, and post-disaster recovery in the Philippines. Her current research is on post-disaster recovery, remittances, social capital, and development in Southeast Asia.

Katherine Velmonte is the Executive Director of A2D Project-Research Group for Alternatives to Development, Inc. A2D is a nonprofit, non-government organization based in Cebu, Philippines, that works toward evidence-based development practice in the areas of inclusive and sustainable development humanitarian and disaster assistance, and disaster risk reduction and climate change adaptation.

Grace Wiebe is a retired Senior Program Manager with World Renew's International Disaster Response Team, having worked predominantly in Asia. She joined World Renew to help manage its response to the Boxing Day Tsunami that devastated Indonesia at the end of 2004. Since then, Grace designed and managed nearly Can$75 million of project implementations for some of World Renew's largest disasters, responding either to conflict, cyclones/typhoons/hurricanes, drought, earthquakes, flooding, or volcanos. These projects were conducted in the Bahamas, Bangladesh, Chile, China, Haiti, India, Indonesia, Kenya, Mexico, Myanmar, Nepal, Nicaragua,

Niger, Pakistan, Philippines, and Sri Lanka. Presently, she continues to support World Renew working on a training manual for the International Disaster Response Team. Previously, she worked as a project manager with a software company and with a local nonprofit focused on poverty alleviation in Canada. Grace studied business and project management at McMaster University, Hamilton, Canada.

Roberto Ariel Abeldaño Zuñiga completed a PhD in Demography at the National University of Córdoba, Argentina, and is currently a tenured professor at the University of Sierra Sur in Oaxaca, Mexico. He has been educated in Argentina, Brazil, and Chile; and has worked as an undergraduate and graduate teacher at public universities in Argentina. He is a researcher accredited by the National System of Researchers of Mexico (Level 2) and belongs to the Researchers Program of Argentina (Category 3). His research interests are related to environmental health, disasters, public health, and mental health in disaster situations. He has been a consultant for United Nations agencies such as UNFPA, UNDP, IOM, and the World Bank.

Acknowledgments

A book has a life of its own, entangled in the lives of contributors written at the most challenging times. We started this project in 2019, and along this collective path of sharing our work came the COVID-19 pandemic since 2020 and, towards the end, the Russian invasion of Ukraine in February 2022. We witnessed the adversities of disasters caused by natural forces, viral transmission, and human acts like war. Writing post-disaster is a challenge in itself; it takes courage and persistence amidst personal and social crises. We are deeply grateful for all contributors who made it through these times.Our journey has been immensely supported by our families and friends whose names are forever etched in our hearts with much gratitude, in Canada, in the Philippines, and elsewhere. They have nurtured our resolve that our lived experiences of disasters be told and be it so for others in island communities.

> From Roxanna: I am endlessly so grateful to my strongest support system and inspiration, Papa Siclo and Mama Sasing, all my siblings (Rodrigo, Rosita, Reynaldo, Redelia, Rita, and Rosario) especially Lot and Jun (my benefactors), and all the members of the big Epe family.
> From Glenda: To my husband, Ike, and my daughters—Charmaine, Czarina, Charelle, Czyna, and Charithe—for all the love and support; to my first grandchild Alyna Avery for the smiles to Nanay; and to our beloved Bichon Shih Tzus, Charly and Niro, for all the joys you bring into the human world.

And, to the staff of Bristol University Press for their patience as we negotiate the unexpected turns in completing this manuscript. *Thank you all!*

Glenda Tibe Bonifacio and Roxanna Balbido Epe

PART I

Disasters in the Philippines: Overview and Responses

1

Introduction

Glenda Tibe Bonifacio and Roxanna Balbido Epe

Disasters shape our realities. We both grew up in the Philippines and experienced the ravage of typhoons, floods, and earthquakes that have shaped our consciousness throughout our lives and continue even after migrating to Canada. Living in different islands in the archipelago was a constant reminder of the beauty and fury of nature and the capacity of our people to survive amidst the risks and impact of disasters. Typhoons, in our memories, simply followed the cycles and are expected in the 'ber' months with strong winds and rains that would cause secondary hazards such as flooding, landslides, health risks, disruption of economic or daily activities, and suspension of classes. The affected communities, local governments, businesses, organizations, and others have to deal with the challenges from the impact of disasters. We listened to our elders about the wrath of nature predicted in the color and movement of the clouds and the anxiety of animals nearby. We all survived the passing typhoons and were seemingly ready to clean up the debris around, bounce back again, and move on. But super typhoon Haiyan changed that.

Haiyan, known as Yolanda in the Philippine Area of Responsibility (PAR), was inconceivable in our living memory. The death and destruction were unprecedented in 2013. This time, we knew almost everyone who lost a family member from the historic storm surge. The immense pain and trauma still linger. Haiyan was the turning point in our attempt to draw more sociological perspectives of disasters in the Philippines, our benchmark of what follows one of the strongest typhoons that made landfall in the world.

Conceptualizing disaster archipelago

Is geography destiny (Gallup, Gaviria, and Lora 2003)? The Philippines is located "curiously enough, on the Eurasian Plate rather than on the adjacent Philippine Plate" (Hinga 2015, 248). The Philippine Sea Plate is "tectonically unusual" because "almost all boundaries are convergent" and the surrounding areas with "very active, producing numerous local earthquakes, tsunamis and volcanic activity" (Vokaty 2014). The Philippines is situated within the Pacific Ring of Fire where most of the world's active above-sea level volcanoes are located and is constantly facing a "variety of environmental hazards"; an archipelago, however, consisting of the "remains of previous eruptions" (Nagle 1998, 27). About 300 of the world's 1,500 most active volcanoes are in the Philippines; 278 are considered dormant and 22 are active (Hilotin and dela Llata 2021). These active volcanoes are found throughout the country that "experiences one large magnitude (7.75 or higher on the Richter Scale) tectonic earthquake every ten years, seven earthquakes of major magnitude (7.0 to 7.4) every ten years, and five earthquakes of moderate magnitude (6.0 to 6.9) every year" (Tope and Nonan-Mercado 2002, 11).

Mount Pinatubo in western Luzon in the Philippines erupted for the first time after 600 years in 1991 and was considered the "second-largest volcanic eruption of this century and by far the largest eruption to affect a densely populated area" in the world (Newhall, Hendley II, and Stauffer 2005). Over 100,000 were displaced and there were about 300 deaths from the heavy ashfalls. Further deaths from diseases at evacuation centers in were recorded in 1991, along with another 72 killed on the volcano's second eruption in 1992 (Encyclopedia Britannica 2021). Since 1572, Taal Volcano has erupted 33 times, most recently on January 12, 2020, displacing over 376,000 people (Aljazeera 2021). Taal Volcano, located south of Manila, is one of the "world's smallest active volcanoes" at 311m but is historically deadly; over 1,300 people died as a result of its eruption in 1911 (CNN 2021).

As of January 2022, the Philippine Institute of Volcanology and Seismology (PHIVOLCS 2018) listed on its website 18 destructive earthquakes since the 1960s: Casiguran (1968 Ms 7.3), Ragay Gulf (1973 Ms 7.0), Moro Gulf (1976 Ms 7.9), Laoag (1983 Ms 6.5), Panay (1990 Ms 7.1), Luzon (1990 Ms 7.8), Mindoro (1994 Ms 7.1), Bayugan (1999 Ms 5.1), Palimbang (2002 Ms 6.8), Masbate (2003 Ms 6.2), Negros oriental (2012 Ms 6.7), Bohol (2013 Ms 7.2), Surigao del Norte (2017 Ms 6.7), Leyte (2017 6.5), Central Luzon (2019 Ms 6.1), Itbayat, Batanes (2019 Ms 5.8), Cotabato (2019 Ms 6.5), and Davao del Sur (2019 Ms 6.9). The Moro Gulf earthquake occurred after midnight on August 17, 1976, followed by 50ft tsunami waves that killed about 8,000 people in Cotabato in Mindanao (Gates and Ritchie 2007; Hamblyn 2014). Known as the "most infamous earthquake in the history of the Philippines" the Luzon earthquake occurred on July 16, 1990, followed by aftershocks for several months with 1,621 deaths and over 3,100 injuries (Gates and Ritchie 2007, 155). Bohol earthquake struck on October 15, 2013, in the central Visayas that "triggered landslides and caused widespread destruction," 200 deaths, and 350,000 displacements (UNICEF 2022). On December 13, 2021, the province of Batangas was hit by an earthquake with greater than 5.0 magnitude that impacted nearby areas in Luzon; the provinces in Luzon experienced six moderate earthquakes within the space of five months in 2021. A moderate earthquake registers a magnitude between 5 and 5.9 on the Richter Scale.

The Philippines is noted as a "disaster archipelago" (Alejandria and Smith 2020) and attracts fervent scholarship both local and international. Academic thesis projects continue to use Haiyan as a case study or point of reference on the impact of global climate change (Santos 2018), vulnerabilities (Noy 2017; Gonzalez Rojas 2019), resilience (Go 2016; Castro 2019), humanitarian assistance (Conrad 2015; van Lint 2016; Junio 2017; Desamero 2018; Evardone 2018), recovery and reconstruction (Arroyo 2019), and remittances (Gazi 2016; Su 2019), among others. These sample works of emerging scholars demonstrate the historic mark left by Haiyan in 2013 in terms of governance, risk management, social protection, international programming, and the built environment. However, in the context of the Asia-Pacific region, Indonesia is another archipelago that experienced an earthquake on July 2, 2013, in the Aceh province of North Sumatra, which resulted in 39 deaths, 2,500 injuries and 50,000 displacements (OCHA 2013). Indonesia and the Philippines are both "prone to natural hazards" (Bagayas 2018). A "disaster archipelago" constructs a country of disasters rendering the visibility of destruction and trauma to respond to and the invisibility of economic and political systems, for example, that contribute to risks and vulnerabilities.

In three consecutive years, from 2016 to 2018, the Philippines ranked third as the most at-risk country globally by the WorldRiskReports (Jesconnek et al 2016, 2017; Kirch et al 2018, 2019) with its geographical and geological factors. In the Disaster Risk Reduction in the Philippines Status Report 2019, about 60% of the "total land area is exposed to multiple hazards, and 74% of the population is susceptible to their impact" with "over 80% of the *natural disasters*" are caused by "hydro-meteorological events including typhoons and floods" (UNDRR 2019, 6, added emphasis). According to Smith (2006), "there is no such thing as a natural disaster. In every phase and aspect of a disaster—causes, vulnerability, preparedness, results and response, and reconstruction—the contours of disaster and the difference between who lives and who dies is to a greater or lesser extent a social calculus." Instead, there are natural hazards that are part of "natural process and event that is a potential threat to human life and property" (Keller and DeVecchio 2016, 5) and sometimes unforeseen. These natural hazards become disasters, when, according to Mizutori (2020), it "affects a human settlement which is not appropriately resourced or organized to withstand the impact and whose population is vulnerable because of poverty, exclusion or socially disadvantaged in some way." In the case of the Philippines, the natural hazards caused by geophysical (earthquakes, landslides, volcanic activity), hydrological (floods), and meteorological (typhoons, storms) (IFRC 2022) result in the loss of lives and the destruction of properties and even communities.

As of the first quarter of 2021, the poverty incidence in the Philippines was 23.7% or 26.14 million and subsistence incidence was 9.9% or 10.94 million (Mapa 2021). Poverty is a risk factor that compounds the impact of disasters. Low-income Filipinos tend to live in marginalized areas in the country, often in slums and along the coastlines or riverbanks susceptible to natural hazards. Poor Filipinos' houses made of light construction materials of cannot withstand typhoons, for example, although concrete houses are also destroyed based on the type and severity of the hazard. Urban congestion in megacities like Manila with over 12 million people visibly marks spaces where rich and poor people live. The population density in the Philippines from 2007 to 2020 was 365.51 people per square kilometer (Statista 2021a); there was an 8% increase in 2020 based on data in 2015 or 337 to 363 persons per square kilometer (PSA 2021). The Philippines ranks 13th in the world based on population with 109,581,085 inhabitants in 2020 (Jaymalin 2017; World Bank 2022).

The built environment has profound implications on the onset of natural hazards and reducing risks. Trees cleared for street widening, mountains flattened to create settlements or housing developments, rivers scraped for gravel, and open-pit mining are examples that we have witnessed in the Philippines. Mountains shield the islands from strong winds, scraped rivers contribute to flooding as well as denuded places. In 2020, the Philippines lost 46.8 Kha of natural forests (Global Forest Watch n.d.). For years, we have observed the rampant destruction of the environment in the name of progress, and sustainability is a contentious political mandate to pursue (Eadie 2016; Fernandez 2021). While geography is a fixed reality of inhabitants, a deep understanding of the relations of people to their environment and effective policies and mechanisms could perhaps control or reduce the impact of disasters. Bradshaw (2013) points out the linkage between development processes and disasters, and those disasters are possible outcomes. Through the years that we have lived in the Philippines, we have witnessed the rise of rapid urbanization without much effective protection for the environment, and hence, disaster awaits. Haiyan as a super typhoon was of a considerable magnitude not experienced in living memory.

Framing cultures and disasters

If "geography is destiny" (Gallup, Gaviria, and Lora 2003), it possibly evokes ways of living in a "disaster archipelago" (Alejandria and Smith 2020) that alludes to some connections between "cultures and disasters" (Bankoff et al 2015; Krüger et al 2015). Indeed, geography is an imprint of nature and seems permanent in the conceptions of a physical place. Landmarks abound where hazards once posed risks to inhabitants; a truism in the histories of civilizations around the world where archeologists have discovered places destroyed by disasters (Gould 2007; Eaton 2015). Geomorphology describes the intricate relations between people and place in the evolution of societies wherein disasters are notably influenced by four factors—"time, space, type of society, and type of event" (Leroy 2020). Our concern is about society, not necessarily about the *type*, but its social constructions, inequalities, inequities, and marginalization that shape diverse experiences of people impacted by disasters or how disasters happen. Typologies construct hierarchies and devalue the unique development of societies. We disengage from labeling societies in the context of decolonizing knowledge.

Societies have varied forms of social organization, cultural beliefs and practices, the evolution of economic and political systems, political culture, and ways of dealing with the environment, to name a few. In *Cultures and Disasters: Understanding Cultural Framings in Disaster Risk Reduction* (Krüger et al 2015), the social constructions of risks and responses of communities towards disasters are intrinsically connected to their cultures. The new language of practical disaster risk reduction (DRR) now recognizes "people's cultural production of risk, and their responses to it" and "why people are affected by hazards and why they do or do not take action to minimize them" (Krüger et al 2015, 1). This work uses an interdisciplinary approach that importantly situates disasters within the purview of various disciplines and requires broadening frames of reference including people's belief systems and actions in DRR (Bankoff et al 2015, 14).

Culture shapes attitudes towards risks from natural hazards, and, as noted by Dalisay and De Guzman (2016, 703), through "generations of engagement with the hazards, people have developed cultural ways of addressing these hazards that reduce risk and vulnerability." One wonders why people tend to return to disaster-prone areas after evacuation or a disaster. In the Philippines, and arguably elsewhere, the relations to the land and conceptions of risks of the surrounding environment form part of the cultural ethos of communities. Traditional ways of coping with disasters demonstrate the complexities of living in island communities, and lessons learned from disasters evolve through time. In their study after super typhoon Haiyan, Dalisay and De Guzman (2016) demonstrate the social and cultural contexts that influenced decisions to evacuate or not. Filipinos in their study from affected areas in Leyte and Samar point to "various elements in culture that shape perceptions of risks" (Dalisay and De Guzman 2016, 709) such as "*sanay na kami diyan* (We are used to it)" (706), the "safe and unsafe spaces" (706) in the community, "[e]vacuation to homes of family members" (707), and "local knowledge" (707).

Communities affected by disasters display an indomitable spirit of resilience. Gaillard and Le Masson (2007, 315) note that "[r]esilient societies are those able to overcome the damage wrought by the occurrence of natural hazards, either by maintaining their pre-disaster social fabric or by accepting marginal or greater change in order to survive." Understanding cultural complexities seem important in exploring ways of living in areas prone to natural hazards such as the Philippines. With about 2,000 out of 7,670 islands inhabited, "ecological diversity and the differential cultural adjustments of particular populations to their effective cultural environment" in the Philippines has resulted in "at least 77 major ethnolinguistic groupings" (NCCA n.d.). Each ethnolinguistic grouping like the Waray in Leyte and Samar islands is socially constructed for their resilient attitudes (e.g., *maisog* or brave) and applied to different contexts. Over 300 years of Spanish colonialism left the Philippines as the only predominantly Catholic country in Asia, and its contemporary expressions are by far not the same as Castilian practice. Filipino folk religiosity is "a syncretic blending of pre-colonial beliefs with the Catholic faith" (Macaranas 2021, 800). How this amalgam of beliefs and practices about the sacred, divine, and disasters provide strength in people's faith of possible outcomes and consequences, even death, expresses unique dimensions of the human spirit.

Disasters before and after super typhoon Haiyan in the Philippines brought destruction, misery, trauma, displacement, death, and all other manifestations of personal and collective response from communities, governments, and civic organizations. Disasters, however, do not define the destiny of Filipinos in the Philippines and those who live in the country amidst onsets of disasters. Despite past and impending catastrophes, the Philippines still ranked as the 8th most favored country to visit in the world in 2019 by the New York-based luxury and travel magazine Condé Nast Traveler Readers' Choice Awards (Rocamora 2019). Popular travel spots in the Philippines include Boracay, the top island destination in Asia, Cebu, and Palawan (Department of Tourism 2019). Tourism is the main driver of island economies in the Philippines yet sustainable practices remain problematic. In 2021, the Philippines ranked "94th or near the tail-end" by Euromonitor's 57 travel indicators from "environmental, social, and economic sustainability; country risk, and sustainable tourism demand, transport and lodging" (De Vera 2021). Island tourism booms lead to preferential treatment in pursuing sustainable economic development; high-traffic areas tend to receive more government support and regulation compared to other islands. For example, Boracay was ordered to close to tourists for six months in 2018 by President Rodrigo Duterte who described it as a "cesspool" (Favila 2018) and appears "ruined by tourism" (Morris 2018). Destruction of the environment contributes to increased natural hazards, and the actors are not only local island inhabitants but visitors from around the world.

Local–global interface

Increasing exposure to natural hazards and disasters before and after Haiyan is not simply an equation of 'acts of nature' in the Philippines but a complex interface of local and global interconnectedness. The global in this context pertains to climate change and its impact as "the country most at risk from climate crisis" in overall natural hazard in 2019 by the Institute for Economics and Peace (Amnesty International UK 2019). These risks include extreme weather due to global warming and rising sea temperatures creating super typhoons like Haiyan, rising sea levels from melting ice caps affecting low and coastal lands, and the destruction of marine life from ocean acidification and pollution (Amnesty International UK 2019). In the age of the Anthropocene, global warming is mainly attributed to human activities since the mid-20th century (Hamilton, Bonneuil, and Gemenne 2015) and the world is at risk. Consequently, "[i]f a natural hazard occurs in a restricted area of the planet, its impact may be felt worldwide due to our current great interconnectedness" (Leroy 2020).

In the Philippines, policies are formulated to protect the environment and people; foster climate change adaptation and disaster risk reduction and management through prevention and mitigation, preparedness, response, rehabilitation, and recovery; and resilience-building toward sustainable development. Laws and implementing guidelines for disaster risk reduction and climate change mitigation have made inroads in the country. In 1941, the National Emergency Commission was established by Executive Order No. 335, the "first landmark," by President Manuel L. Quezon (Domingo and Manejar 2018, 16). This was followed by the creation of the National Civil Defense Administration in 1954 by Republic Act 1190 under President Ramon Magsaysay. The National Disaster Coordinating Council was created together with the Office of Civil Defense by Presidential Decree No. 1566 by President Ferdinand Marcos (Domingo and Manejar 2018, 16) in 1978 "to strengthen Philippine disaster control" (Rey 2015). Under the Local Government Code of 1991 or Republic Act 7160 under President Corazon Aquino, local government units can access 5% of estimated revenues for "unforeseen expenditures such as the occurrence of calamities" but only if the president of the country declares the area "in a state of calamity" (Rey 2015). At present, the Philippines has a National Disaster Risk Reduction and Management Council (NDRRMC), established by Republic Act 10121 in 2009 under President Gloria Macapagal Arroyo, composed of diverse sectors "responsible for ensuring the protection and welfare of the people during disasters or emergencies" (United Nations n.d.). Republic Act 9729 or the Climate Change Act of 2009 created the Climate Change Commission Philippines as the "lead agency for the coordination, monitoring and evaluation of programs and action plans" on climate change (NICCDIES n.d.). The Philippine Disaster Reduction and Management Act or Republic Act 10121 passed in 2010 "shifted the policy environment and the way the country deals with disasters from mere response to preparedness" (Grantham Research Institute on Climate Change and the Environment n.d.). This Act was considered a "landmark legislation for Philippine disaster risk-reduction and management" that replaced Presidential Decree 1566 after 32 years (Patajo-Kapunan 2019). As of January 2022, there is pending legislation in the upper House of Senate, House Bill No. 5989 which was passed by the lower House of Representatives in September 2020, that proposes to create the Department of Disaster Resilience by mid-2022 and replace NDRRMC (*Philippine Daily Inquirer* 2021).

The policy ideal, however, is not substantiated by robust implementation mechanisms in a postcolonial country with fewer resources or, perhaps a lack of political will by the local and national leadership on sustained disaster preparedness. Government expenditures related to disaster risk reduction amounted to PHP20.6 billion from 2009 to 2017, with the highest value in 2013 after super typhoon Haiyan (Statista 2021b). In the study of disaster preparedness and local governance Domingo and Manejar (2018, 17–18) note that "[m]unicipal mayors shared disappointment when even after a year, they have not received a single centavo from the national government" after Haiyan. Three years after Haiyan struck in Tacloban City, our field research notes reveal that many survivors were still housed in temporary shelters and the promised monetary assistance was not fully distributed. From 2000 to 2016, there were over 23,000 deaths and about 125 million people affected by disasters in the Philippines at a cost of US$20 billion, or an average yearly cost of US$1.2 billion (Jha et al 2018). However, the administration of President Rodrigo Roa Duterte (2016–2022), with his strong political will (Lopez 2021), has significantly enhanced the policy agenda, investments, and governance on disaster risk reduction with the aim of effective programming, resource mobilization, service delivery, and environmental protection (Rappler 2017; Gutierrez 2019).

Economically developing countries such as the Philippines experience more profound impact on the challenges posed by global climate change due to its lack of financial capacity to respond to disasters. International humanitarian assistance is often relied upon, including from former colonial powers like the United States. The US Agency for International Development (USAID), for example, has "responded to more than 40 disasters" since 1990 (USAID 2022). The Philippines was the first US colony that was ceded by Spain in 1898 for US$20 million, and America took direct control for nearly 50 years (Hincks 2016). In response to Haiyan, the US government provided more than US$37 million through USAID and the Department of Defense (White House 2013). Spain considers the Philippines as a "Partner Country (i.e., of highest priority) in the Asia-Pacific region by the current Master Plan of the Spanish Cooperation Agency"; the only European Union member with a "permanent humanitarian aid team" and the largest donor since 2007 (LaMoncloa 2013). The 16th-century colonial power that gave the Philippines its name began its official development assistance only in the 1980s by the Spanish International Cooperation Agency, and humanitarian assistance is "relatively small" and "among the smallest of the Development Aid Committee (DAC) members of the OECD" (Humanitarian Practice Network 2001). In the aftermath of Yolanda (Haiyan), the Spanish government sent two aircraft with supplies to Cebu to support the Spanish Red Cross, Action Against Hunger, an additional €200,000 through the International Federation of Red Cross and Red Crescent Societies (IFRC), Spanish Technical Cooperation Office, and the Spanish Embassy in the Philippines (Gobierno de España 2013). Lastly, Japan occupied the Philippines during the Second World War and, since liberation, continues to foster "*kizuna* or bonds of friendship" (Fernandez 2013). The Japan International Cooperation Agency (JICA) provided about US$610,000 worth of relief supplies immediately after Haiyan, sent medical teams, and technical experts on Haiyan-related oil spills (Fernandez 2013). Japan has been the Philippines' largest source of official development assistance (ODA) for the last 20 years amounting to US$14.139 billion or 72% of foreign aid (Business World 2021).

In the aftermath of super typhoon Haiyan, individuals, non-government organizations, international organizations, civil society, and other countries responded with compassion after constant reporting from international media outlets. Conflicting messages from a national television channel aligned with the government added media attention to the slow response in Tacloban City and other affected areas in the Visayas (Dadis 2013; de Jesus 2013; Romero 2013). The United Nations Office for Coordination (UNOCHA) reported a total of US$865,151,866 (PHP41.8 billion) for Yolanda (Haiyan) survivors; the top donor was the United Kingdom at US$122,743,593 and the smallest came from Cyprus at US$25,221 (Gavilan 2016). As of October 7, 2016, of the former colonial powers in the Philippines, the US was ranked 3rd highest foreign donor for Yolanda survivors with US$90,585,530; Japan was ranked 6th with US$63,328,022, and Spain was ranked 25th with US$4,171,931 (ABS-CBN 2016). Based on the same period of the foreign donors in 2016, private individuals and organizations ranked 2nd with US$98,589,169 out of 80 in the list (ABS-CBN 2016).

Humanitarian assistance is a critical support for emergency response, rehabilitation, and reconstruction. But this comes with the accountability of a shared future before and after the onset of disasters. The Philippines has long been connected to global trade with the Manila Galleon Trade since the 16th century (Fish 2011; Giraldez 2015). In 2020, the US was the top foreign investor in the Philippines with PHP35.4 billion, followed by China, the United Kingdom, Singapore, Japan, Netherlands, and Taiwan among 20 countries in the list (Statista 2021c). The sectors receiving the most foreign direct investments in 2021 were "information and communication, electricity, gas, steam and air conditioning supply, manufacturing, and administrative and support service activities" (InCorp Philippines 2021). These investments should consider the impact on the environment and contribute to disaster risks reduction and mitigation. In manufacturing, local groups have advocated for the reduction, if not abolition, of plastic packaging, particularly single-use plastic sachets that pollute seawater (Sarmiento 2018). Plastic waste contributes to the clogging of drainage which causes floods during typhoons. After a four-year ban by President Rodrigo Duterte, open-pit mining for "copper, gold, silver and complex ores" was reinstated by a Department Administrative Order 2021–40 in December 2021 and was lauded by foreign businesses (Talavera 2022). Open-pit mining has a grave impact on the environment, such as "deforestation, soil erosion, destruction of the natural landscape, landslides, water-level reduction and surface water depletion through dewatering, as well as water and soil pollution from mining waste and tailing dust" (Park et al 2019, 248). In these examples, the risks to natural hazards are compounded by business activities funded by foreign investors. Disasters are waiting to happen with wanton disregard of the environment. The connections between the local and the global in reducing the likelihood of disasters are paramount in recognizing a shared sustainable future.

Organization of the book

Intersectional and multidisciplinary perspectives foster critical insights, for example, on human adaptation, risk mitigation, government response, humanitarian assistance, and community rehabilitation. In this collection, we present the disasters in the Philippines in two parts: Part I provides an overview and responses on disasters pre-Haiyan and Part II on the Haiyan aftermath. Part I has five chapters related to the socio-economic impact of disasters, use of information and communication technology (ICT), food and nutrition security, disability, and children and education. Part II has nine chapters organized into three themes: women and empowerment, economics and inclusion, and communities and resilience.

In Part I, Chapter 2, Roberto Ariel Abeldaño Zuñiga characterizes the patterns in the occurrence of disasters, associated deaths, population affected, and economic damages in the Philippines over the period 2000–2019. It is a descriptive study based on secondary data from the International Disaster Database (EM-DAT) of the Center for Disaster Epidemiology Research (CRED) of the School of Public Health, Catholic University of Louvain, Belgium. This chapter outlines the direct and secondary effects of disasters, vulnerability, and strategies for disaster risk reduction in the Philippines. Chapter 3 by Xavier Venn Asuncion presents a study on 'climate e-governance' which aims to contribute to this new area by looking into the state of ICT use, or what and how ICTs are being deployed for monitoring of climate change and impacts and adaptation and disaster management. The study is focused on local government units (LGUs) in Metro Manila, Philippines, which is among the countries in the world with high climate risk vulnerability and slow e-governance progress. Chapter 4 by Erlidia Llamas-Clark discusses the implications of gender roles to the food and nutrition security of children and households in extreme weather events. Using a mixed-methods approach, this chapter highlights the pre- and post-disaster coping strategies of women engaged in unpaid and paid work and the need for immediate and long-term support during disasters. Chapter 5 by Roxanna Balbido Epe presents a follow-up study on integrated child-centered disaster risk reduction management programs and activities of the Department of Education, local government units, and selected communities affected by the Bohol earthquake. The collaboration and synergies of these sectors contribute to resilience building, good governance, and children's empowerment.

Part II of this book focuses on the aftermath of Haiyan. The sociological dimensions of the impact of the super typhoon include women's participation, economic recovery, and community resilience. Chapter 6 by Anita Cular demonstrates the case of women in Sta. Rita, Samar, using an ethnographic study as a microcosm of the country's struggle with development in the aftermath of the disaster. Women in the community demonstrated diverse forms of agency in rebuilding their lives locally instead of migrating to other areas after Yolanda. Chapter 7 by Rowena Guiang and Ervina Espina look at relations of disaster events and role transitions among low-income rural women. The shift in women's traditional gender roles rendered them visible in comparison with their past experiences of invisibility and marginalization. Women's participation and engagement in disaster recovery efforts highlighted their capability to be partners in community building. Chapter 8 by Dhino Geges discusses how women's participation through social enterprises became valuable in enhancing resilience in island communities of Concepcion, Iloilo. Using a pragmatic worldview and pluralistic approach, this chapter explores women's lived experiences of collective work, leadership, and reproductive and productive roles in disaster recovery. Women's empowerment is crucial in fostering transactional and transformational development processes in communities impacted by disasters. Chapter 9 by Shiely Anne Julian and Francine Blaise Loja uses counterfactual analysis to evaluate the effect and the actual event if Yolanda had not happened. This is an important consideration to determine the subsequent economic losses post-Yolanda to the productive physical and human capital in Region VIII (Leyte and Samar) using Gross Regional Domestic Product per capita in constant 2000 prices. Chapter 10 by Glenda Tibe Bonifacio underscores the relations between gender and labor market regimes that developed post-Yolanda in Leyte. Based on observational fieldwork and interviews, this chapter explores the interplay of local and international actors in creating these labor regimes and their impact on the labor participation of local populations. Chapter 11 by Kaira Zoe Alburo-Cañete, Corazon Clarin, Marivic Codiñera, and Katherine Velmonte aims to contribute to the discourse on inclusive DRR in the Philippines. Based on a case study in Cebu with a multistakeholder advocacy coalition of organizations of persons with disability (OPDs), this chapter sets out the key lessons for disability inclusion as one of the pioneering initiatives in the country. Chapter 12 by Dennis Gupa examines sea rituals, climate change, and applied theater in Guian, Eastern Samar. An autoethnographic work probes deeper nuances of cultural signs in the lifeways of people and their ecological knowledge. The use of applied theater enables community mobilization in crafting theatrical pieces intended as a contemporary performative tool in unpacking ecological dilemmas of a fishing village. Chapter 13 by Yvonne Su, Ladylyn Mangada, and Ara Joy Pacoma interrogates the Filipino practice of *bayanihan* (mutual help) in post-disaster recovery. Drawing from two separate but overlapping case study research projects in the province of Leyte after Haiyan and their "insider-outsider" status, this chapter posits the view that while *bayanihan* was once a principle that was believed to be upheld by the whole community, its contemporary expression is often on a much smaller scale from neighbor to neighbor and only for a brief period during crises. Chapter 14 by Grace Wiebe describes the response of World Renew (WR), an international Christian-based humanitarian organization, to achieve community resilience and the behavioral challenges it overcame as it transitioned from emergency, early recovery, rehabilitation, and reconstruction in Iloilo from 2013 to 2019. WR highlights the fisherfolk and farmers' efforts to adapt to the effects of climate change; the economic recovery activities that supported men and women in the community; and how the promulgation of the Canadian government's Feminist International Assistance Policy mid-stream resulted in project plans becoming more gender transformative. Chapter 15 by Maria Cecilia Medina elucidates the role of religion in the public sphere in Leyte after super typhoon Yolanda. This chapter gives attention to how Catholic religious leaders and lay leaders have facilitated and assisted in service delivery in support of relief and rehabilitation efforts going beyond their traditional religious roles. As part of civil society, religious leaders also performed functions of building community, advocacy, and protection of citizens' welfare, monitoring for accountability in the use of state resources and improving interventions for the disaster.

These chapters subscribe analyses to applicable international frameworks such as the United Nations 2030 Agenda for Sustainable Development, Sendai Framework for Disaster Risk Reduction (2015–2030), and the Hyogo Framework for Action (2005–2015). A conclusion is provided on the post-Haiyan Philippines and beyond to situate, again, how disasters in the Philippines are lived realities of island states that need to be recognized globally as an imprint of what could unfold in other areas in the world. These realities are, however, not exactly the same anywhere. Some similarities of experiences could perhaps raise a global consciousness on how our lives are interconnected by systems and forces that impact the environment then, now, and tomorrow.

References

ABS-CBN. (2016) "How much Yolanda aid did PH get from US, EU?," *ABS-CBN News*, October 8. https://news.abs-cbn.com/focus/10/08/16/how-much-yolanda-aid-did-ph-get-from-us-eu [Accessed January 28, 2022].

Alejandria, M. C. P. and W. Smith (eds) (2020) *Disaster Archipelago: Locating Vulnerability and Resilience in the Philippines*, Lanham, MD: Lexington Books.

Amnesty International UK. (2019) "Philippines country most at risk from climate crisis". https://www.amnesty.org.uk/philippines-country-most-risk-climate-crisis [Accessed January 26, 2022].

Arroyo, I. (2019) "User involvement in housing recovery: Cases from Haiyan affected areas in the Philippines," PhD Thesis, Department of Architecture and Built Environment, Housing, Development and Management, Lund University, Sweden. https://portal.research.lu.se/files/63093007/Ivette_Arroyo_doctoral_thesis_kappa.pdf [Accessed January 2, 2022].

Aurelio, M., S. D. Catugas, J. A. Escudero, A. M. F. Lagmay, and G. A. Tapang. (2021) "Luzon, Philippines, sees sixth strong earthquake in five months," *Temblor*, December 21. https://temblor.net/earthquake-insights/luzon-philippines-sees-sixth-strong-earthquake-in-five-months-13571/ [Accessed January 22, 2022].

Aljazeera. 2021. "Philippine scientists warn of another eruption at Taal Volcano," *News*, July 4. https://www.aljazeera.com/news/2021/7/4/philippine-scientists-warn-of-another-eruption-at-taal-volcano [Accessed January 22, 2021].

Bagayas, S. (2018) "Why Philippines and Indonesia are prone to natural hazards," Rappler, October 18. https://www.rappler.com/moveph/214623-why-philippines-indonesia-prone-natural-hazards/ [Accessed January 23, 2022].

Bankoff, G., T. Cannon, F. Krüger, and E. L. F. Schipper (2015) "Introduction: Exploring the links between cultures and disasters," in F. Krüger, G. Bankoff, T. Cannon, B. Orlowski, and E. L. F. Schipper (eds) *Cultures and Disasters: Understanding Cultural Framings in Disaster Risk Reduction*, London: Routledge, pp 1–16.

Bradshaw, S. (2013) *Gender, Development and Disasters*, Cheltenham, UK: Edward Elgar.

Business World. (2021) "Japan top ODA source over last 20 years, providing over $14B," August 10. https://www.bworldonline.com/japan-top-oda-source-over-last-20-years-providing-over-14b/ [Accessed January 27, 2022].

Castro, M. T. (2019) "Strengthening capacities towards a resilient future: The case of sexual and gender minorities in Tacloban City, Philippines after the 2013 typhoon Haiyan," Masters thesis, Development Studies, School of Geography, Environment and Earth Sciences, Victoria University of Wellington, New Zealand. https://doi.org/10.26686/wgtn.17142425.v1 [Accessed January 2, 2022].

CNN. (2021) "Philippines raises Taal Volcano danger level as thousands evacuate," July 2. https://www.cnn.com/2021/07/01/asia/mass-evacuation-philippines-volcano-intl-hnk/index.html [Accessed January 22, 2022].

Conrad, A. E. (2015) "Expect the unexpected: An autoethnography of typhoon "Yolanda" (Haiyan)," Masters thesis, Department of Sociology and Anthropology, Illinois State University, USA. https://ir.library.illinoisstate.edu/etd/439 [Accessed January 2, 2022].

Dadis, Z. (2013) "Anderson Cooper vs. Korina Sanchez: Cooper's response and Sanchez's alleged suspension," *When in Manila*, November 21. https://www.wheninmanila.com/anderson-cooper-vs-korina-sanchez-coopers-response-and-sanchezs-alleged-suspension/ [Accessed January 27, 2022].

Dalisay, S. N. and M. De Guzman (2016) "Risk and culture: The case of typhoon Haiyan in the Philippines," *Disaster Prevention and Management* 25(5): 701–714. DOI:10.1108/DPM-05-2016-0097

De Jesus, T. V. (2013) "Anderson Cooper's report Irks Korina Sanchez," Inquirer.Net, November 14. https://entertainment.inquirer.net/120857/anderson-coopers-report-irks-korina-sanchez [Accessed January 27, 2022].

Department of Tourism. (2019) "DOT celebrates PHL's ranking in CNT world top countries," Republic of the Philippines, October 25. http://www.tourism.gov.ph/news_features/phsrankingincntworldtopcountries.aspx [Accessed January 26, 2022].

Desamero, J. D. (2018) "Typhoon Haiyan recovery assistance shelter project compliance to the Philippine Red Cross Shelter Settlements and Implementation Guidelines," MBA thesis, School of Graduate Studies, Central Philippine University, Philippines. https://hdl.handle.net/20.500.12852/327 [Accessed January 2, 2022].

De Vera, B. O. (2021) "PH ranks poorly in tourism sustainability, resilience indices," Inquirer.Net, March 15. https://business.inquirer.net/319499/ph-ranks-poorly-in-tourism-sustainability-resilience-indices [Accessed January 26, 2022].

Domingo, S. N., and A. J. A. Manejar (2018) "Disaster preparedness and local governance in the Philippines". Discussion Paper Series No. 2018–52, Philippine Institute for Development Studies, Quezon City, Philippines. https://pidswebs.pids.gov.ph/CDN/PUBLICATIONS/pidsdps1852.pdf [Accessed January 26, 2022].

Eadie, P. (2016) "The Philippines, environmental politics and the challenges ahead," *Ballots & Bullets*, School of Politics & International Relations, University of Nottingham, April 18. http://nottspolitics.org/2016/04/18/the-philippines-environmental-politics-and-the-challenges-ahead/ [Accessed January 24, 2022].

Eaton, G. (2015) *A History of Civilization in 50 Disasters*, Thomaston, ME: Tilbury House Publishers.

Encyclopedia Britannica. (2021) "Mount Pinatubo," September 13. https://www.britannica.com/place/Mount-Pinatubo [Accessed January 22, 2022].

Evardone, R. N. L. (2018) "Participation of Youth Volunteers during Typhoon Haiyan Emergency Response," Masters thesis, International Studies, College of Liberal Arts, De La Salle University, Philippines. https://animorepository.dlsu.edu.ph/etdm_intlstud/22/ [Accessed January 2, 2022].

Favila, A. (2018) "Philippines closes popular tourist island after president calls it a 'cesspool'," *Global News*, April 5. https://globalnews.ca/news/4125090/philippines-boracay-tourism-pollution/ [Accessed January 26, 2022].

Fernandez, H. A. (2021) "Five years after Duterte: Has sustainability in the Philippines won or lost?," *Eco-Business*, July 27. https://www.eco-business.com/news/five-years-of-duterte-has-sustainability-in-the-philippines-won-or-lost/ [Accessed January 24, 2022].

Fernandez, M. (2013) "From emergency support to a full-scale recovery: JICA's Comprehensive Response in the Aftermath of Typhoon Haiyan," JICA USA November/December, Japan International Cooperation Agency, https://www.jica.go.jp/usa/english/office/others/newsletter/2013/1311_12_01.html [Accessed January 27, 2022].

Fish, S. (2011) *The Manila-Acapulco Galleons: The Treasure Ships of the Pacific*, Milton Keynes, UK: Author House.

Gaillard, J. C. and V. Le Masson (2007) "Traditional societies' response to volcanic hazards in the Philippines: Implications for community-based disaster recovery," *Mountain Research and Development* 27(4): 313–317.

Gallup, J. L., A. Gaviria, and E. Lora (2003) *Is Geography Destiny: Lessons from Latin America*, Washington, DC: Stanford University Press and The World Bank.

Gates, A. E. and D. Ritchie (2007) *The Encyclopedia of Earthquakes and Volcanoes*, 3rd ed., New York: Facts on File Inc.

Gavilan, J. (2016) "What typhoon Yolanda foreign aid looks like without US, EU and UN," *Rappler*, October 9. https://www.rappler.com/newsbreak/iq/148685-united-nations-european-union-united-states-foreign-aid-philippines-typhoon-yolanda/ [Accessed January 27, 2022].

Gazi, N. (2016) "Bridging the Gap between relief and early recovery: The role of cash transfers in rebuilding livelihoods: A case study of post-typhoon Haiyan recovery in the Philippines," Masters thesis, International Development Studies, University of Birmingham, UK. https://doi.org/10.31237/osf.io/b9wrx [Accessed January 24, 2022].

Giraldez, A. (2015) *The Age of Trade: The Manila Galleons and the Dawn of the Global Economy*, Lanham, MD: Rowman & Littlefield.

Global Forest Watch (n.d.). "Philippines," https://www.globalforestwatch.org/dashboards/country/PHL/ [Accessed January 24, 2022].

Go, C. O. (2016) "Kababayen-an han Karak-an (women of storm surges): A feminist ethnographic research on waray women survivors of super typhoon Haiyan," Masters thesis, Gender, Race, Sexuality and Social Justice, Faculty of Arts, The University of British Columbia, Canada. http://hdl.handle.net/2429/57666 [Accessed January 24, 2022].

Gobierno de España. (2013) "Spanish cooperation responds to the humanitarian emergency caused by typhoon 'Yolanda' in the Philippines," press release 211, November 11. http://www.exteriores.gob.es/Portal/en/SalaDePrensa/NotasdePrensa/Paginas/Articulos/20131111_NOTA211.aspx [Accessed January 27, 2022].

Gonzalez Rojas, A. (2019) "Vulnerability assessment using time series mapping: A case study of typhoon Haiyan in Tacloban City, Philippines," Masters thesis, Harvard University, USA. http://nrs.harvard.edu/urn-3:HUL.InstRepos:42004236 [Accessed January 24, 2022].

Gould, R. (2007) *Disaster Archeology*, Salt Lake City, UT: University of Utah Press.

Grantham Research Institute on Climate Change and the Environment. (2010) "Philippine Disaster Reduction and Management Act (RA 10121)," https://climate-laws.org/geographies/philippines/laws/philippine [Accessed January 26, 2022].

Gutierrez, J. (2019) "Gina Lopez, who led crackdown on mines in the Philippines, dies at 65," *The New York Times*, August 19. https://www.nytimes.com/2019/08/19/obituaries/gina-lopez-philippines-dies.html [Accessed February 2, 2022].

Hamblyn, R. (2014) *Tsunami: Nature and Culture*, London, UK: Reaktion.

Hamilton, C., C. Bonneuil, and F. Gemenne (eds) (2015) *The Anthropocene and the Global Environmental Crisis: Rethinking Modernity in a New Epoch*, New York: Routledge.

Hilotin, J. and S. dela Llata (2021) "Deadliest earthquakes in the Philippines: What you need to know," *Gulf News*, February 7. https://gulfnews.com/special-reports/deadliest-earthquakes-in-the-philippines-what-you-need-to-know-1.1612705914449 [Accessed January 22, 2022].

Hincks, J. (2016) "A brief history of US–Philippine relations," *Time*, October 26. https://time.com/4543996/history-of-us-philippine-relations/ [Accessed January 27, 2022].

Hinga, B. D. R. (2015) *Ring of Fire: An Encyclopedia of the Pacific Rim's Earthquakes, Tsunamis, and Volcanoes*, Santa Barbara, CA: ABC CLIO.

Humanitarian Practice Network (HPN). (2001) "A critical view of Spanish humanitarian aid," April. https://odihpn.org/magazine/a-critical-view-of-spanish-humanitarian-aid/ [Accessed January 27, 2022].

IFRC (International Federation of Red Cross and Red Crescent Societies). (2022) "What is a disaster?" https://www.ifrc.org [Accessed January 24, 2022].

InCorp Philippines. (2021) "What to know about foreign direct investments in the Philippines for 2021," Business in the Philippines, February 17. https://kittelsoncarpo.com/foreign-direct- investments-in-the-philippines/ [Accessed January 28, 2022].

Jaymalin, M. (2017) "Philippines 13th most populated country," *PhilStar Global*, July 27. https://www.philstar.com/headlines/2017/07/27/1722195/philippines-13th-most-populated- country [Accessed January 24, 2022].

Jeschonnek, L., P. Mucke, J. Walter, and L. Kirch (eds) (2016) *WorldRiskReport 2016*, Berlin and Bonn, Germany: Bündnis Entwicklung Hilft (Alliance Development Works) and United Nations University – Institute for Environment and Human Security (UNU-EHS). https://reliefweb.int/sites/reliefweb.int/files/resources/WorldRiskReport2016_small.pdf [Accessed September 17, 2021].

Jeschonnek, L., L. Kirch, and P. Mucke (eds) (2017) *WorldRiskReport Analysis and Prospects 2017*, Berlin, Germany: Bündnis Entwicklung Hilft. https://reliefweb.int/sites/reliefweb.int/files/resources/WRR_2017_E2.pdf [Accessed September 17, 2021].

Jha, S., A. Martinez, P. Quising, Z. Ardaniel, and L. Wang (2018) "Natural disasters, public spending, and creative destruction: A case study of the Philippines," ADBI Working Paper Series No. 817, Asian Development Bank Institute, March. https://www.adb.org/sites/default/files/publication/408351/adbi-wp817.pdf [Accessed January 26, 2022].

Junio, A. (2017) "After the wrath of typhoon Haiyan: Unveiling representation of vulnerabilities as beneficiaries of the emergency shelter assistance for survivors in the Philippines," Masters thesis, Development Studies, Human Rights, Gender and Conflict Studies, International Institute of Social Studies, Erasmus University, Netherlands. https://thesis.eur.nl/pub/41655 [Accessed January 24, 2022].

Keller, E. A. and D. E. DeVecchio (2016) *Natural Hazards: Earth's Processes as Hazards, Disasters, and Catastrophes*, 4th ed., London and New York: Routledge.

Kirch, L., P. Mucke, and J. Walter (eds) (2018) *WorldRiskReport 2018*, Berlin and Bochum, Germany: Bündnis Entwicklung Hilft and Ruhr University Bochum – Institute for International Law of Peace and Armed Conflict (IFHV). https://reliefweb.int/sites/reliefweb.int/files/resources/WorldRiskReport-2018.pdf [Accessed September 17, 2021].

Kirch, L., P. Mucke, and J. Walter (eds) (2019) *WorldRiskReport 2019*, Berlin and Bochum, Germany: Bündnis Entwicklung Hilft – Gemeinsam für Menschen in Not e.V. and Institute for International Law of Peace and Armed Conflict (IFHV) Ruhr University Bochum (RUB. https://reliefweb.int/sites/reliefweb.int/files/resources/WorldRiskReport-019_Online_english.pdf [Accessed September 17, 2021].

La Moncloa. (2013) "Spanish cooperation agency responds to humanitarian emergency caused by typhoon Yolanda in Philippines," November 11. https://www.lamoncloa.gob.es/lang/en/gobierno/news/Paginas/2013/20131111_Philippines.aspx [Accessed January 27, 2022].

Leroy, S. A. G. (2020) "Natural hazards, landscapes and civilizations," *Reference Module in Earth Systems and Environmental Sciences*, July 30. DOI: 10.1016/B978-0-12-818234-5.00003-1 [Accessed January 23, 2022].

Lopez, A. (2021) "Strong political will is Duterte's biggest legacy: Corvera," *Philippine News Agency*, September 30, https://www.pna.gov.ph/articles/1155205 [Accessed February 2, 2022].

Macaranas, J. R. G. (2021) "Understanding folk religiosity in the Philippines," *Religions* 12(10): 800. https://doi.org/10.3390/rel12100800 [Accessed January 26, 2022].

Mapa, D. S. (2021) "Proportion of poor Filipinos registered at 23.7 percent in the First semester of 2021," Philippine Statistics Authority, December 17. https://psa.gov.ph/content/proportion-poor-filipinos-registered-237-percent-first-semester-2021 [Accessed January 24, 2022].

Mizutori, M. (2020) "Time to say goodbye to 'natural' disasters," *PreventionWeb*, UNDRR, July 16. https://www.preventionweb.net/blog/time-say-goodbye-natural-disasters [Accessed January 23, 2022].

Morris, H. (2018) "What's happening in Boracay, the island paradise ruined by tourism?," *The Telegraph*, September 13. https://www.telegraph.co.uk/travel/destinations/asia/philippines/articles/boracay-closure-when-will-isl and-reopen/ [Accessed January 26, 2022].

NCAA (National Commission for Culture and the Arts) (n.d.) "Ethnic differentiation," Republic of the Philippines. https://ncca.gov.ph/about-culture-and-arts/culture-profile/glimpses-peoples-of-the-philippines/eth nic-differentiation/ [Accessed January 26, 2022].

NICCDIES (National Integrated Climate Change Database and Information Exchange) (n.d.) "Climate reports," https://niccdies.climate.gov.ph/clim ate-reports [Accessed January 26, 2022).

Nagle, G. (1998) *Hazards*, Cheltenham, UK: Nelson Thornes.

Newhall, C., J. W. Hendley II, and P. H. Stauffer (2005) "The cataclysmic 1991 eruption of Mount Pinatubo, Philippines," US Geological Survey Fact Sheet 113–197. https://pubs.usgs.gov/fs/1997/fs113-97/ [Accessed January 15, 2022].

Noy, I. (2017) "Assessing the vulnerability and resilience of the Philippines to disasters," PhD thesis, School of Economics and Finance, Victoria University of Wellington, New Zealand, http://creativecommons.org/licenses/by-nc-sa/3.0/nz/ [Accessed January 24, 2022].

OCHA (UN Office for the Coordination of Humanitarian Affairs). (2013) "Asia-Pacific Region 9–15 July, 2013, Natural Disasters and Other Events Being Monitored by the OCHA Regional Office for the Asia-Pacific". *Reliefweb*, July 15, https://reliefweb.int/report/india/asia-pacific-reg ion-9-15-july-2013-natural-disasters-and-other-events-being-monitored [Accessed January 23, 2022].

Park, J., E. Kwon, E. Chung, H. Kim, B. Battogtokh, and N. C. Woo (2019) "Environmental sustainability of open pit coal mining practices at Baganuur, Mongolia," *Sustainability* 12: 248. doi:10.3390/su12010248 [Accessed January 26, 2022].

Patajo-Kapunan, L. (2019) "Department of disaster resilience," *Business Mirror*, April 29, https://businessmirror.com.ph/2019/04/29/departm ent-of-disaster-resilience/ [Accessed January 26, 2022].

Philippine Daily Inquirer. 2021. "Proposed department for disaster response pushed". December 26, https://newsinfo.inquirer.net/1532213/proposed-department-for-disaster-response-pushed [Accessed January 26, 2022].

Philippine Institute of Volcanology and Seismology (PHIVOLCS). (2018) "Destructive earthquakes in the Philippines," https://www.phivolcs.dost.gov.ph/index.php/earthquake/destructive-earthquake-of-the-philippines [Accessed January 22, 2022].

PSA (Philippine Statistics Authority). (2021) "Highlights of the population density of the Philippines 2020 Census Population and Housing (2020 CPH)," July 23. https://psa.gov.ph/content/highlights-population-density-philippines-2020-census-population-and-housing-2020-cph [Accessed January 24, 2022].

Rappler. (2017) " 'I would rather follow Gina Lopez,' Declares Duterte," March 11, https://www.rappler.com/nation/163910-duterte-support-gina-lopez-after-ca-grilling/ [Accessed February 2, 2022].

Rey. A. (2015) "Timeline: PH policies on climate change and disaster management". *Rappler*, November 1, https://www.rappler.com/moveph/111307-climate-change-disaster-management-policies-philippines/ [Accessed January 26, 2022].

Rocamora, J. A. L. (2019) "PH ranks 8th among travelers' favorite countries in the world," *Philippine News Agency*, October 23. https://www.pna.gov.ph/articles/1083943 [Accessed January 26, 2022].

Romero, A. (2013) "CNN's Anderson Cooper tells Korina: Go to Tacloban," *The Philippine Star*, November 15, https://www.philstar.com/headlines/2013/11/15/1257065/cnns-anderson-cooper-tells-korina-go-tacloban [Accessed January 27, 2022].

Santos, C. T. (2018) "The impact of tropical cyclone Hayan in the Philippines: Contribution of spatial planning to enhance adaptation in the city of Tacloban," PhD thesis, Climate Change and Sustainable Development Policies, Faculty of Science, University of Lisbon, Portugal. https://repositorio.ul.pt/bitstream/10451/35141/1/ulsd732258_td_Carlos_Santos.pdf [Accessed January 26, 2022].

Sarmiento, B. S. (2018) "Philippines: Local groups demand companies to pay for pollution caused by plastics," Business & Human Rights Resource Centre, August 12, https://www.business-humanrights.org/en/latest-news/philippines-local-groups-demand-companies-to-pay-for-pollution-caused-by-plastics/ [Accessed January 28, 2022].

Smith, N. (2006) "There's no such thing as a natural disaster," *items, Insights from the Social Sciences*, Social Science Research Council, June 11. https://items.ssrc.org/understanding-katrina/theres-no-such-thing-as-a-natural-disaster/ [Accessed January 23, 2022].

Statista. (2021a) "Population density of the Philippines from 2007 to 2020," Statista Research Department, November 18, https://www.statista.com/statistics/778517/philippines-population-density/ [Accessed January 24, 2022].

Statista. (2021b) "Expenditures of the government on disaster risk reduction in the Philippines from 2009 to 2017," Statista Research Department, December 16, https://www.statista.com/statistics/1090427/philippines-government-expenditures-disaster-risk-reduction/ [Accessed January 26, 2022].

Statista. (2021c) "Leading foreign investors in the Philippines in 2020, by value of investment," Statista Research Department, September 10, https://www.statista.com/statistics/1091420/philippines-leading-foreign-investors-by-investment/ [Accessed January 28, 2022].

Su, Y. (2019) "Remittances after disasters: Case study of Tacloban City, Philippines after typhoon Haiyan," PhD thesis, Department of Political Science, University of Guelph, Canada. http://hdl.handle.net/10214/17475 [Accessed January 24, 2022].

Talavera, C. (2022) "Foreign businesses laud lifting of ban on open-pit mining," *The Philippine Star*, January 6, https://www.business-humanrights.org/en/latest-news/philippines-local-groups-demand-companies-to-pay-for-pollution-caused-by-plastics/ [Accessed January 28, 2022].

Tope, L. R. and D. P. Nonan-Mercado (2002) *Cultures of the World: Philippines*, Tarrytown, NY: Marshall Cavendish Benchmark.

UNDRR. (2019) "Disaster risk reduction in the Philippines: Status report 2019," Bangkok, Thailand, United Nations Office for Disaster Risk Reduction (UNDRR), Regional Office for Asia and the Pacific. https://www.unisdr.org/files/68265_682308philippinesdrmstatusreport.pdf [Accessed August 30, 2021].

UNICEF. (2022) "2013 Philippines earthquake," https://www.unicefusa.org/mission/emergencies/earthquakes/philippines-2013 [Accessed January 22, 2022].

United Nations. (n.d.) "Philippines National Disaster Risk Reduction and Management Council (NDRRMC)," Office for Outer Space Affairs, UN-SPIDER Knowledge Portal, https://www.un-spider.org/philippines-national-disaster-risk-reduction-and-management-council-ndrrmc [Accessed January 26, 2022].

USAID. (2022) "Philippines," https://www.usaid.gov/humanitarian-assistance/philippines [Accessed January 26, 2022].

Van Lint, S. (2016) "Sense-making of the Netherlands Red Cross priority index model: Case: Typhoon Haiyan, Philippines," Masters thesis, MSc International Development Studies, Sociology-Disaster and Conflict Studies, Wageningen University, Netherlands, https://edepot.wur.nl/421814 [Accessed January 24, 2022].

Vokaty, M. (2014) "The Philippine Sea Plate," https://tectonicsofasia.weebly.com/philippine-sea-plate.html [Accessed January 15, 2022].

White House. (2013) "Fact sheet: U.S. response to typhoon Haiyan," Office of the Press Secretary, November 19, https://obamawhitehouse.archives.gov/the-press-office/2013/11/19/fact-sheet-us-response-typhoon-haiyan [Accessed January 27, 2022].

World Bank. (2022) "Population, total–Philippines," https://data.worldbank.org/indicator/SP.POP.TOTL?locations=PH [Accessed January 24, 2022].

2

Socio-Economic Impact of Disasters in the Philippines: The First Two Decades of the 21st Century

Roberto Ariel Abeldaño Zuñiga

Introduction

The Philippines is an island country in Southeast Asia surrounded by the Pacific Ocean. An archipelago made up of 7,107 islands, the Philippines' location in the Pacific Ring of Fire make it a country prone to earthquakes, while its tropical climate makes it vulnerable to typhoons (Alcayna et al et al 2016). In 2017, the Philippines had an estimated population of 102 million inhabitants, which placed it as the 12th most populous country in the world (World Bank 2019). Its biodiversity is one of the world's richest, and its growing industrialization process has led to the replacement of agricultural activity by the manufacture of industrialized products and services (World Bank 2019). However, the Philippines faces numerous challenges linked to, among other things, the lack of access to education, health services, and human development for large sectors of its population (United Nations Development Program 2019, 76).

The Philippines is one of the top three countries with the highest exposure to disaster risks (UN Office for Disaster Risk Reduction 2019, 6), among which typhoon Haiyan stood out in November of 2013. The Haiyan disaster resulted in 7,354 deaths, affected 16,106,870 people (requiring immediate assistance), and caused an estimated economic loss of more than US$10 billion (Centre for Research on the Epidemiology of Disasters 2019). There are also earthquakes and permanent volcanic activity in the Philippines due to the presence of 53 volcanoes, including the Mayón Volcano, which is one of the most active in the world (Centre for Research on the Epidemiology of Disasters 2019). These major disasters have cumulative impacts on the population, not only in terms of deaths but also in their social and economic effects.

In economic terms, the average annual loss due to multiple risks for the Philippines is estimated at over US$7 billion, which is equivalent to 69% of the country's social spending in 2011 (Alcayna et al 2016, 2). Although the Philippines recorded a GDP of US$330.91 billion that placed it in 39th position in the world in 2018 (World Bank 2019), the United Nations Human Development Report placed it in the 106th place in 2018, with a Human Development Index (HDI) of 0,712 (United Nations Development Program 2019, 301). Despite it being considered a high HDI, there are deep inequalities in its population; for example, 21.6% of its population lived in poverty in 2017 (United Nations Development Program 2019, 321).

Disasters seriously compromise the sustainable development of nations; they cause loss of lives and properties that hit less economically developed countries harder, particularly in countries where there are deep socio-economic inequalities (Caballeros Otero and Zapata Martí 1994; Adger et al 2003; Swart et al 2003; Daoud et al 2016). Hence, the United Nations has outlined the Sustainable Development Goals (SDGs) that operate in line with pre-existing international treaties on disaster risk reduction. One of the goals set in the SDGs involves supporting the most vulnerable regions to adapt to climate change by integrating disaster risk reduction measures into national policies and strategies, and limiting the increase in global average temperature to 2°C above pre-industrial levels (United Nations 2015). This highlights the complex network of relationships between climate change, disasters, and sustainable development in less economically developed countries. Some authors have developed the idea that disasters affect the development of nations in two ways: a disaster can lead to poverty, and that poverty generates conditions of vulnerability in dealing with disasters (Lopez-Calva et al 2009; Daoud et al 2016). The complexity of these two-way relationships creates a framework that becomes difficult to measure empirically. For the Philippines, this is particularly important since its geographical characteristics make it very prone to frequent high-intensity disasters, mainly from hydrometeorological hazards.

This chapter aims to provide an overview of the socio-economic impact of disasters in the Philippines in the first two decades of the 21st century and to document examples of disaster risk reduction in the country. In the following sections, estimates of the socio-economic impact of disasters, vulnerability to disasters, and strategies for disaster risk reduction in the Philippines are discussed.

Estimates of the socio-economic impact of disasters in the Philippines

Major disasters cause considerable loss of life, affect large numbers of the population (due to injuries and losses of homes), a large number of economic losses, and population displacement when families lose their homes (Abeldaño Zúñiga and González Villoria 2018). Focusing on an economic point of view, damage caused by disasters can be summarized by the following three categories: direct damages, indirect damages, and secondary effects of disasters (Economic Commission for Latin America and the Caribbean 2003; Hochrainer 2006). *Direct damages* refer to capital assets and direct assets of individuals, families, and institutions that are affected immediately at the time of the disaster. *Indirect damages* are a result of the cessation of household income and income of public or private institutions; the interruption in the production of goods and services due to a chain of effects; and the extraordinary expenses that are intended to provide immediate assistance and emergency (provisional) services to the affected population. *Secondary effects of disasters* are those that affect the macroeconomy, such as the increased rates of inflation, the fall in economic growth, the imbalances in the balance of payments, the flight of investment capital, the increase in public spending, the decrease in international reserves, the loss of agricultural land, and the increase in poverty and indicators of inequality, among others. Some indirect costs like damage to cultural heritage cannot be measured in monetary terms.

The Centre for Research on the Epidemiology of Disasters (CRED) records the impact of disasters on the population based on the following variables: deaths, missing people, and total affected population. The CRED records the total estimated damages of economic impact using the current value of the US dollar. This variable expresses the value of all economic damages and losses directly or indirectly related to the disaster (Centre for Research on the Epidemiology of Disasters 2019). In the Philippines, typhoon Haiyan caused a large number of human losses and economic damage in 2013. It also caused persistent damage to basic health services infrastructure, resulting in the disappearance of important infrastructure segments (hospitals) in the affected communities in the islands of Leyte and Bohol, among others. Indirect effects are added to the infrastructure losses, which are the costs that the government faces to prevent consequent health effects, such as epidemics that arose after the disaster (Murray et al 2013; Shilkofski et al 2016; Gil Cuesta et al 2019).

From 2000 to 2019, a total of 384 events of diverse origins were recorded in the Emergency Events Database (EM-DAT) of the Université Catholique de Louvain (Belgium). Disasters are classified as climate-related, which includes meteorological, hydrological, climatological, and biological hazards;

and non-climate-related disasters, including geophysical and technological hazards (Table 2.1).

Of those 384 recorded events, 72.7% were climate-related hazards (caused by meteorological, hydrological, climatological, or biological hazards) and the remaining 27.3% were events of geophysical or technological origin. The proportion of climate-related events has been increasing over the years (see Figure 2.1). This trend has also been observed for several years in other countries (Abeldaño Zúñiga and González Villoria 2018; Abeldaño Zuñiga 2019).

In the first two decades of the 21st century, the Philippines has faced, on average, 19 disaster situations per year. In 2011, it recorded the highest number with 41 events, followed by 2009 with a total of 29 events (see Figure 2.1) (Centre for Research on the Epidemiology of Disasters 2019). One of the worst years in the history of disasters for the Philippines occurred in 2013 with typhoon Haiyan (Yolanda), although there were also 12 other climate-related events and 3 non-climate-related events during that year (Centre for Research on the Epidemiology of Disasters 2019).

Table 2.1: Hazard classifications based on their origin

Hazard groups		Hazard subgroups
Climate-related	Meteorological	Storms
		Cyclones and hurricanes
		Extreme temperatures
	Hydrological	Flood
		Landslide
		Wave action
	Climatological	Drought
		Wildfire
	Biological	Infectious diseases (epidemics)
Non-climate-related	Geophysical	Earthquakes
		Volcanic activity
		Mass movement
	Technological incidents	Industrial
		Transport
		Miscellaneous

Source: Centre for Research on the Epidemiology of Disasters (2019).

Figure 2.1: Recorded disasters in the Philippines in the period 2000–2019

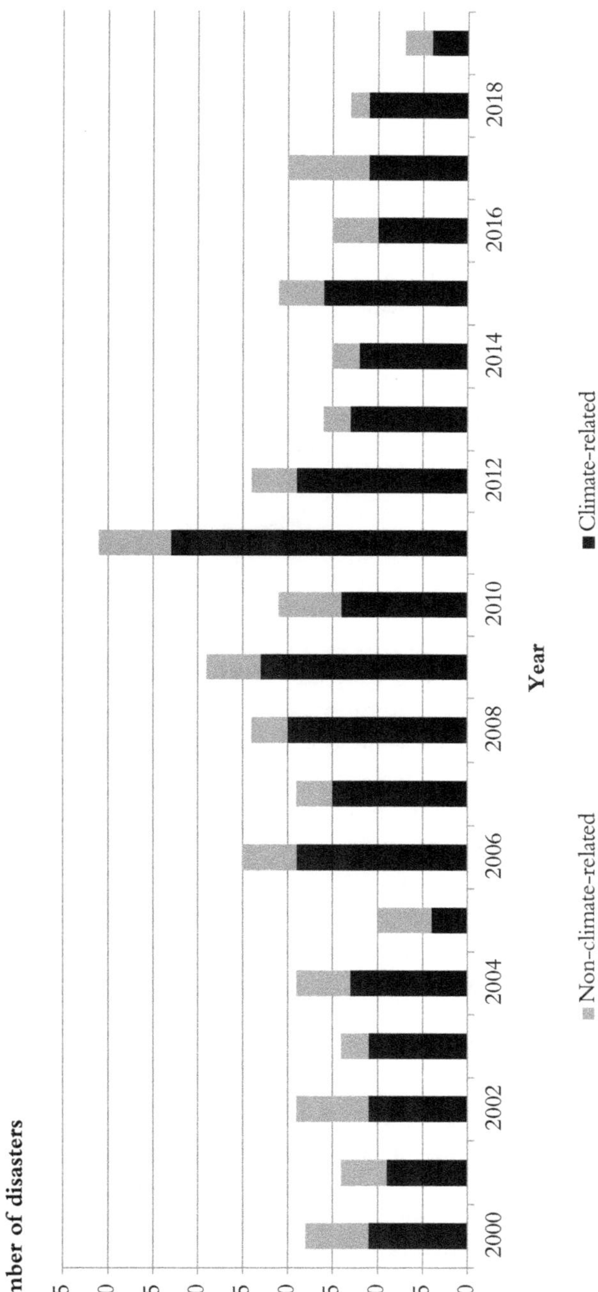

Source: Centre for Research on the Epidemiology of Disasters (2019).

Direct effects of disasters

Table 2.2 shows the direct effects of disasters on the population over the last two decades in the Philippines. An estimated 88.2% of the total 28,086 disaster-related deaths over 20 years were caused by climate-related events, and 11.8% of the remaining deaths were caused by non-climate-related events. The number of persons affected by disasters is most alarming. With a cumulative total of over 142 million people affected during the period (see Table 2.2) in a country with a total population of 102 million, the impact of disasters is recurring in some sectors with the most vulnerable population.

Disasters are hitting the most vulnerable households repeatedly. In this context, the poorest populations in the rural areas of the Philippines witness typhoons, floods, and earthquakes repeatedly destroying their homes, causing unhealthy conditions, and ruining crops and fishing boats that affect livelihood and well-being, thereby eliminating opportunities to escape poverty. These result in a recurring cycle of vulnerability and poverty (Daoud, Halleröd and Guha-Sapir 2016; Gil Cuesta et al 2019). While the Philippines is a country with a great wealth of biodiversity and vast resources to enable its population to recover, the frequency of disasters results in a constant rebuilding of the lives among the people.

Between 2000 and 2019, a total of 142,739,817 people were affected by a disaster situation; 96.9% were affected by a climate-related event (see Table 2.2), indicating that Filipinos are highly exposed to vulnerabilities and risks of climate change. The reported data of the fateful year 2013 for the Philippines was 7,909 deaths (95.1% of them caused by climate-related disasters), and more than 25 million were injured.

From 2010 to 2019, three typhoons (Haiyan, Bopha, and Washi) were among the 10 disasters that caused the highest number of deaths in the Philippines, each exceeding a thousand. Figure 2.2 shows the events according to their origin: technological, biological, hydrological, geophysical, and climatological. This figure highlighted the participation of hydrological hazards (that is, floods, landslides, and wave action) that represented 27.9% of the total, and those of meteorological origin (storms, extreme temperatures, cyclones, and hurricanes) that represented 40.6% of the total.

Epidemics have been described as a recurring post-disaster problem in the Philippines and other less economically developed countries (Adeagbo et al 2016; Alcayna et al 2016; Shilkofski et al 2016; Lagmay et al 2017; Gil Cuesta et al 2019; Murray et al 2019). This is also shown in Figure 2.2, where the situations of biological disasters caused by infectious disease epidemics represent 6.5% (mainly in 2013). Among the epidemics that have affected the Philippines in the last 20 years are acute diarrheal syndrome, dengue, cholera, and measles (Centre for Research on the Epidemiology

Table 2.2: Total deaths and total affected people by year and type of hazard in the Philippines in the period 2000–2019

Year	Total deaths					Total affected				
	Climate-related hazards		Non-climate-related hazards		Total	Climate-related hazards		Non-climate-related hazards		Total
	n	%	n	%	n (100%)	n	%	n	%	n (100%)
2000	737	67.6	354	32.4	1,091	6,356,576	98.9	68,473	1.1	6,425,049
2001	630	85.4	108	14.6	738	3,541,737	97.7	83,272	2.3	3,625,009
2002	305	57.9	222	42.1	527	1,137,761	93.9	73,527	6.1	1,211,288
2003	352	71.8	138	28.2	490	687,761	99.9	203	0.1	687,964
2004	1,950	91.5	180	8.5	2,130	3,263,076	99.4	21,108	0.6	3,284,184
2005	39	33.6	77	66.4	116	213,057	97.8	4,750	2.2	217,807
2006	2,984	96.9	97	3.1	3,081	8,568,968	99.1	76,492	0.9	8,645,460
2007	129	66.2	66	33.8	195	2,009,056	98.6	28,187	1.4	2,037,243
2008	959	51.2	914	48.8	1,873	8,459,896	99.9	5,174	0.1	8,465,070
2009	1,307	93.0	99	7.0	1,406	13,304,955	99.6	48,486	0.4	13,353,441
2010	1,113	92.4	92	7.6	1,205	5,567,346	99.7	16,195	0.3	5,583,541
2011	1,989	97.4	53	2.6	2,042	11,695,988	99.5	63,997	0.5	11,759,985
2012	2,301	93.7	156	6.3	2,457	12,178,289	97.2	353,160	2.8	12,531,449
2013	7,520	95.1	389	4.9	7,909	22,444,909	87.4	3,222,873	12.6	25,667,782
2014	331	82.1	72	17.9	403	13,214,113	99.5	61,158	0.5	13,275,271

(continued)

Table 2.2: Total deaths and total affected people by year and type of hazard in the Philippines in the period 2000–2019 (continued)

Year	Total deaths					Total affected				
	Climate-related hazards		Non-climate-related hazards		Total	Climate-related hazards		Non-climate-related hazards		Total
	n	%	n	%	n (100%)	n	%	n	%	n (100%)
2015	201	60.2	133	39.8	334	4,019,201	98.6	57,645	1.4	4,076,846
2016	79	88.8	10	11.2	89	5,534,610	99.8	9,202	0.2	5,543,812
2017	222	67.1	109	32.9	331	4,808,048	98.5	73,738	1.5	4,881,786
2018	832	98.8	10	1.2	842	11,196,885	99.2	86,056	0.8	11,282,941
2019	789	95.4	38	4.6	827	158,757	86.3	25,132	13.7	183,889
Total	24,769	88.2	3,317	11.8	28,086	138,360,989	96.9	4,378,828	3.1	142,739,817

Source: Centre for Research on the Epidemiology of Disasters (2019).

of Disasters 2019). All these diseases are related to the current patterns of climate change in less economically developed countries (Yang et al 2014; Metcalf et al 2017).

The Philippines is included in the area formed by the Pacific Ring of Fire, which is characterized by a concentration of some of the most important subduction zones in the world, causing intense seismic and volcanic activity in the area. However, the incidence of events of geophysical origin is relatively low with 3.1% of the total of events recorded by the CRED in the period 2000–2019 (see Figure 2.2).

Economic damages were estimated at a constant price based on the year 2010. In the CRED databases, economic damages caused by disasters are expressed in US$ at current value. To express them in constant values (and make temporary comparisons), these values were deflated using 2010 as the base year, and as a deflator the Philippine Consumer Price Index (CPI) published by the World Bank (2019). Between 2000 and 2019, the total economic damages amounted to a total of US$35,605,243,000, in which US$18,229,233,000 was recorded in 2013 and US$11,051,278,000 in 2015 (see Figure 2.2). The annual economic loss due to disasters was on average US$1,780,262,000. The GDP per capita at constant prices (based on the year 2010) for the year 2000 was US$1,600, which steadily increased to reach US$3,021 in 2018 (World Bank 2019).

Figure 2.1 shows the incidence of disasters hazard origin and economic damages, by year, between 2000 and 2019 in the Philippines. Here, the five events that caused the most significant economic damage in the last two decades were typhoon Haiyan (Yolanda) in 2013, typhoon Chan-Hom in 2015, the floods in August of 2013, typhoon Koppu (Lando) in 2015, and typhoon Bopha in 2012 (Centre for Research on the Epidemiology of Disasters 2019).

Secondary effects of disasters

Forced displacement of the population is another secondary effect of disasters. Globally, the phenomenon of forced displacement due to disasters is increasing, partly due to the increase in disasters related to climate change. Kälin (2010) points out that in 2008, there were 36 million people displaced within their own countries due to a sudden-onset disaster. The Philippines is no stranger to this trend since in recent decades there have been major disaster situations that have caused large numbers of internal displacements. The Internal Displacement Monitoring Centre (IDMC, Norway 2019), data reveals that between the years 2008 and 2018 there has been a cumulative total of 39,469,642 internal population displacements due to disasters, although the data are probably underestimated due to the lack of registration of displacements in some rural zones (see Figure 2.3).

Figure 2.2: Number of disasters by hazard origin and economic damages by disasters in the Philippines in the period 2000–2019

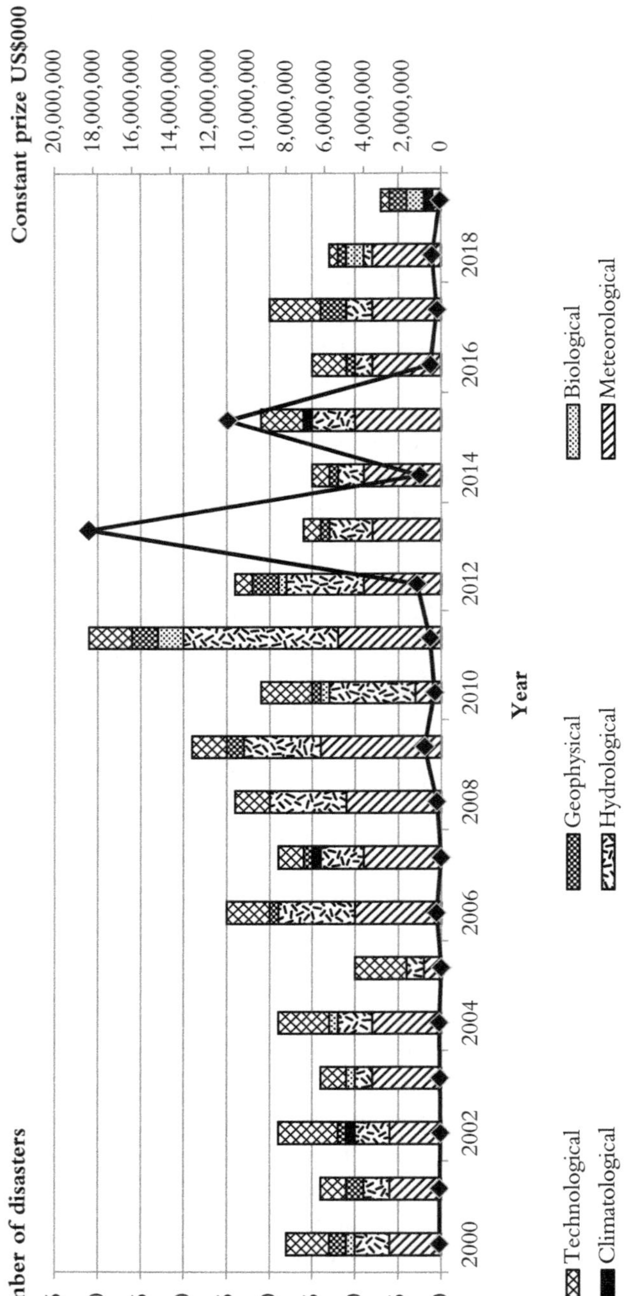

Source: Centre for Research on the Epidemiology of Disasters (2019).

Figure 2.3: Number of internally displaced people by disasters in the Philippines in the period 2008–2018

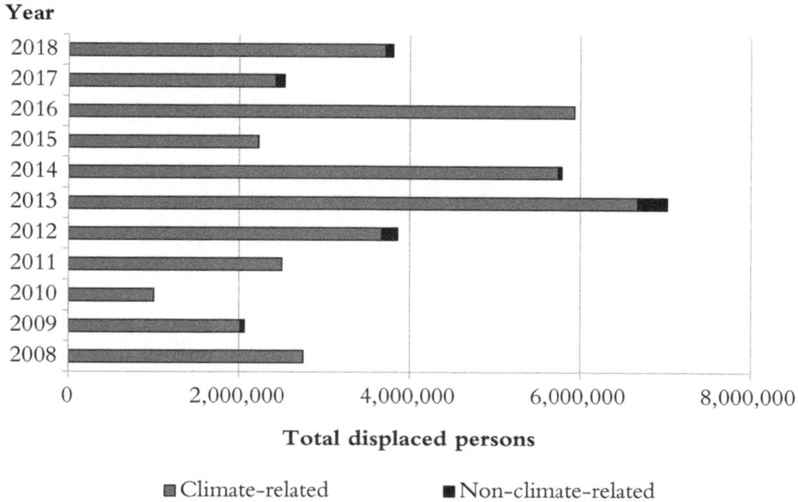

Source: Internal Displacement Monitoring Centre (2019).

During the period 2008–2018, disasters have displaced many people, equivalent to one-third of the total population in the Philippines. Every year there has been an average of 3.5 million internally displaced Filipinos. The number of displaced persons in the Philippines in a year is equivalent to the total annual internal displacements due to disasters in all 33 countries of Latin America and the Caribbean (Abeldaño Zuñiga and Fanta Garrido 2020). Between 2008 and 2018, 97.9% of the displacements registered in the IDMC Global Database were caused by climate-related events, among which were the 7 million people displaced in 2013 and the 5.9 million people displaced in 2016 (see Figure 2.3).

Vulnerability to disasters in the Philippines

Once the social and economic impact of disasters on the Philippines is analyzed, one wonders why the population in this country is repeatedly affected by disasters. This section analyzes socio-economic variables that make up the social vulnerability of the population when facing a disaster situation.

The concept of social vulnerability implies that a disaster is a situation that cannot be considered as 'natural,' in part because its effects can be counteracted by strong institutional mechanisms, such as good governance. A threat of natural origin becomes a disaster situation when it interacts with a population that is vulnerable when confronting that threat. For example, if a storm strikes a well-built house, the storm probably does not result in a disaster situation, but if instead, the storm strikes a home without a roof, the family facing that storm will experience a real disaster during the days that follow.

Daoud et al (2016) argue that the effects of disasters on communities can contribute to worsening poverty, but life in conditions of poverty can also increase vulnerability to natural hazards, as shown in the example above. Assuming that governance is synonymous with efficiency and quality in decision-making interventions by the state, Oliver-Smith et al (2017) argue that good governance could reverse or prevent the negative impacts of disasters on populations. On the contrary, Daoud et al (2016) argue that the lack of governance has a negative effect, probably over-aggregated, on the effects of disasters on different dimensions of the population such as health, education, and mortality, among others. Thus, a third element can be integrated into the cycle of disasters and poverty: governance in terms of disaster risk management.

Climate change trends in the Philippines show that average temperatures in the Southeast Pacific have increased approximately 1°C since 1970, with an average rate of 0.3°C per decade, although temperatures appear to be rising more rapidly in the southern area of the archipelago (World Bank Climate Knowledge Portal 2019). In the islands of the archipelago, the average annual temperature increased by 0.14°C between 1971 and 2000. Recent evidence suggests a tendency to more humid conditions since the number of rainy days in the Philippines has increased since the 1990s, as has the annual variability of the beginning of the rainy season. The number of Category 4 and 5 storms in the Pacific region has doubled since the 1970s, which means that the frequency of cyclones that have entered its territory has also increased (World Bank Climate Knowledge Portal 2019). The carbon dioxide (CO_2) emissions have increased from 73,307ktn in 2000 to 105,654ktn in 2014 (see Table 2.3).

Problems related to economic development are a crucial point in the vulnerability analysis of the Philippine population. It is known that at least 1,000 of the 7,107 islands that make up the archipelago are uninhabitable due to their geographical formations. According to World Bank (2019) data, more than 60% of the population lives in coastal areas, and 5% of the total population are in areas where the elevation is less than 5m. This is an important dimension of vulnerability, taking into account that typhoon Haiyan registered winds of more than 300kph, raising waves to 15m high.

Regarding the socio-economic variables that can be used as a proxy for social vulnerability, the Philippines Human Development Index[1] between 2000 and 2017 placed this country in an average human development rating between 0.50 and 0.70 (United Nations Development Program 2019, 305). However, poverty affected one in four people in 2003, although it slowly

Table 2.3: Socio-economic indicators in the Philippines in the period 2000–2018

Year	HDI	GDP per capita (US$)	Inflation (%)	Poverty (%)	Forestlands (%)	Rainfall (mm)	Average annual temperature (°C)	CO_2 emissions (ktn)
2000	0.624	1607	5.7	ND	23.6	3013.8	25.9	73,307
2001	0.628	1619	5.5	ND	23.6	2964.1	26.0	71,052
2002	0.633	1643	4.2	ND	23.6	2352.0	25.9	71,338
2003	0.637	1690	3.2	24.9	23.7	2701.7	25.8	71,426
2004	0.647	1767	5.5	ND	23.7	2311.6	25.9	74,066
2005	0.650	1817	5.8	ND	23.7	2436.5	26.0	74,832
2006	0.651	1878	4.9	26.6	23.6	2413.3	26.0	67,693
2007	0.657	1969	3.1	ND	23.4	2594.2	26.0	72,170
2008	0.661	2017	7.5	ND	23.3	3112.4	25.6	78,859
2009	0.659	2007	2.8	26.3	23.1	2777.7	25.6	77,568
2010	0.665	2124	4.2	ND	22.9	2430.4	25.9	84,869
2011	0.670	2165	4.0	ND	23.7	3213.6	25.7	85,496
2012	0.677	2271	2.0	25.2	24.6	2994.4	26.1	91,206
2013	0.685	2390	2.0	ND	25.4	2592.2	26.1	98,129
2014	0.689	2496	3.2	ND	26.2	2374.1	25.9	105,654
2015	0.693	2605	-0.6	21.6	27.0	2281.3	26.1	ND
2016	0.696	2743	1.7	ND	27.8	2432.7	26.5	ND
2017	0.699	2884	2.3	ND	ND	ND	ND	ND
2018	ND	3022	3.8	ND	ND	ND	ND	ND

ND: No data available.

Source: World Bank (2019) and United Nations Development Program (2019).

decreased by 21.6% in 2015 (World Bank 2019). Although GDP almost doubled from 2000 to 2018, the other indicators infer that there are still deep social inequalities in the Philippines, since the Gini index[2] reflects a value above 40% (World Bank 2019), which is considered as an unequal distribution of income. Consequently, social inequality here is considered a situation of social vulnerability in the face of disasters.

Inflation in the Philippines is relatively low, although it is observed that it increased by 1.2 percentage points in 2014 (World Bank 2019), compared to 2013 when typhoon Haiyan hit. Regarding social vulnerability, development, governance and disasters, Oliver-Smith (2017) notes that disasters are problems generated by *unresolved problems of development*, such as lack of social investment by governments, environmental degradation, and the unequal distribution of wealth, among other dimensions; in other words, he attributes causality to asymmetric development patterns.

Strategies for disaster risk reduction in the Philippines

The World Bank has estimated that close to 1 million Filipinos fell into poverty in 2013 after the passage of typhoon Haiyan, which destroyed more than 1 million homes and undermined the national economy, causing losses of US$12.9 billion (Hallegatte et al 2017). If the effects of disasters intensify poverty, and at the same time poverty intensifies vulnerability to cope with disasters, the comprehensive disaster risk reduction strategies focused on governance must be linked to poverty reduction.

A World Bank global report (Hallegatte et al 2017) proposes a set of resilience policies and strategies that would help the poor cope with the consequences of extreme climate-related events. These strategies include early warning systems, increased access to personal banking services, insurance policies, and social protection systems related to conditional cash transfer systems and national impact public works programs, which could prepare people to respond with a better level of insurance and to recover from crises. The report suggests that national governments must make essential investments in infrastructure, dikes, and other means to control water levels, and develop appropriate land use policies and construction standards. This implies that in the Philippines, the resilience strategies should aim to protect the poorest and most vulnerable citizens since the population is highly vulnerable to the effects of disasters. The World Bank report (Hallegatte et al 2017) also highlights the efficiency of risk transfer mechanisms through the insurance of household assets.

Regarding agricultural and land use issues, which are the assets most affected by climate-related disasters, insurance strategies should involve crops, fields, homes, and movable property such as agricultural machinery, among others. In that sense, the Food and Agriculture Organization (FAO) has also suggested strategies to improve land tenure regimens and insurance in rural areas with agricultural activity (Eleazar 2010).

Although the transfer of risk through insurance is included in the local disaster risk management legislation dating from 2010 in the Philippines, it is clear that the implementation of national legislation at the local level has not yet reached a degree of adequate concreteness. There are international initiatives led by the European Union that have resulted in at least 1,500 households in the Philippines contracting micro-insurance paid in part with community savings funds (Barbin, Lomboy, and Soriano 2002; European Civil Protection and Humanitarian Aid Operations 2019).

Disaster risk reduction strategies

Disaster risk reduction and climate action policies have been adopted as mandatory functions of all levels of governance in the Philippines. The institutionalization of disaster risk management has become a permanent element of official governance. This has been done in consideration of the global political frameworks for disaster risk management and climate change: Sendai Framework for Disaster Risk Reduction, Sustainable Development Goals, and the Paris Climate Agreement (UN Office for Disaster Risk Reduction 2019).

The UNDRR 2019 Report cites that the Philippines has made significant progress in collecting complete and updated risk data using different technological tools, including GIS, remote sensing, early warning systems, and computer simulations. They have also collected data on population vulnerability through different national agencies and international cooperation. At the national level, technical agencies have been put in place for disaster risk management in various fields such as the Philippine Institute of Volcanology and Seismology (PHIVOLCS), the Philippines Atmospheric, Geophysical and Astronomical Services Administration (PAGASA), the National Mapping and Resource Information Authority (NAMRIA), the Mines and Geosciences Bureau (MGB), and the UP Resilience Institute.

The aforementioned agencies have technical expertise that complements each other when conducting multiple risk assessments. Performing risk assessments for volcanic eruptions, earthquakes, and tsunamis are the main technical services of PHIVOLCS. It is also in charge of earthquake impact assessments, as well as the impact of strong winds and floods in some provinces. This service, in turn, requires contributions from PAGASA, responsible for data collection, modeling and monitoring of typhoons, and other hydrometeorological hazards, along with climate projections (UN Office for Disaster Risk Reduction 2019).

The Manila Observatory is a research institute in atmospheric and earth sciences with monitoring work relevant to sustainable development and poverty reduction (Manila Observatory 2019). As stated on their website, the Manila Observatory "advocates a science-based approach to the areas of climate change and disaster science" aiming to disaster risk reduction (Manila Observatory 2019). The main monitoring areas of this observatory are remote sensing and ground-based environmental observation and monitoring; climate change, variability, and extremes; disaster risk and sustainable development; and public health and human vulnerability. Within those areas, the following elements are specifically monitored: air quality; regional climate and extreme weather events; upper atmosphere and solid earth dynamics for disaster applications; and integrated risk analysis for disasters and climate change.

The case of Albay province

A model to highlight in this chapter is the organization of Albay province in the southern Philippines for local management of disaster risks. Beginning in the year 2006, when typhoon Durian hit Albay province, the government began to organize a multisectoral system for disaster risk management that involved local government and health agencies (Province of Albay 2019; Provincial Government of Albay 2019; UN Office for Disaster Risk Reduction 2019).

In 2006, typhoon Reming devastated the province of Albay, located in the Bícol region on the island of Luzon, and affected more than 2.5 million people (Centre for Research on the Epidemiology of Disasters 2019). Daraga and Legazpi (the capital of Albay) suffered significant damage due to their lack of preparation to face a disaster of this magnitude. From 2006, the local government of Albay took measures in this regard, resolving to educate its citizens in reducing disaster risks, and preparing to face new threats of natural origin. Thus, was created the first Local Office for Disaster Risk Management in the country, the Albay Public Safety and Emergency Management Office (APSEMO) (Province of Albay 2019; Provincial Government of Albay 2019; UN Office for Disaster Risk Reduction 2019).

The objectives of creating APSEMO were: 1) to establish a proactive, permanent, comprehensive, rational system of disaster management in Albay to reduce and mitigate the effects of disasters, and 2) to institutionalize the Disaster Management office of Albay (now APSEMO) as an independent department of the Albay provincial government and serves as the technical and administrative arm of the Provincial Disaster Coordination Council (PDCC) (Province of Albay 2019; Provincial Government of Albay 2019; UN Office for Disaster Risk Reduction 2019). The office was established with the following concrete actions: 1) to formulate the Albay Disaster Management Manual, of which the Action Plans for types of threats are part; 2) to establish of the Disaster Operations Center; 3) to establish a good working relationships with PDCC and Alert Agencies (PHIVOLCS, PAGASA); 4) to develop alert criteria, communication protocol, and evacuation procedures; 5) to develop the Barangay Disaster Preparedness and Response Manual; 6) to prepare risk maps; and 7) to increase public awareness of the nature and effects of hazards through seminars and workshops (Province of Albay 2019; Provincial Government of Albay 2019; UN Office for Disaster Risk Reduction 2019).

One of the actions carried out by APSEMO, which became well known worldwide, was the establishment and institutionalization of an early warning system, a communication protocol, and preventive evacuation procedures in Albay. They also addressed issues related to health and resource management. The early warning system was one of the first to respond when typhoon Haiyan made landfall. Because of the impact of their achievements in disaster risk reduction and management, the office became an awarded team at the national level and a model of good disaster risk reduction practices at the international level. Albay's office states that "so far they have achieved zero casualties in disasters related to typhoons and volcanic eruptions" (Provincial Government of Albay 2019).

An early warning system in disaster situations operates from the generation and dissemination of warning information promptly to allow people, communities, and organizations exposed to a hazard to prepare and act appropriately and in advance. For example, in March 2011, when an earthquake of magnitude 9.0 shook the coast of Japan, more than 120,000 people living in an exposed coastal community in the Philippines were alerted to a possible tsunami. In that event, early warning systems implanted in the region allowed that, in the first five minutes after the earthquake, more than 108,000 people from 150 coastal districts were evacuated as a precaution, since they received a message on their cellphones alerting them to a possible tsunami (PNUD 2013).

Conclusions

This chapter analyzed the socio-economic impacts of disasters in the Philippines over the past 20 years. It can be concluded that the disasters that have caused the greatest toll on human life have been caused by climate-related events, and typhoons represent the most important danger in the Philippines. Another impact highlighted is population displacement, since each year the average volume of internal displacement caused by disasters in the Philippines is 3.5 million people. Based on figures about the direct effects of disasters, there are recurring impacts for the most vulnerable communities in the country. This implies not only the involvement at the individual level with injuries, but also the recurrent loss of family assets such as homes, crops, and personal property. At the community level, this also has serious implications, as described throughout the chapter, in terms of the recurrence of epidemics and nutritional alterations affecting collective health (Eleazar 2010; PNUD 2013; UN Office for Disaster Risk Reduction 2019).

Climate-related disasters have caused the greatest volume of economic losses to the Philippines over the past two decades. Disasters have mostly affected the poorest, and a cyclical relationship is woven between poverty—vulnerability—disasters (Daoud et al 2016). In the Philippines, it is also clear that economic development patterns that generate social inequalities make the poorest communities more vulnerable.

Given this panorama of the incidence of disasters in the Philippines, some recommendations for disaster risk management, and to strengthen resilience, can be focused to facilitate the access of the poorest population to the mechanisms of economic risk transfer through universal insurance, increase investment in national and local infrastructure, to expand the coverage of early warning systems to protect the population in coastal areas, the creation and maintenance of technical-scientific agencies for permanent monitoring of threats. Observatories and local disaster risk inventories have shown great potential for the production and analysis of data and their subsequent linkage to the generation of evidence-based risk management policies (Moscoso 2011; United Nations Office for Disaster Risk Reduction 2018). In this regard, the United Nations has supported these local technical initiatives in various parts of the world, including the Philippines, but as it is a territory made up of thousands of islands, it is advisable to establish local community observatories.

The future outlook in terms of disasters is not encouraging. Enrico Quarantelli (1999) suggests that since the end of the 1990s, the global scenario would be one of greater frequency and severity of disasters. Regardless of the agents involved, there is an increase both in the frequency and the force of threats. As this has been documented since the 1990s, national and local governments must make their best efforts to achieve the goals of the United Nations Sustainable Development Agenda, since, as described in this chapter, disasters have a social origin in the unsolved problems of asymmetric development.

From a governance perspective, the extension of models like Albay's is crucial in the Philippines for disaster risk reduction and community resilience building. Disasters will continue to affect the Philippines, and it is necessary to extend good practices in disaster risk management.

Notes

1 The Human Development Index is a social measurement carried out by the United Nations since 1990. It is composed of variables related to life expectancy at birth, access to education, and Gross Domestic Product per capita. HDI levels between 0.51 and 0.70

are considered as medium human development, while levels between 0.71 and 0,80 are considered as high human development (United Nations Development Program 2019.

2 The Gini index is a measure of social inequality used to measure inequality in monetary income within a country. The coefficient can take a value between 0 and 1, where zero is perfect equality, and 1 is the maximum inequality.

References

Abeldaño Zuñiga, R. A. (2019) "El Cambio Climático en la Incidencia de Desastres en América Latina y El Caribe," in D. A. Fabre Platas, I. Ortiz Sánchez, and G. Busso (eds) *Agua. Territorialidades y Dimensiones de Análisis*, 1st ed., ch. 33. Veracruz, Mexico: Universidad Veracruzana IIESES-UV (México), Universidad Nacional de Río Cuarto – UNRC (Argentina), Universidad de Granada – UGR (España), Red Latinoamericana de Estudios sobre Vulnerabilidades – ALAP.

Abeldaño Zuñiga, R. A. and J. Fanta Garrido (2020) "Internal displacement due to disasters in Latin America and the Caribbean," in W. Leal Filho (ed) *Climate Change, Hazards and Adaptation Options: Handling the Impacts of a Changing Climate*, Cham: Springer, pp 389–409. doi:10.1007/978-3-030-37425-9_21

Abeldaño Zúñiga, R. A. and A. M. González Villoria. (2018) "Desastres en México de 1900 a 2016: Patrones de Ocurrencia, Población Afectada y Daños Económicos," *Revista Panamericana de Salud Pública* 42(e55): 1–8. doi:10.26633/RPSP.2018.55.

Adeagbo, A., A. Daramola, A. Carim-Sanni, C. Akujobi, and C. Ukpong (2016) "Effects of natural disasters on social and economic well being: A study in Nigeria," *International Journal of Disaster Risk Reduction* 17: 1–12. doi:10.1016/j.ijdrr.2016.03.006

Adger, N., S. Huq, K. Brown, D. Conway, and M. Hulme (2003) "Adaptation to climate change in the developing world," *Progress in Development Studies* 3(3): 179–195. doi:10.1191/1464993403ps060oa

Alcayna, T., V. Bollettino, P. Dy, and P. Vinck (2016) "Resilience and disaster trends in the Philippines: Opportunities for national and local capacity building," *PLoS Currents* 8 (Disasters): 1–10. doi:10.1371/currents.dis.4a0bc960866e53bd6357ac135d740846

Barbin, E. A., C. Lomboy, and E. S. Soriano (2002) "A field study of microinsurance in the Philippines". Working paper No. 30. Employment Sector. Geneva, Switzerland. https://www.ilo.org/wcmsp5/groups/public/---ed_emp/documents/publication/wcms_117980.pdf [Accessed January 2, 2022].

Caballeros Otero, R. and R. Zapata Martí (1994) "The impacts of natural disasters on developing economies: Implications for the international development and disaster community," in World Bank (ed) *Economic Aspects of Natural Disaster. Reduction for Sustainable Development. World Conference on Natural Disaster Reduction*, Yokohama, Japan: National Academy of Science, pp 11–15. doi:10.17226/6425

Centre for Research on the Epidemiology of Disasters. (2019) "The International Disaster Database (EM-DAT)." Université Catholique de Louvain, Belgium. http://www.emdat.be [Accessed December 6, 2019].

Daoud, A., B. Halleröd, and D. Guha-Sapir (2016) "What is the association between absolute child poverty, poor governance, and natural disasters? A global comparison of some of the realities of climate change," *PLoS ONE* 11(4): 1–20. doi:10.1371/journal.pone.0153296

Economic Commission for Latin America and the Caribbean (2003) *Handbook for Estimating the Socio-Economic and Environmental Effects of Disasters*, Santiago, Chile: ECLAC/United Nations.

Eleazar, L. (2010) "Land tenure and natural disasters: Philippines," in A. Herrera Garibay, P. De Wit, L. Eleazar, F. Bucheli, S. Norfolk, R. Sánchez Mena, and S. Shafi (eds) *Land Tenure and Natural Disasters. Addressing Land Tenure in Countries Prone to Natural Disasters*, 1st ed. Rome: Food and Agriculture Organization. https://www.fao.org/3/I1855E/i1855e.pdf [Accessed January 2, 2022].

European Civil Protection and Humanitarian Aid Operations. (2019) "Philippines factsheet". https://ec.europa.eu/echo/where/asia-and-pacific/philippines_en [Accessed January 2, 2022].

Gil Cuesta, J., J. A. Frank van Loenhout, M. L. de Lara Banquesio, M. Mustaffa, and D. Guha-Sapir (2019) "Medical consultations after typhoon Haiyan in a field hospital in the Philippines," *Disaster Medicine and Public Health Preparedness*, November, 1–5. doi:10.1017/dmp.2019.99

Hallegatte, S., A. Vogt-Schilb, M. Bangalore, and J. Rozenberg (2017) *Unbreakable: Building the Resilience of the Poor in the Face of Natural Disasters*, Washington DC: World Bank Publications, The World Bank Group. doi:10.1596/978-1-4648-1003-9

Hochrainer, S. (2006) *Macroeconomic Risk Management against Natural Disasters*, Wiesbaden: Deutscher Universitats-Verlag (DUV). doi:10.1007/978-3-8350-9441-3

Internal Displacement Monitoring Centre. (2019) "Global Internal Displacement Database". http://www.internal-displacement.org/database/ [Accessed January 2, 2022].

Kälin, W. (2010) "Conceptualising climate-induced displacement," in J. McAdam (ed) *Climate Change and Displacement: Multidisciplinary Perspectives*, 1st ed, Oxford, United Kingdom: Hart Publishing, pp 81–104.

Lagmay, Alfredo Mahar Francisco, Bernard Alan Racoma, Ken Adrian Aracan, Jenalyn Alconis-Ayco, and Ivan Lester Saddi (2017) "Disseminating near-real-time hazards information and flood maps in the Philippines through Web-GIS," *Journal of Environmental Sciences* 59 (September): 13–23. doi:10.1016/j.jes.2017.03.014

Lopez-Calva, L. F. and E. Ortiz-Juarez (2009) "Evidence and policy lessons on the links between disaster risk and poverty in Latin America: Methodology and summary of country studies," MDG-01-2009. New York: Research for Public Policy, MDGs and Poverty. https://mpra.ub.uni-muenchen.de/18342/1/01_RPPLAC_POV.pdf [Accessed January 2, 2022].

Manila Observatory. (2019) "Manila observatory," http://vm.observatory.ph/. [Accessed November 27, 2021].

Metcalf, C. J. E., K. S. Walter, A. Wesolowski, C. O. Buckee, E. Shevliakova, A. J. Tatem, W. R. Boos, D. M. Weinberger, and V. E. Pitzer (2017) "Identifying climate drivers of infectious disease dynamics: Recent advances and challenges ahead," *Proceedings of the Royal Society B: Biological Sciences* 284(1860): 20170901. doi:10.1098/rspb.2017.0901

Moscoso, J. (2011) "La Investigación de Frontera en Humanidades y Ciencias Sociales," *Lychnos* 5: 80–84. http://www.fgcsic.es/lychnos/es_es/articulos/investigacion_de_frontera_en_humanidades_y_ciencias_sociales [Accessed January 2, 2022].

Murray, K. O., N. T. Castillo-Carandang, A. M. Mandalakas, A. T. Cruz, L. M. Leining, and S. R. Gatchalian (2019) "Prevalence of tuberculosis in children after natural disasters, Bohol, Philippines," *Emerging Infectious Diseases* 25(10): 1884–1892. doi:10.3201/eid2510.190619

Murray, L. K, I. Familiar, S. Skavenski, E. Jere, J. Cohen, M. Imasiku, J. Mayeya, J. K. Bass, and P. Bolton (2013) "An evaluation of trauma focused cognitive behavioral therapy for children in Zambia," *Child Abuse & Neglect* 37(12): 1175–1185. http://www.ncbi.nlm.nih.gov/pmc/articles/PMC3823750/?tool=pubmed [Accessed January 2, 2022].

Oliver-Smith, A., I. Alcántara-Ayala, I. Burton, and A. Lavell (2017) "The social construction of disaster risk: Seeking root causes," *International Journal of Disaster Risk Reduction* 22 (June): 469–474. doi:10.1016/j.ijdrr.2016.10.006

Programa de las Naciones Unidas para el Desarrollo (PNUD) (2013) *Protección Del Desarrollo Contra Los Desastres. Apoyo Del PNUD Al Marco de Acción de Hyogo*. 1st ed. New York: Programa de las Naciones Unidas para el Desarrollo. https://www.undp.org/sites/g/files/zskgke326/files/publications/es/UNDPHyogo-BCPR_Final_ES.pdf [Accessed January 2, 2022].

Province of Albay. (2019) "Albay Provincial Disaster Risk Reduction and Management Council," http://albay.gov.ph/category/pdrrmc/page/2/ [Accessed December 6, 2019].

Provincial Government of Albay. (2019) "Albay Public Safety and Emergency Management Office," http://www.csc.gov.ph/profile-of-awardees/308-albay-public-safety-and-emergency-management-office.html [Accessed December 6, 2019].

Quarantelli, E. (1999) "Implications for programmes and policies from future disaster trends," *Risk Management* 1(1): 9–19. doi:10.1057/palgrave.rm.8240011

Shilkofski, N., M. Agueh, M. Fonseka, A. Tan, and J. Cembrano (2016) "Pediatric emergency care in disaster-affected areas: A firsthand perspective after typhoons Bopha and Haiyan in the Philippines," *Journal of Pediatric Intensive Care* 6 (1): 19–27. doi:10.1055/s-0036-1584910

Swart, R., J. Robinson, and S. Cohen (2003) "Climate change and sustainable development: Expanding the options," *Climate Policy* 3(Suppl 1): 19–40. doi:10.1016/j.clipol.2003.10.010

UN Office for Disaster Risk Reduction. (2019) "Disaster risk reduction in the Philippines". Bangkok, Thailand. https://reliefweb.int/report/philippines/disaster-risk-reduction-philippines-status-report-july-2019 [Accessed January 2, 2022].

United Nations. (2015) "Sustainable Development Goals," http://www.un.org/sustainabledevelopment/sustainable-development-goals/ [Accessed January 2, 2022].

United Nations Development Program. (2019) "Human development report 2019. Beyond income, beyond averages, beyond today: Inequalities in human development in the 21st century," New York: UNDP, http://report.hdr.undp.org/ [Accessed January 2, 2022].

United Nations Office for Disaster Risk Reduction. (2018) "Words into action guidelines. Implementation guide for local disaster risk reduction and resilience strategies. A companion for implementing the Sendai Framework target E Santa Fe, Argentina: UNDRR, https://www.preventionweb.net/files/57399_drrresiliencepublicreview.pdf [Accessed January 2, 2022].

World Bank. (2019) "World Bank Open Data," https://data.worldbank.org/ [Accessed December 6, 2019].

World Bank Climate Knowledge Portal. (2019) "World Bank Climate Change Knowledge Portal for Global Climate Data and Information," https://climateknowledgeportal.worldbank.org/country/philippines/climate-data-historical [Accessed December 5, 2019].

Yang, Q., C. Fu, N. Wang, Z. Dong, W. Hu, and M. Wang (2014) "The effects of weather conditions on measles incidence in Guangzhou, Southern China," *Human Vaccines & Immunotherapeutics* 10(4): 1104–1110, doi:10.4161/hv.27826

3

State of ICT-Use within the Local Climate E-Governance in Metro Manila, Philippines

Xavier Venn A. Asuncion

Introduction

Globally, the effects of climate change have increasingly affected various populations (Tollin et al 2017). These effects occur in various forms—from sea level rise, melting of ice sheets to increasing frequency and intensity of weather events such as rainfall, extreme heat waves, and droughts (European Commission, n.d.). Although there are no exceptions when it comes to the adverse impacts of climate change, the extent of risk among countries varies due to differences in their 'vulnerability' which, as defined by the Intergovernmental Panel on Climate Change (IPCC), is "the extent to which climate change may damage or harm a system... (it) depends not only on a system's sensitivity, but also on its ability to adapt to new climatic conditions" (Olmos 2001, 2). This indicates, as conceptualized by Ospina and Heeks (2010), that vulnerability has two facets—one that is outside of a system which is the source of shocks, and another that is inside the system which is the subject of the shocks and variations. This is crucial in coming up with adaptation measures as it shows that a country's vulnerability depends not only on its exposure to climate risks but also on its adaptive capacity or ability to address and manage risks. As such, developing countries are usually more vulnerable to climate risks because of constraints in their social, technological, and financial resources (United Nations 2007).

A tool that is now being used to enhance adaptive capacity for effective climate change adaptation and risk management are information and communication technologies (ICTs) or technologies with the capability to collect, store, edit, and transfer data and information in various forms (Kundishora 2010). The recent growth in the application of ICTs for development processes has resulted in the use of ICTs for climate change action (Karanasios 2011). The continued growth of ICT use for tackling climate change-relevant concerns is due to the capacity of ICTs to generate and disseminate information; to facilitate coordination of actors in and beyond government; to make public and private services more efficient (World Bank 2012). This is because ICTs enable the facilitation of a great deal of information which is necessary to effectively formulate and implement adaptation-related decisions (Dinshaw et al 2012). General usage of ICTs in the context of climate change includes generation, organization and communication of climate change-related information; development of information systems; development of early warning and hazard risk information systems, and development of ICT mapping tools (World Bank 2012).

This chapter is based on a study about how local government units (LGUs) in Metro Manila, Philippines—a region vulnerable to climate risks (Yusuf and Francisco 2009)—use ICTs for local climate change action which, using the basic conceptualization of e-governance by Kumar, Kumar, and Kumar (2014), is referred to in this chapter as 'local climate e-governance.' The study explored the different types of ICTs being utilized for local climate e-governance; the roles that ICTs play in helping localities adapt to, and manage, climate risks; and the issues and barriers to ICT use within the context of local climate action in the region. It was based on a survey among 46 respondents from all LGUs in Metro Manila—Caloocan, Makati, Malabon, Mandaluyong, Manila, Marikina, Muntinlupa, Navotas, Parañaque, Pasay, Pasig, Pateros, Quezon City, San Juan, Taguig, and Valenzuela. This survey is referred to in this chapter as the 'Metro Manila survey'. The study also included a case study of Marikina, one of the cities in Metro Manila that is most vulnerable to climate risks, through key informant interviews (KIIs) with their six LGU officers from different departments who are involved in the formulation and implementation of their local climate change action programs, activities, and projects. This case study is referred to in this chapter as the 'Marikina case study.'

The first section of this chapter presents the types of ICTs and how they are being used by LGUs in Metro Manila for local climate e-governance. The second section discusses the different processes involved in using ICTs to address and manage climate risks. The third section summarizes the different issues/barriers to ICT adoption in Metro Manila, and the last section provides general recommendations to address these challenges.

ICT-use for local climate e-governance in Metro Manila, Philippines

Based on the responses from the Metro Manila Survey, the ICTs that most LGUs in Metro Manila use for performing climate change action tasks are social media; wireless broadband technologies; geographical information systems (GIS); and mobile technologies. Many of the LGUs, as suggested from the results of the Metro Manila Survey, use these ICTs to a high extent as almost half of the respondents, 45–48%, believe that the climate change action tasks can only be achieved using it. The rest of them mostly believe that local climate action can be done without using ICTs but using them helps in accomplishing climate change action tasks more efficiently and effectively. How the different ICTs are used by the LGUs are discussed below.

Social media

Social media was the most selected ICT for local climate e-governance from the Metro Manila survey. According to Alampay and Delos Santos (2016), the use of social media among LGUs in Metro Manila started around 2008 and peaked from 2012 onwards. Out of the 16 LGUs they monitored, all of them have Facebook accounts while 15 have Twitter accounts. One of the early adopters of social media is Marikina. It started as a personal account of an employee but because it eventually exceeded the 5,000 limit of friends, a page for Marikina was created (Alampay et al 2018). One of the key informants (KIs) from the Marikina case study also shared that the employees' use of social media back then, when it was not yet adopted for official use, was done secretly, suggesting that the use of social media in the workplace used to be an unacceptable practice. The City Administrator (CA) eventually thought of utilizing social media instead for the benefit of the LGU since its use among the members of the LGU seemed unavoidable. Generally, social media is used for disseminating information that focuses on disaster preparedness, such as updates on typhoons, water levels, flooding, earthquakes, traffic and vehicular accidents, and power and water utility interruptions (Alampay and Delos Santos 2016). As such, social media activities among Metro Manila LGUs increase when hazard occurrences are at their peak.

Wireless broadband technologies

Many respondents from the survey also selected wireless broadband technologies as one of the ICTs being used for local climate action. Internet connectivity in the Philippines, however, is relatively inferior in terms of overall performance. A report by the Department of Information and Communications Technology (DICT 2017) showed that compared to neighboring ASEAN countries such as Indonesia, Malaysia, Singapore, and Thailand, the internet performance in the Philippines is still weak in terms of affordability, availability, and speed. Siar (2005), in her assessment of local e-governance in the Philippines, also attributes the slow progress of e-governance in the Philippines to the poor quality of internet connectivity in the country. As such, despite the high diffusion of broadband technologies among Metro Manila LGUs, the quality of the connection is most likely poor. Even in Marikina LGU, for example, internet connection often slows down or drops out. Moreover, since the internet connection in the LGU is centralized, if their broadband system goes down, the whole LGU goes offline. In such cases, since online access is crucial to the execution of many of their tasks, the employees and officials are forced to use their mobile data. Aside from a centralized broadband system in Marikina, pocket WiFi is also used by field officers, which allows them to communicate and transmit data to the main office even when on the ground.

GIS

Many LGUs in the Philippines have already been using GIS for mapping to support decision-making (Tongco 2011). The Housing and Land Use Regulatory Board (HLURB), a national agency in the Philippines, even created a set of guidelines for GIS use by LGUs in 2007 called the GIS Cookbook, which describes how to create GIS-based land use plans. A report by Aban (2016) also found that GIS, along with other ICT-enabled services, is already being used at different levels of local government in the Philippines. GIS software used includes ArcGIS as well as other free and open-source software such as QGIS. Marikina also uses GIS, specifically ArcGIS, to a great extent. They have already created many thematic maps that are available online through a website called One-Map that Marikina designed. The maps of Marikina are integrated into the plans of the LGU, especially to their Comprehensive Land-Use Plan (CLUP) which is mandated by the Local Government Code of the Philippines.

Mobile technologies

The literature on the use of mobile technologies for local e-governance is limited; but based on the Metro Manila survey, two noteworthy mobile-based applications were specified: the Batingaw and Makatizen apps. Batingaw was created by the National Disaster Risk Reduction and Management Council (NDRRMC) of the Philippines (Batingaw n.d.). The term *batingaw* means siren or alarm, which somehow encapsulates the main objective of the application which is to help the public prepare for emergencies and disasters by providing them access to relevant information through government and NGO websites, disaster or emergency-related Twitter feeds, and different hazards and safety tips. Digital tools such as sirens, flashlights, compasses, strobes, an e-library, and FM/AM radio are also available in the app (Waidyanatha 2015). The application is available for free download at the Google Play store, making it accessible to smartphone users. Another mobile-based app is the Makatizen app, used by Makati. It is not specifically designed for climate action purposes, but it offers different types of information about Makati, which may include news, updates and announcements about impending disasters, and climate change-related programs, projects, and activities (SOCA 2018). The Makatizen app is also available for free download at the Google Play store (Makatizen app n.d). For Marikina, the LGU personnel and officials use mobile technologies to communicate amongst themselves and with the public. Mobile devices such as cellphones allow them, for instance, to reach people from other departments, especially in Marikina where offices are not necessarily positioned close to one another. Some offices such as the Disaster Risk Reduction and Management Office (DRRMO) also use cellphones as one of the channels for the public to reach them in cases of emergency.

Aside from the above-mentioned ICTs, around 19–30% of the Metro Manila survey respondents said that they also use other web-based applications. Many of these applications are locally operated, such as the Nationwide Operational Assessment of Hazards (NOAH n.d.). NOAH is based in the University of the Philippines (UP) and some of the features of its web-based application include measurement of linear distance and area of the polygon on map, and provision of risk- and hazard-related data and information. Another web-based application being used is the Met-Hydro Decision Support Infosys (MDSI), which is operated by the Philippine Atmospheric, Geophysical and Astronomical Services Administration (PAGASA), an attached agency of the Department of Science and Technology (DOST). Like the NOAH, the MDSI also provides hazard-related information produced mainly from PAGASA's radar dopplers, automatic rain gauges (ARG), and automatic weather stations (AWS) that are deployed all over the country allowing the measurement of barometric pressure, atmospheric pressure, air temperature, relative humidity, wind speed and direction from remote areas using meteorological sensors (PAGASA Met-Hydro Decision Support Infosys, n.d).

ICT application processes for local climate e-governance: Marikina case study

This section outlines and discusses the different processes that enable the achievement of climate change adaptation and risk management

Figure 3.1: ICT application processes for local climate e-governance in Marikina

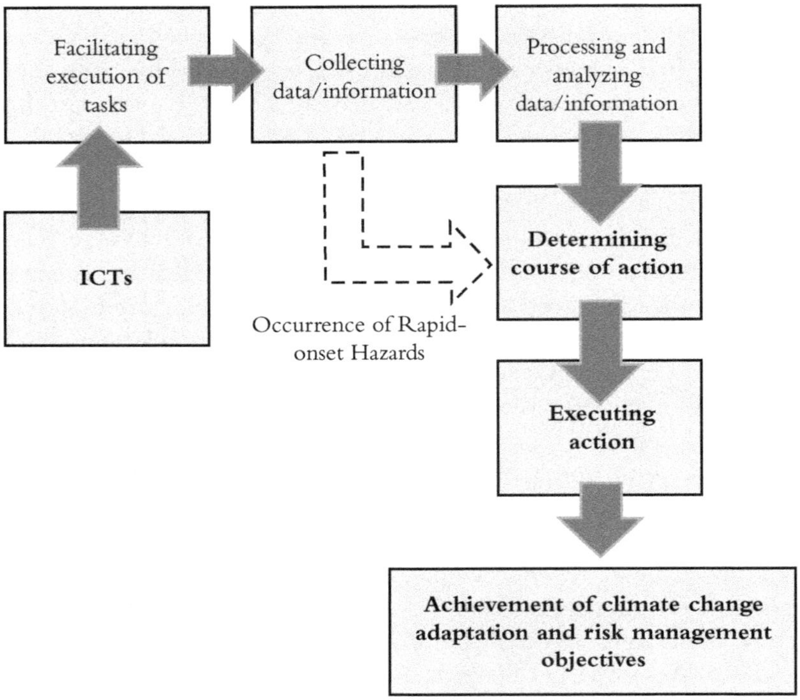

objectives using ICTs as derived from the Marikina case study. Three main processes were identified: 1) facilitating execution of tasks; 2) collecting data/information, and 3) processing and analyzing data/information (see Figure 3.1).

Facilitating execution of tasks

This process is about how ICTs enable an environment that allows for the facilitation of executing day-to-day and other important tasks including those related to climate action. In Marikina's case, this is mainly manifested through the capacity of ICTs to provide internet connectivity and two-way communication channels.

Most, if not all, climate e-governance tasks require internet connectivity, which is made available using wireless broadband technologies, specifically through WiFi. Marikina makes use of environment- or weather-related information from the NOAH's web-based application as well as from the social media pages of relevant government agencies such as PAGASA and the Department of Environment and Natural Resources (DENR), to decide on future actions—especially when there are threats of impending hazards. Access to these web applications and pages is only possible through an internet connection. The air quality monitoring stations owned by the DENR deployed in Marikina also make use of internet connection to transmit data. The data that DENR collects from these stations are analyzed, processed, and posted on their social media accounts and website, which the LGU uses as a reference to determine whether or not they need to implement air quality-related interventions. Marikina's field officers are also equipped with pocket WiFi devices, which are mainly used to access online communication channels that enable fast transmission of data/information from the field to the main office. This allows for immediate analysis of the data/information as well as for quick decision making. Internet connectivity is also especially crucial concerning Marikina City's disaster management operations. The DRRMO views the CCTV footage using the web and mobile applications, allowing them to monitor the water level of the Marikina River in real time. As such, they can determine whether the water in the Marikina River reaches a level that requires immediate attention. The CCTV footage is also streamed live on Marikina City's website and social media accounts, allowing the residents to know and prepare before a flood would occur. This whole operation, from real-time monitoring to streaming of the CCTV footage, requires internet connectivity.

The use of two-way communication channels is also evidently crucial in the conduct of many tasks within the LGU, and these channels are mainly provided for by ICTs. This occurs both internally and externally. Internal communication involves communication between members of the LGU while external communication is between the LGU and the public or other institutions, such as the national government and other LGUs. In Marikina, internal two-way communication occurs, for instance, when field officers transmit data/information from the field to the people in the main office. Field officers mainly use radio as their main channel for communicating with people in the main office; but they also make use of calling, texting, or chatting through cellphones or tablets. All departments are equipped with two-way radio devices, and many members of the LGU are also equipped with walkie-talkies, from field personnel and project engineers up to the mayor. Coordination among different departments is also facilitated via two-way communication channels, especially in Marikina's case where the department offices are not necessarily close to one another. This is mostly done using telephones, online messaging applications such as Facebook messenger, texting, and e-mail. One of the key informants from Marikina's Management Information Systems and Call Center (MISCC) even shared that they have a private Facebook group where they can call the attention of other departments to address specific requests. The informant added that aside from Facebook, they also use a software called 'compliance slip' that allows them to send and track requests to other departments. This software is used by all departments in Marikina. Through the compliance slip, the different departments within the LGU can send and receive requests to and from each other. They can also track the status of these requests. This software fosters accountability and promotes timeliness in terms of accomplishing deliverables. Another example of internal communication in action is during typhoon occurrences when the mayor sends directives via text or messenger to the CA, which are then cascaded to the different departments.

Concerning external communication, this usually involves communication between the LGU and the public, or between the LGU and other institutions. This is exemplified in various ways. For one, nearby LGUs can inform Marikina through text or Facebook whenever they experience heavy rainfall. ICTs have allowed them to better communicate and coordinate with other LGUs. The KI from Marikina's City Planning and Development Office (CPDO) shared that, after typhoon Ondoy (2009), a network of LGUs in Metro Manila was formed where member LGUs alert and coordinate with each other whenever a hazard occurs. Their main mode of communication is through text or Facebook. This is important since Marikina is a catch basin for its nearby cities and towns. This kind of coordination allows Marikina to implement appropriate measures before floodwaters from nearby areas, for example, flow into their city.

Marikina also uses ICTs to coordinate or receive data from the national government. Large-scale and more advanced ICTs for hazard monitoring and data collection are owned and operated by national government agencies such as the NDRRMC, PAGASA, and DENR. As such, Marikina relies on the data/information sent to them by these agencies. Marikina's DRRMO, for instance, relies on PAGASA's weather forecasts sent through email, fax, or even through a Viber group (a mobile-based messaging application), for the whole Metro Manila DRRMC in which the Marikina LGU is represented. As mentioned, the DENR also sends daily emails to Marikina's City Environment Management Office (CEMO) regarding air quality data and information from their monitoring stations. In the same way, other institutions, such as the Korea International Cooperation Agency (KOICA) and the Japan International Cooperation Agency (JICA), which own water-level sensors, and the Metro Manila Development Authority (MMDA) that operates the Effective Flood Control Operating System (EFCOS), send Marikina the weather and hazard-related data that are crucial to the local climate action activities of the LGUs.

External communication with the public usually happens using social media. The residents of Marikina usually message the LGU through a public, closed Facebook group, "Marikina News," currently has over 300,000 members,[1] which is over 66% of the city's population.[2] While it may not be validated whether all members are actual residents of Marikina, the number is still suggestive of the Facebook group's reach. There are other avenues for the public to reach Marikina, such as calling them through their telephone lines, but this rarely occurs as most of the time they are reached through social media or email. Aside from Marikina News, there are also other Facebook pages that the public can message to communicate with the LGU such as the Marikina PIO (@MarikinaPIO) which is the main Facebook page of Marikina handled by their public information office (PIO) with over 220,000 followers.[3] Whenever the page receives a message from the public, the information is forwarded to the relevant departments or offices. Another popular Marikina Facebook page is Marikina City Rescue 161 (@MarikinaRescue161), which is the official page of the LGU for disseminating disaster-related information in Marikina. It has almost 52,000 followers and is also utilized by the DRRMO as a channel for the public to reach them, especially at times of disasters when residents need assistance or rescuing. Marikina also makes use of a data-tracking system (DTS) which serves as the counterpart of the 'compliance slip' software for public communication. It allows residents with any kind of transaction with the Marikina LGU to track the status or check updates of their requests, concerns, or inquiries. This is done through a tracking number.

Collecting data/information

When internet connection and communication channels are already in place, concerning climate e-governance, ICTs play their next major role through the collection of data and information. In Marikina's case, this is possible because of the monitoring devices, access to relevant information online, and public communication.

Among the departments interviewed, CEMO and DRRMO are the main handlers of the monitoring devices deployed in Marikina. For CEMO, the main monitoring devices they use are air quality monitoring stations. However, as mentioned, these stations are owned by the DENR. Nevertheless, the CEMO gets daily data and information from the DENR extracted from the monitoring stations. If further data is needed, they would have to send a formal request to the DENR. Marikina has two real-time and two manual air quality monitoring stations. On the one hand, the real-time stations produce data that can be immediately extracted. The manual monitoring stations, on the other hand, have filters that must be replaced every six days and it takes a day for the data to be available. These stations monitor particulate matter PM 10 and PM 2.5. Data from these are posted on the DENR's social media accounts, such as Twitter. The post makes use of process or analyzes the data where the qualities of air on different areas where they have monitoring stations are already tagged as good, moderate, unhealthy for sensitive groups, unhealthy, very unhealthy, and hazardous according to the Air Quality Index or AQI in compliance with RA 8749 or the Philippine Clean Air Act of 1999 as well as with the standards set by the Department of Environment & Natural Resources (DENR).

Another monitoring method extensively used in Marikina is CCTV. In terms of climate e-governance, CCTV is used to monitor flood and water-level situations. CCTV is mainly located along entry and exit points, major intersections, and within flood-prone areas in Marikina. The CCTV is linked to the server in their command center and the footage can be viewed live online through Marikina's website or their Facebook page. One of the CCTV locations is especially used to monitor the water level in the Marikina River where a gauge is located. This allows both the DRRMO and the public to be aware when the water level reaches alarming heights. Marikina implements evacuation protocols in this regard. When the water reaches 15, 16, and 18m, first, second, and third alarms are triggered respectively. For the first alarm, the sirens would go off for a minute, which would serve as a signal for the residents to stay alert and to prepare for possible evacuation. For the second alarm, the sirens are sounded for 2 minutes, which indicates that the residents are already advised to evacuate, while for the third and final alarm, the sirens are sounded for 5 minutes, and the residents are forced to evacuate. The sirens in Marikina are still manually operated but can be triggered remotely from the command center. It is as such that the officers in the DRRMO are required to always stay alert and to regularly monitor the CCTV footage even when they are not in the command center. This can be done using mobile applications where they just need to input the internet protocol (IP) address used by their CCTV to be able to view the footage. In situations where there is an occurrence of rapid-onset hazards such as flooding, the data/information collected does not go through the analysis or processing stage. In this instance, actions are already predetermined once data/information is collected from the ground.

Marikina also releases social media warnings showing the level of alarm, what it signifies, the volume of rainfall count at selected flood-prone areas, and the water level itself. It also shows the emergency contact numbers of the LGU. The data for the rainfall count comes from a public regional institution—the Metro Manila Development Authority (MMDA). The MMDA operates EFCOS, the equipment funded by JICA for monitoring rainfall volume and analyzing whether flooding will be imminent. The MMDA could transmit data from the EFCOS to Marikina's command center on an hourly basis during weather disturbances. To help the public interpret the rainfall data, they also constantly share the rainfall warning guide regularly posted by PAGASA.

Marikina is also able to collect climate change-related data and information from the residents themselves through their grievances, concerns, and messages, in general, addressed to the LGU through social media, telephone lines, and e-mail. Again, the most popular option among residents to reach the LGU is social media. Looking into the Marikina News Facebook group provides an overview of the different kinds of information collected from the residents. This was done by using the name of the departments as keywords as well as other relevant terms to search over the content of the posts of the group. Examples of concerns that came out from this simple analysis were broken drains, local policy violations, rescue needs, evacuation conditions, broken CCTV, and uncovered manholes.

Processing and analyzing data/information

After data and information are collected, they are analyzed and processed to guide decision-making. The most evident use of ICTs for processing and analyzing data and information for local climate action in Marikina is making maps such as hazard and risk maps. The MISCC is the main department in charge of map-making. The main bases for the maps are historical data and the key software used is ArcGIS. When flooding occurs, for instance, the engineering department surveys Marikina to determine the level of flooding within different areas, then the data are given to the MISCC to overlay the data to their base map. Aside from the survey, they also get data from satellite images. Sometimes they get maps from external sources, such as the Marine and Geosciences Bureau (MGB), a national government agency; and the Greater Metro Manila Area Risk Assessment Project (GMMA-RAP), a three-year program which started in 2014 and was funded by AusAID through the UNDP. Different infrastructure types are also plotted on the Marikina LGU's maps. For instance, if new buildings are established in Marikina after securing permits from the LGU, they plot them on their maps right away. Note that what MISCC uses is a thematic map where they have a base map and they overlay data to the maps according to different themes or data types.

To facilitate map creation in MISCC, they also designed an online program called 'One-Map'. This software allows for centralized creation and revision of their maps. For example, if someone from the department is adding or revising the data of their thematic maps, others who are accessing One-Map can monitor the changes being made. One-Map is also available for public online viewing. As such, anyone can see and even download the different maps that Marikina has produced. At present, there are currently four categories of thematic maps that can be viewed in One-Map: environment, hazards, land-use, and settlement. Different infrastructures can also be located on their maps, such as government offices, hospitals, health centers, police stations, fire stations, public schools, private schools, churches, heritage sites, cemeteries, transportation terminals, CCTV live feed locations, bike lanes, transport routes, street drives, barangay boundaries, buildings, roads, and rivers. Distances and areas within Marikina can also be measured using One-Map. It also connects to Google Maps if the user requires a street view of a location on the map.

Marikina also conducted a type of mapping that includes social and other vulnerability factors through the Community-Based Monitoring System (CBMS). The CBMS is a survey that tries to capture the different facets of poverty using different indicators (Reyes et al 2017). Some maps produced from CBMS that could aid significantly in LGU's climate action activities include a proportion of households living in makeshift housing, households who are informal settlers, households without access to safe water supply, households with income below the poverty threshold, and so on. All the data are processed from the mapping are integrated into Marikina's CLUP, as well as other as other plans that involve the creation of programs, projects, and activities.

Processing of data can also take the form of validation. For instance, when the LGU receives a complaint or grievance from the public on social media, they visit the area where the complaint relates to in order to determine whether it is valid or not. The field officers usually take pictures of the area if the complaint involves structural concerns, which are then sent to the main office for evaluation. It is then that a course of action is determined and executed. During heavy rainfall, the DRRMO also analyzes data from the EFCOS. They can calculate the probability that rainfall would lead to heavy flooding. Through this analysis, they can also determine whether they will need to advise the residents to evacuate even when water levels in the Marikina River have not reached an alarming level.

The role of ICTs becomes very minimal after the analysis and processing of data and information, at least in Marikina's case. Courses of action, for example, are decided upon Local Development Council (LDC) meetings and the actions themselves usually do not involve the use of ICTs. For instance, when a resident complains about clogged drainage, ICTs are involved from when the resident communicated with the LGU for particular concerns, to its processing and validation, but the actual de-clogging does not involve ICTs.

Issues/barriers to ICT use and adoption

This section presents a summary of the different issues/barriers to ICT use and adoption from the results of the Metro Manila survey and Marikina case study which used the framework of Cucciniello et al (2015) that was derived from a systematic literature review to determine barriers to ICT adoption and diffusion in the public sector. The framework identifies nine types of factor that could hamper ICT adoption and diffusion in the public sector:

1. *Inter-institutional dynamics*: inter-institutional networks or the lack thereof do/does not encourage ICT adoption
2. *Legal factors*: strict legal requirements
3. *Political factors*: political attitude of relevant actors against innovations
4. *Economic factors*: budget constraints
5. *Social factors*: absence of public pressure to drive ICT adoption
6. *Demographic factors*: small demographic context does not encourage innovation
7. *Technological factors*: lack of ICT infrastructure support/low public ICT readiness/threats to information security/technological complexity
8. *Organizational factors*: bureaucracy and organizational structures/arrangements prevent innovation
9. *Individual factors*: lack of ICT readiness among individuals in the system

Among these factors, the Metro Manila survey respondents selected technological and economic factors to be the biggest barrier to ICT use/adoption. This suggests that lack of infrastructure support; low public ICT readiness; threats to information security; technological complexity; and budget constraints are the main hindrances to the diffusion of ICTs among LGUs in Metro Manila. Technological factors as one of the main barriers to ICT diffusion is also reflected in the recommendations provided by the Metro Manila respondents to strengthen ICT use for local climate e-governance, specifically within the aspect of technological complexity. Their comments, as provided below, suggest that their respective LGUs have limited ICT equipment as well as human resources with adequate skills and knowledge to operate and manage ICTs in terms of DRR-CCA:

> The LGU is in great need of ICT officers and specialist. As of the moment, DRR-CCA is about to be mainstreamed here in the city, The DRRM office has no technical personnel such as GIS specialist because of lack of budget for personnel services. Although budget of ICT equipment is being prioritized in the DRRM fund, we still need to wait for the office ratification to hire people. The office greatly relies on consultants (EMI and UP Planades) and other offices such as City Planning Office (Land Use and Zoning) and Information Technology and Development Office—Special Operation Support Staff, DRRMO, Quezon City
>
> Issue policies mandating a team of IT skilled employees with envi[ronmental] and urban planning technical employees to work on the ICT program for CC [Climate Change]—Project Development Officer, CPDO, Marikina
>
> I think there is a need to strengthen the mainstreaming of the usage of ICT in terms of local climate governance. With that said, there should

> be more equipment and training made available to the offices and personnel concerned, like our local DRRMO and the other member offices of the Protective Services sub-cluster...—Administrative Officer, CPDO, Muntinlupa

Budget constraints could also lead to technological issues. The lack of ICT specialists, for instance, could be attributed to budget constraints. According to the key informant from Marikina's CPDO, good ICT professionals are not encouraged to work for the government since they can get jobs with better pay in the private sector. Another technological issue that stems from budget constraints is the lack of ICT infrastructure. For instance, concerning internet connectivity, the KI from Marikina's CEMO shared that they are not able to improve the quality of their connection because, to do so, they would have to install another antenna which the LGU cannot afford.

Budget constraints also overlap with issues on legal restrictions and inter-institutional dynamics. The KI from Marikina's DRRMO said that they were being held back by the funding restrictions under RA 10121 or the Philippine Disaster Risk Reduction and Management Act of 2010. Under this law, the DRRMO gets a 5% allocation from the LGU's total budget, which is called the Local Disaster Risk Reduction and Management Fund (LDRRMF). About 30% of this fund should be set aside as Quick Response Fund (QRF) or funds that can be used only when the locality is in a 'state of calamity' where relief and recovery programs and activities are necessary. The remaining 70% can be used for pre-disaster preparedness activities. Acquisition of internet services, however, is not considered by the law as a pre-disaster activity. Instead, they get funding from the general fund which is what is used for paying for general utilities, which is smaller than what they could have obtained from the LDRRMF. For inter-institutional dynamics, many of the weather and climate-related monitoring ICTs are owned and operated at the national level. As such, the quality, as well as the timeliness of data received by the LGU, depends on the efficiency of concerned national government agencies. This suggests that leadership changes affect the dynamics between national-local governments for data sharing. Although the LGU also wants to purchase their equipment and facilities, many of these are beyond the financial capacity of the LGU.

Recommendations

With the current state of ICT diffusion and use among Metro Manila LGUs, the majority of the Metro Manila survey respondents said that ICTs are not necessary to accomplish the different adaptation entry points provided by World Bank (2012)—addressing drivers of vulnerability; building response capacity of local systems; reducing and managing risks; and confronting climate change—but ICTs help with achieving these objectives more efficiently and effectively. In general, the majority of the respondents also said that they barely achieve these objectives. These are also reflected in a report by the Adaptation Knowledge Platform (AKP 2012), which highlighted that many LGUs in the Philippines still do not have the technical capacity to undertake climate change adaptation interventions. More specifically, the report found that LGUs are weak in three areas: 1) assessing a situation and creating a vision and mandate, 2) formulating policy and strategy, and 3) monitoring and evaluation. The report recommended that LGUs should have access to more data and information such as climate projections, multitemporal and multispatial hazards, biophysical and social vulnerabilities, and future climate impacts. Given its nature, ICTs could contribute significantly to these issues and recommendations from the report. All these suggest is that at present, the degree and state of ICT diffusion within the local climate e-governance among Metro Manila LGUs is still inadequate when it comes to performing climate change adaptation and risk management measures. This means that Metro Manila LGUs may need to adopt and maximize more types of ICTs on top of those that they already use. This, however, would require the government, both at the national and local levels, to invest more in improving ICT support infrastructures and facilities, such as improving internet connectivity.

Generally, the findings from the Metro Manila survey and Marikina case study also suggest that ICT use is not yet integrated within the overall climate change action framework of the government. The KI from Marikina's DRRMO shared that there are still no rules and regulations regarding ICT use for local climate action. The guidebooks, for instance, on the conduct of Climate and Disaster Risk Assessment (CDRA) as well as on the formulation of Local Climate Change Action Plans (LCCAP) do not also provide for what and how ICTs can be utilized to collect, process, and analyze climate-relevant data. The use of ICTs can be especially useful in this regard as many developing countries such as the Philippines do not have enough available climate risk-related data which limits the development of Disaster Risk Reduction—Climate Change Adaptation policies (GFDRR 2016). This disintegration of the framework also results in a gap between the capacity and the responsibility of LGUs for climate e-governance. As exemplified in the Marikina case study, there are climate change action tasks that require the use of certain ICTs that they do not have. This is also supported by the report of AKP (2012) where they found that LGUs lacked the technical capacity to fulfill their mandate under the Local Government Code. The current institutional setup restricts ICT diffusion for climate e-governance, which in turn hampers the capacity of LGUs to serve as frontline agencies for climate action. To maximize ICT use for local climate action, there is a need to integrate ICTs within the climate change action framework both at the national and local levels. If the LGUs are able to operate their monitoring stations, they will have faster access to relevant climate-related data and be able to get access to raw data that they could analyze and use for their own respective local needs. By integrating ICTs into the climate change action framework, the distribution of ICTs may be synchronized within the dynamics of the national and the LGUs in terms of climate action. Doing so could also result in the integration of ICTs into other climate change-related policies such as RA 10121. Policies, especially those that have funding implications, should recognize ICTs as integral to climate action. This will allow LGUs to access more funds and resources, such as the 70% under the LDRRMF, to acquire ICTs that can be used for local climate e-governance.

Notes

1 Checked at around 9.00pm, August 18, 2018, Netherlands.

2 Based on the 2015 census (PSA 2016).

3 Checked around 8.40pm, August 17, 2018, Netherlands.

References

Aban, J. E. (2016) "Philippines: Enhancing community resource mapping through GIS," Technical Assistance Consultant's Report. Asian Development Bank (ADB), Manila http://www.iapad.org/wp-content/uploads/2016/01/47156-001-tacr-01.pdf [Accessed April 13, 2018].

Adaptation Knowledge Platform (AKP). (2012) "Scoping assessment on climate change adaptation in the Philippines," Regional Resource Centre for Asia and the Pacific, Bangkok http://www.rrcap.ait.asia/Publications/AKP-book_CRphilippines-DAC-V1.pdf [Accessed April 13, 2018].

Alampay, E. and M. Delos Santos (2016) "Information, cities and disaster mitigation: A look at social media use by cities in Metro Manila, Philippines," CPRSouth. http://www.cprsouth.org/wp-content/uploads/2015/08/07082015-CPRsouth2015_InformationCitiesandDisasterMitigation_PP47.pdf [Accessed April 20, 2018].

Alampay, E., X. V. Asuncion, and M. Delos Santos (2018) "Management of social media for disaster risk reduction and mitigation in Philippine local government units," in Atreyi Kankanhalli, Adegboyega Ojo, and Delfina Soares (eds) *Proceedings of the 11th International Conference on Theory and Practice of Electronic Governance (ICEGOV '18)*, New York: ACM, pp 183–190. https://doi.org/10.1145/3209415.3209452

Batingaw. (n.d.) https://play.google.com/store/apps/details?id=com.batingaw.tudlo&hl=en [Accessed April 13, 2018].

Climate Change Commission. (n.d.) "National climate change action plan 2011–2028," http://extwprlegs1.fao.org/docs/pdf/phi152934.pdf [Accessed April 20, 2018].

Cucciniello, M., G. Nasi, V. Mele, G. Valotti, and R. Bazurli (2015) "Determinants and barriers of adoption, diffusion and upscaling of ICT-driven social innovation in the public sector: A systematic review," LIPSE Working Papers (No. 6). Erasmus University Rotterdam.

DICT. (2017) "National broadband plan: Building infostructures for a digital nation," DICT. Quezon City http://www.dict.gov.ph/wp-content/uploads/2017/09/2017.08.09-National-Broadband-Plan.pdf [Accessed April 13, 2018].

Dinshaw, A., D. Aarjan, and H. Mcgray. (2012) "Information for climate change adaptation: Lessons and needs in South Asia," working paper, World Resources Institute, Washington DC https://www.wri.org/sites/default/files/pdf/climate_change_adaptation_lessons_south_asia.pdf [Accessed April 19, 2018].

European Commission. (n.d.) "Climate change consequences," https://ec.europa.eu/clima/change/consequences_en [Accessed April 19, 2018].

Global Facility for Disaster Reduction and Recovery (GFDRR). (2016) "Feature story: Bridging critical gaps in actionable climate and disaster risk information," https://www.gfdrr.org/en/feature-story/bridging-critical-gaps-actionable-climate-and-disaster-risk-information [Accessed April 19, 2018].

Karanasios, S. (2011) *New & Emergent ICTs and Climate Change in Developing Countries*. Center for Development Informatics. Institute for Development Policy and Management, SED. University of Manchester. Available at: http://www.niccd.org/wp-content/uploads/2017/11/KaranasiosClimateChangeEmergentICTs.pdf [Accessed April 13, 2018].

Kumar, P., D. Kumar, and N. Kumar (2014) "E-Governance in India: Definitions, challenges and solutions," *International Journal of Computer Applications* (0975–8887) 101(16): 6–8. https://arxiv.org/pdf/1411.1876.pdf [Accessed April 13, 2018].

Kundishora, S. M. (2010) "The role of information and communication technology in enhancing local economic development and poverty reduction," World Bank https://siteresources.worldbank.org/CMUDLP/Resources/Role_ICT_paper.pdf [Accessed April 19, 2018].

Makatizen App. (n.d.) https://play.google.com/store/apps/details?id=project.smsgt.makaapp&hl=en [Accessed April 13, 2018].

NOAH. (n.d.) Nationwide Operational Assessment of Hazards, https://center.noah.up.edu.ph/the-center/ [Accessed April 13, 2018].

Olmos, S. (2001) "Vulnerability and adaptation to climate change: Concepts, issues, assessment methods," Climate Change Knowledge Network https://pdfs.semanticscholar.org/902b/ffcf082c044cdd76918974be9571f0140a5b.pdf [Accessed April 13, 2018].

Ospina, A. V. and R. Heeks (2010) "Linking ICTs and climate change adaptation: A conceptual framework for e-resilience and e-adaptation," International Development Research Centre (IDRC), Manchester https://www.preventionweb.net/files/14900_ConceptualPaper1.pdf [Accessed April 13, 2018].

PAGASA Met-Hydro Decision Support Infosys. (n.d.) Philippine Atmospheric, Geophysical and Astronomical Services Administration (PAGASA). https://v2.meteopilipinas.gov.ph/home [Accessed April 19, 2018].

PSA (Philippines Statistics Authority) (2016) Philippine population density (based on the 2015 Census of Population), http://psa.gov.ph/content/philippine-population-density-based-2015-census-population [Accessed April 13, 2018].

Reyes, C., A. Mandap, J. Qulitis, S. Calubayan, J. Naviong, R. Audar, V. Biñas, E. Bacud, and J. Moreto (2017) *The Many Faces of Poverty: Volume 8*, Manila: De La Salle University. Available at: https://www.pep-net.org/sites/pep-net.org/files/CBMS/Publications/City_of_Marikina_2015_0.pdf

Siar, S. V. (2005) "E-governance at the local government level in the Philippines: An assessment of city government websites," *Philippine Journal of Development* 32(2). Philippine Institute for Development Studies. https://core.ac.uk/download/pdf/6370626.pdf [Accessed April 13, 2018].

SOCA (State of the City Address) (2018) Makati City Government. http://www.makati.gov.ph/portal/mayors_corner/view.jsp?id=1073 [Accessed April 19, 2018].

Tollin, N., J. Hamhaber, S. Grafakos, S. Lwasa, and J. Morato (2017) *Sustainable Urbanization in the Paris Agreement: Comparative Review of Nationally Determined Contributions for Urban Content*, Nairobi: UN-Habitat. https://unhabitat.org/books/sustainable-urbanization-in-the-paris-agreement/ [Accessed April 13, 2018].

Tongco, A. (2011) "Developing a GIS for rural villages in the Philippines," *IAMURE: International Association of Multidisciplinary Research Journal* 1(1), https://ejournals.ph/article.php?id=653 [Accessed April 19, 2018].

United Nations. (2007) "Climate change: Impacts, vulnerabilities and adaptation in developing countries," https://unfccc.int/resource/docs/publications/impacts.pdf [Accessed April 19, 2018].

Waidyanatha, N. (2015) "Philippines kickoff workshop and meetings: 'CAP on a Map' for improving institutional responsiveness," workshop report, Sahana Foundation https://sahanafoundation.org/wp-content/uploads/2015/06/workshop_report_ph-CAP_final.pdf [Accessed April 19, 2018].

World Bank. (2012) "ICTs for climate change adaptation in Africa," World Bank http://siteresources.worldbank.org/EXTINFORMATIONANDCOMMUNICATIONANDTECHNOLOGIES/Resources/282822-1346223280837/ClimateChangeAdaptation.pdf [Accessed April 13, 2018].

Yusuf, A. and H. Francisco (2009) "Climate change vulnerability mapping for Southeast Asia," EEPSEA Special and Technical Paper tp200901s1, Economy and Environment Program for Southeast Asia (EEPSEA), https://idl-bnc-idrc.dspacedirect.org/handle/10625/46380 [Accessed April 13, 2018].

4

Food and Nutrition Security and Children's Health Challenges in Extreme Weather Disasters in the Philippines: Understanding the Implications of Gender Roles

Erlidia F. Llamas-Clark

Introduction

While several studies have emphasized the vulnerability and helplessness of women during disasters due to sociocultural norms (Ikeda 1995; Enarson and Morrow 1998; Fothergill 1998; Rashid 2002; Enarson and Meyreles 2004; Islam 2011), research in this area remains limited. This chapter presents the implication of gender roles to children's food and nutrition security in extreme weather events (EWEs).[1] More specifically, it links three factors: EWEs, maternal care, and child nutrition preservation in the face of a flooding disaster, using the 2009 floods in the Philippines as a case study. It provides a detailed insight into how women used critical caring roles to the advantage of the family during this calamity and contributes to building the argument that maternal care strongly contributes to child nutrition preservation in normal pre-crisis conditions and more so during EWEs and disasters.

On September 26, 2009, tropical storm Ketsana (Ondoy) brought an unusually heavy volume of rain in 12 hours causing widespread flooding in 25 provinces of Luzon, including Manila (World Bank 2011). A week later, typhoon Parma (Pepeng) brought more rain (> 1,000mm) over the already-flooded areas (World Bank 2011). Thousands of households were caught unprepared in the torrential rain and floodwaters. Agriculture was heavily affected, threatening food security across the entire country.

Undernutrition and climate

Malnutrition is a broad term used to describe the state of a body unable to get the appropriate amounts of nutrients, vitamins, and minerals needed to maintain the health of tissues and enable proper functioning of organs.[2] There are two main types of malnutrition: undernutrition (used here synonymously with protein-energy malnutrition), which is a result of inadequate intake of calories from proteins, vitamins, and minerals; and over-nutrition, which is a dietary imbalance or excess as a result of eating large amounts of food, unhealthy food choices, lack of physical activity, and unnecessary use of vitamins, minerals, and nutrients.

Climate is defined as the average weather pattern for a particular area usually over a period of 30 years. The climate has a complex system that brings about the average weather and variations in conditions (IPCC 2007). Human activities depend on the climate system. In the first instance, food production systems are influenced by the availability of crops and water, which are dependent on the climate and weather conditions. At the societal/community level, government policies and institutional regulations are partly influenced by weather/climate variables. At the household level, access to food is influenced by food prices, the financial capability of the household members, and access to credit, insurance, and support from government and business institutions. Food access and safety, access to water resources, and food preferences also influence nutrition outcomes.

Climate change as a factor in undernutrition

Global environmental change is increasingly acknowledged as one of the biggest challenges faced by humankind. There is a growing concern that undernutrition and health will be increasingly threatened by future climate change (Parry and Rosenzweig 1994; Parry, Rosenzweig, and Livermore 2005; McMichael et al 2007; Schmidhuber and Tubiello 2007; Nelson et al 2009). One of the contributors to the global burden of disease and premature deaths is climate change and its effects (Ezzati et al 2002; Palmer and Raisanen 2002 Campbell-Lendrum and Woodruff 2007). Millions of people are at risk of increasing shortages of water and food in many parts of the world that have experienced the direct impacts of EWEs—heat, storms, and flooding disasters. "By the 2080s, up to 170 million additional people will be at risk of hunger due to climate change" (Parry, Rosenzweig, and Livermore 2005, 2136).

Climate change is predicted to intensify and increase the frequency of EWEs, magnifying the risk of disasters (IPCC 2012). However, there is still uncertainty and ongoing debate about whether global warming increases incidence of tropical cyclones due to other complex factors that can alter variables. The current challenge presented by EWEs is their interface with the human population that can potentially lead to a disaster event, and how preventive and adaptation measures can be put in place (Malilay 1997; Ebi and Schmier 2005; Noji 2005; O'Brien et al 2006; Sauerborn and Ebi 2012).

Nutrition in EWEs/disasters

Nutrition in emergencies focuses on safeguarding the nutritional status of vulnerable populations to prevent disease, malnutrition and even death (Davis 1996; Kim 2012). A substantial body of literature concerns nutrition in complex emergencies (Bagchi et al 2004; White 2005). Natural disasters could also potentially increase the occurrence of complex emergencies, mass movements, and nutrition crises.

In disasters, malnutrition is not just a consequence of a lack of food. Lack of food, loss of shelter and clothing, destruction of crops and livestock, and disruption of the routine way of life of people occur in calamities. Apart from food unavailability, a rise in infectious diseases or the development of outbreaks, inadequate care and poor feeding practices, inadequate access to health care services, and hazardous environments promote malnutrition in calamities (Johns Hopkins and the International Federation of Red Cross and Red Crescent Societies 2008). Acute malnutrition is the main concern during disasters. However, because of prevailing micronutrient deficiencies and chronic malnutrition in areas of crises, emergency interventions are more likely (Webb 2009).

There are some difficulties in the implementation of nutrition interventions during disasters because of the challenging environment in these situations, lack of coordinated efforts with different sectors in relief, recovery, and rehabilitation, as well as the lack of agreed treatment protocols in a nutrition crisis (Young 1999). Household and community preparedness are advocated for better outcomes (Delica 1993; Babu and Mthindi 1995; Abarquez and Murshed 2005; Bevaola and Alam 2012; Levac, Toal-Sullivan and O'Sullivan 2012). The minimum standards for addressing these issues have recently been addressed by the Sphere Project, and its use is encouraged by the global community (Young et al 2004).

Child undernutrition

Child undernutrition occurs as a result of many unfavorable biophysical, social, and economic conditions. It is caused not only by inadequate dietary intake and disease but is also a result of insufficient food access, poor maternal and childcare practices, lack of access to water and health care services, and inadequate sanitation. These generally accepted underlying and immediate causes of malnutrition are also relevant in the development of undernutrition in EWE/disaster situations. This is apparent in the UNICEF Conceptual Framework of Food and Nutrition Security (1990, 1998), where the concepts of nutritional status and food security overlap.

Operational framework

Based on a review of the literature, I developed the EWEs, Food Security, and Population Health Operational Framework (Figure 4.1) to guide the conceptual and contextual understanding of food insecurity during disasters. In this framework, the consequences of a disaster such as food insecurity, and negative population and economic impacts are set out.

Boxes A and B indicate that the Philippines' climate and weather indicators are showing some changes such as increasing temperature, but no significant changes to trends in rainfall. The number and frequency of cyclones have not changed but there are suggestions of increased cyclone clustering. EWEs can lead to increasing incidence of floods that impact local conditions due to land use change, loss of forest and vegetation, urbanization of low-lying areas, and constriction of waterways.

Box C shows that EWEs have direct and indirect impacts on food security. EWEs can suddenly decrease food production areas, which can threaten food production, availability, and supply. This can increase food insecurity at the household level. Because of decreases in food supply and food access due to loss or lack of income, a decrease in dietary intake happens. As a result, there can be some changes in food preferences and the way food is allocated to household members.

Box D illustrates that changes in climate and local weather conditions can have direct and indirect health effects on the community and households. The direct effects are injuries, deaths, initial hunger, infectious diseases, and mental stress. These effects can compromise the health, socio-economic status, and food security of the household, particularly if the head of the household is also affected. The indirect long-term health effects are changing incidence of illnesses and undernutrition, influenced by the household's socio-economic and nutritional status.

Box E shows that environment-related problems that cause flooding can magnify the impacts of the EWEs and have direct and indirect socio-economic impacts on the community and households. EWEs can destroy infrastructures such as housing, hospitals, and communication facilities, and disrupt water supply and social services that are essential in maintaining health. As a result of EWEs, households can experience a decrease in income and reduced access to health and other essential services.

Figure 4.1: Developed EWEs, food security and population health operational framework

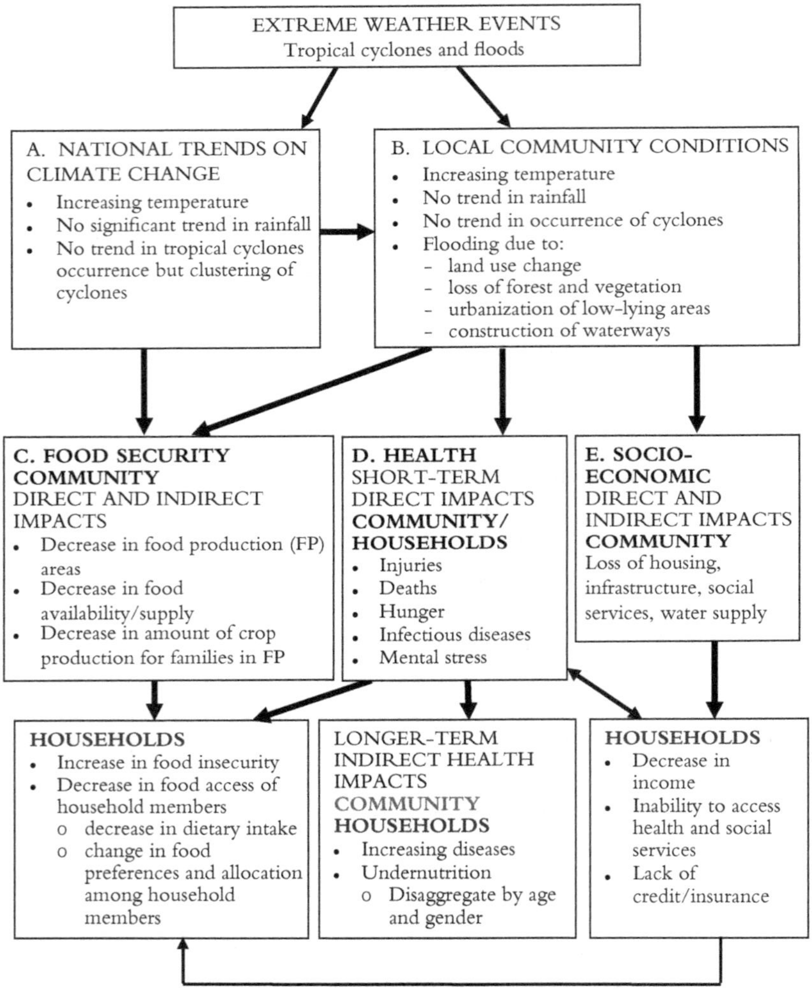

Method

The methodology was mainly driven by the research question, "how did the mothers and the community preserve Child Nutrition Security (CNS) during and after the EWEs?" This complex question is best addressed qualitatively, warranting a deeper understanding of participants' accounts of their experiences. To demonstrate division of labor and gender roles operating in disaster situations, it is necessary to understand the "lived experiences" of mothers and the communities who were exposed to, and affected by, the 2009 EWEs in the Philippines. To accomplish this, I lived among the participants in their community for almost 12 months of fieldwork to experience their culture and the activities of their daily lives to gather direct observations among the people and build a contextual understanding of the central role of the mothers as caregivers.

Several research strategies were used in gathering data from ethnographic, phenomenological, and gender perspectives. Several survey questionnaires were used to gather information about the determinants of CNS and Household Food Security (HFS) which utilized quantitative methodology. In-depth interviews were undertaken with the objective of better understanding the child, maternal, and household factors, and other determinants affecting CNS in the flood-exposed villages of Laguna. Sampling in the qualitative phase was based on the results of the quantitative phase.

The pool of mothers of either healthy or malnourished children from six municipalities was sampled. The purposive sampling technique was undertaken by selecting willing participants targeted from the study sites. The local health workers approached the mothers for initial screening with written and verbal information about the project. I conducted 20 in-depth interviews of mothers of children, with some categorized as healthy and others as malnourished, who were initially surveyed in various research sites. Pseudonyms were used in quoting responses in this chapter.

Characteristics of women participants, children, and households

At the time of the in-depth interviews, most of the women participants (14) were married and stated their religion as Catholic (15) and were of Tagalog ethnicity (19). Nine of the women were aged between 20 and 34 and ten were aged between 34 and 44. Half of the women had attained a high school or vocational education, while four had a college degree or higher. Most respondents were able to read (17) or partly read (2). Nine of the women reported their employment as home duties, five were engaged in agriculture, and four in the service/manufacturing sector. In the 12 months before the survey, 11 were engaged in some form of employment.

Characteristics of the children

Of the surveyed children of the women participants, 11 were male and nine were female. At the time of the in-depth interviews, 11 were less than 5 years of age, while 9 were older. All were born full-term, the majority (15) had been breastfed, while 18 had been immunized. In terms of birth order, six were the second child, five were the fourth child and four were the fifth child. Ten of the children had three or more siblings, while seven had one sibling. Most children (13) had some form of illness at the time of the survey. About 40% of the children were underweight and stunted.

Characteristics of households

The household income for eight of the women respondents was between PHP5,000–9,999 pesos per month, while another eight had a household income between PHP10,000–19,999 pesos per month. Half of the women respondents owned or rented their homes, while the other half shared a home with other people. All households had a toilet facility. The majority (15) had no electricity, with 13 having access to a water source that was outside the premises.

Flood exposure

The women respondents described their exposure to the flood at varying levels: to the knee (7), to the thigh (5), and chest or higher (4). They described their duration of exposure as anywhere between a few hours to up to two months.

Understanding the causes of flood

In addressing the flood disaster issue, how people understand the cause of this calamity must be determined. The affected women and their households viewed the three main causes of EWEs and the consequent flood disaster as due to 1) natural causes, 2) man-made occurrence, and 3) supernatural, religious/spiritual.

Some described the Ketsana/Parma flood disaster as an unavoidable consequence of many natural environmental factors. To understand the causes of flood according to the women who experienced it, I asked, "In your view, what was the cause of the flood?" One of the women said:

> "[…] nature. Sometimes, you do not know because the typhoons come suddenly. Even if the direct hit is not here it just comes." (Kyla)

Of those who perceived the disaster as a "man-made occurrence," the majority recognized that negative human behavior against nature or *kalikasan* was the main cause that brought them this calamity. In the traditional Filipino worldview, nature is understood to be the sum of all things in the environment. It has a life-giving essence (*likas*), which holds the universe together and "provides solidarity the world needs to exist and support human life" (Landa-Jocano 2001, 16). Human beings are perceived as part of nature and nature affects people's lives. Deforestation and excessive logging, quarrying of the river, poor waste disposal, and lack of environmental sanitation leading to clogging of waterways, and poor coordination of dam management staff with local government placed the population at grave risk.

> "Because of the garbage and destroyed trees, such that water is not anymore absorbed when it rains and then we get flooded. Do not cut down the trees. Also, the garbage should not be thrown into the river so that the waterways will not clog and disrupt the flow of the waterways." (Marie)

For some, while they understood the role of nature and humans in the disaster, they also associated this event with supernatural, religious/spiritual undertones. This is consistent with the teleological dimension of the traditional Filipino worldview. The teleological dimension is defined as the "strong tendency to attribute certain occurrences in life to be due to the workings of the environmental spirits and psychic forces" (Landa-Jocano 2001, 145). As such, the role of supernatural powers such as good or malevolent gods, spirits, and life force in the affairs of human beings is acknowledged (Landa-Jocano 2001; Bankoff 2004). One woman equated this to biblical pronouncements of the end of the world and the lack of faith and prayers of many people which brings these calamities.

> "They said it is because there are many bad people already hence the end of the world is near, and these calamities come ... Maybe because humans have become evil. Because we lack faith in God, lack of prayers. Maybe we need to strengthen our faith in God, and he will listen to us and those calamities need not happen." (Kyla)

Another woman takes a more personal view of the floods and questions God about the disaster happening at the time of their downtrodden condition.

> "It is like a trial. I said, God why is it like this? It is already flooding. It is at this time that we are in strife and I was in tears already." (Pamela)

Understanding the disaster phenomenon is an important step in setting up any post-disaster intervention. Being aware of the participants' perception of the causes of flood presents an opportunity to address these perceptions in the implementation of effective policies and programs.

Information and warning

All the women interviewed agreed that their 2009 typhoon-flood experience was extraordinary. They were accustomed to the seasonal occurrence of typhoons and did not see much threat. Before the actual floods, there was media coverage of the developing weather disturbance. The tropical storm bulletins were periodically released by the weather bureau and most of the women interviewed had watched the television, listened to the radio, and were provided information through local government officials. Nevertheless, they were caught unawares by the sudden and unexpected severity of the flooding that happened in the early hours of the morning.

As the flood came, 14 of them were caught in their homes and did not have a chance to move out. The three who felt that their homes were in danger due to house construction using light materials moved to higher ground and stayed with relatives or friends within their community. The other three stayed in the village hall and schools that were used as evacuation centers.

Anxiety and fear of loss of loved ones, home, and possessions

All the women respondents felt fear and stress at the height of the flood experience, not just for themselves but also about losing their loved ones, especially their children. Most were at home in conditions of anxiety and fear and later were traumatized by the unforgettable experiences of survival they had to endure just to get themselves and their children to safety.

> "[...] three children ... I carried on my shoulders, we went through floodwaters, and my children were already stiff as the water rose to the chest. That is what I will never forget." (Janice)

Most of the women experienced many inconveniences that added burden to their feelings of uncertainty post-flood. Two women had just recently delivered a child and one was pregnant in the third trimester during the floods, presenting challenges to mobility and requiring extra effort to ensure the safety of themselves and their children.

Next to losing family members, their greatest fear was losing their home and possessions. They felt they needed to save their few possessions from being washed out by floodwaters or from being looted, and because of this, many were reluctant to move to evacuation centers or temporary shelters. Some respondents returned to their homes within three days of the floodwaters receding to find if they had lost their personal effects, household appliances, and utilities, and at least two women lost their entire house. They had feelings of despair and devastation due to the loss of their possessions and homes. They also felt *kawawa* (bottom of the pile and filled with self-pity) when they saw that their meager possessions such as kitchen utensils, washbasins, and clothes had gone.

Different mothering experiences

Normal daily routines before the flood

All the respondent-mothers performed the child-rearing practices that were expected of them, including food preparation. All of them placed a strong emphasis on maintaining their children's cleanliness and most recognized the importance of good sanitation and hygiene practices. Respondent-mothers with school-aged children would prepare and send their children to school, do housekeeping chores, and engage in some form of income-generating activities such as food vending, ironing, or laundry work. These mothers were able to work with the help of older children or relatives who looked after the younger children.

Controlling children's activities during the flood

Three mothers noted they specifically prevented their children from being exposed to floodwater to prevent sickness. Marie reported that she did not allow them to wade and bathe in the floodwaters. After the flood, another mother, Olivia, said she provided her children with fresh water when going to school and instructed them, "Do not buy the cold refreshments because the water is dirty". One family who was confined in their house for nearly two months due to floodwaters said they practiced strict toileting instructions for their young children to prevent them from getting illnesses, with the father or older sibling carrying them to the toilet.

Long-term experiences concerning food and nutrition

For many of those who were exposed to the floods for long periods, the common themes about their flood experience included food scarcity and insecurity, loss of livelihood opportunities, and the daily struggle due to financial constraints. Even for those who were exposed to the flood only for a few days, the availability of food, and loss of income, were issues that still needed to be addressed.

Food scarcity and insecurity

Most of the women respondents found it difficult to procure food after the flood because it was scarce, and they became dependent on rations and the generosity of their neighbors and relatives.

> "It was so hard. There was no way to get food." (Pamela)

> "For us the challenge was food. Because we did not earn a lot, also I had a lot of children, seven. Instead of having three meals, I had only two meals." (Sarah)

As the flood duration lengthened, the women felt the pressure of their inability to earn their livelihoods.

> "[...] there was a lack of food, because my husband's job was to drive a tricycle and my in-law sells fruits. At that time, you cannot work and sell because of flood ... the children, that time, were feeling the pinch." (Pamela)

Food provision and allocation issues: 'Hating kapatid' *(divide like siblings) but not really ...*

'*Hating kapatid*' is an expression alluding to equal sharing. Within the household, the entire family usually eats together, ensuring everyone would get some share of the food. During the flood, food was still shared among household members; but the children would eat first, then the adults would eat together. In instances when food was scarce, households would divide the portions:

> "When it comes to allocation, we budget it and divide. If we have dried fish, we get one each, that is enough. We just have one each." (Olivia)

Faye, a mother of three children, saved money by buying already-cooked food. She divided two dishes between five of them to get by. When probed about not cooking meals, she said, she thought she would be able to stretch their food money as the cooked food was considered cheap.

Mothers skipping meals for the sake of the children

In the homes, women were the ones who allocated food. Children were given preference over adults, and mothers would find ways to cope with food insecurity. In some extreme conditions of scarcity, women sacrificed their food consumption for the sake of their children. One mother revealed:

> "Sometimes, I would end up hungry because of skipping meals. I do not eat breakfast and it is already late. I would look after them first. I eat last. Sometimes, I feel lazy to eat. ... There is none already. Sometimes I was not able to find anything to eat so sometimes I don't ... (eat anymore)." (Sarah)

Another mother, Bernadette, would forgo eating because she was living with her in-laws and not contributing financially for her children. She felt *hiya*—or coyness—to partake much because she could not contribute to the household's food or money. According to a 1994 study, cultural belief has a strong influence on this practice, with the argument that because she is not contributing to income and stays home (but does not discern the value of her unpaid work in the household), it is alright for her to forgo food, rather than her husband who has to work or the child who still needs to grow (Datinguinoo 1999).

Resourcefulness to make ends meet

After the floods, women contributed to household incomes where possible. Nita, a mother of five, decided to sell food because her husband had become unemployed. She was able to solicit some help from her sibling who provided capital, and she engaged in food vending.

Additional burdens affecting child nutrition and health

Access to local health services

Most of the children of these mothers were assessed to be healthy. However, at some point within two years of the flood, many children had some form of common illness. Three mothers said their children were sickly after the flooding. Nearly all mothers used the services of their local health centers, ensuring that their children's immunizations were complete and monthly weighing was done, and to obtain medicines for common ailments. Although the local health center could not handle some of their health problems, they continued to use these facilities due to the costs associated with transportation and accessing more distant hospitals and clinics. Several medical missions came to the villages after the flood which provided an opportunity to avail of free consultations and medicines that were lacking in the local village clinics.

Employment

In the Philippines, most of the poor are involved in the unskilled and informal sector, which is often characterized by low wages. In the flooded areas, most of the husbands were employed on contracts and their jobs were not secure. They were daily wage laborers, helpers, waiters, or dependent on demand for their skills, such as drivers, electricians, and plumbers. Unemployment was common during and after the flood for a few months and most husbands lost their livelihood and jobs.

Two spouses working in other places sent less money home as the floods also affected their contract work. One husband, who was working overseas, sent more money to the family to make sure they were comfortable after the flooding disaster. During the flood, the husbands tried to earn an income doing alternative jobs such as becoming helpers in the clean-up and operating boat rides on inundated roads. However, for some, it was hard to get alternative employment. The loss of a steady income and the lack of opportunity to secure alternative employment placed more pressure on the household budget. This in turn made it more difficult for a family to purchase sufficient food supplies to maintain their children's nutrition and health.

Rebuilding amidst financial constraints

All families reported some financial difficulty and concern, particularly during the floods. Women tried to budget but the floods placed an additional burden on their finances. While the prices of basic food items such as rice, sugar, coffee, and cooking oil were regulated by the national government, fish and vegetable prices increased due to flood effects. Most had previously lived on their daily wage, bought their food according to their earnings, and did not have any food stocks when the disaster came. During and after the flood, mothers had to do very strict budgeting to afford food, be able to send their children to school, and buy their daily non-food needs.

> "That time, I was pregnant, so I only cared for my children. My husband tried so hard to drive the tricycle even if there is flooding. He goes out at dawn then goes home only at night so we can have additional income for the children." (Beth, a mother of six)

Another respondent said her husband had to do odd jobs during the flood to tide over their family.

> "It was hard especially when (aid) was not there. We were looking hard for provisions. I took up laundry, after I would buy food in the market such as a kilo of rice, fish, cooking oil, then go home. The next day, my husband has to find some opportunities again. How will we get by?" (Carla, a mother of two)

In this flooding incident, some women were determined and fortunate enough to transform their domestic skills into income-generating enterprises (for example, doing laundry, cooking food for others) or engaging in micro-enterprises (such as a small home shop *sari-sari* store) with the financial help of relatives, thereby ensuring their children's nutritional needs.

Lack of sustainable livelihood support and financial constraints

A few women expressed their desire for support for livelihood opportunities. A number tried training seminars offered for livelihood programs such as soap and dishwashing liquid making. However, no further micro-credit and financing assistance were offered to them, so their learned skills went unused. The opportunities provided by local government post-disaster were not sustained.

Keeping your head above water: Coping strategies

Coping strategies are "social responses used to offset threats to a household's food and economic resources in times of hardship" (Open University UK 2011, 9). Coping strategies consist of food (alteration of consumption patterns) or non-food (change of income-earning or food production patterns) strategies (Maxwell et al 2003). People affected by the flood used several approaches to temper their hunger and food insecurity. Some of the households were in different stages of their food insecurity experiences while others had to forgo meals, used alternative or cheaper foods, converted assets to cash, or engaged in income-generating enterprises to ensure their children and entire household could eat and prevent illness during and after the Ketsana/Parma disaster.

Food coping strategies

Participation in local nutrition programs as a coping mechanism

Pre-flood, local governments had feeding programs in place targeting all children within the age group 0–72 months, and undernourished children. These programs provided not only food for needy children but an opportunity for children and their mothers to socialize, assist in the local community, and even partake in the meal. Sending the children to the feeding program was by itself a coping strategy for some households.

During the floods, the feeding program became irregular, and in some cases, completely stopped due to food rationing and budget constraints. This left some families struggling for adequate food supplies. While food aid temporarily filled the gap, it was not a long-term solution and children were affected again when the aid stopped.

Faye, a mother of three, and Lanie, a mother of 12, said they took advantage of the village feeding program to assist them before and after the flood. They relied on the program because they lacked food on most days, so it assured their children of food for the day.

Non-food coping strategies

Income-related coping

While many of the women in this group were unemployed, they had engaged in some occasional income-earning activities such as selling vegetables, food snacks, and other merchandise in the markets. However, when markets became inaccessible in some areas, some women sought to sell at alternative marketplaces within their village, or from home or from peddling on the roadside. One woman resorted to selling some goods given by relatives from overseas so her children could have some milk and food, or would loan from people she knew. One of the women used an extreme and rare form of coping mechanism, which is suggestive of severe food insecurity. She used a 'distress strategy' by sending her older children (from the first partner) to her brother in the nearby province of Cavite, admitting she could no longer afford to feed and send all her children to school.

Social networks as safety nets

The term *bayanihan* (spirit of cooperation/communal unity such as "toiling on another's behalf and assuming another's burden") refers to a collective action of community members to achieve a shared goal (Enriquez 1976). *Pakikipagkapwa* (shared identity) is a Filipino core value that obliges people to treat one another as fellow human beings (*kapwa*) and therefore as equals (Enriquez 1976). In the Philippines, social networks (immediate families, neighbors, and friends) can function as safety nets in times of crisis. One's religious or social groups also provide support. NGOs, private companies, and individual citizens or families offer additional help.

During the flood, mothers and their households experienced how the community exercised this Filipino spirits of *bayanihan*, *pakikipagkapwa* and *pakikisama* (sense of shared community). In times of crises, there is a collective exercise of *pakikiisa* (oneness), *pagtulong* (help), and *pagdamay* (empathy). Neighbors are expected to help not only in day-to-day work but also in times of great need (Landa-Jocano 2001). This complex concept of *pakikiramay* goes beyond 'unity' or 'togetherness' and is culture-specific (Bankoff 2007, 3). Based on these Filipino values, people act accordingly in times of challenging situations and this flood disaster was no exception.

Within the communities affected by the floods, the mothers related how the neighborhood expressed generosity and concern for one another. Some families who were better off offered their home to immediate and extended families that lacked financial resources, while others helped to ensure the safety of neighbors. There was also support in the form of food assistance and other resources such as clothes and some building materials to repair or rebuild damaged house structures. Others made sure their next of kin or neighbor was notified about relief efforts and helped in packing and distributing relief packs. Several health providers mobilized their professional organizations to provide free medical services. Many donated money or in-kind assistance to institutions that handled the coordination of assistance with the local government.

Family members, who were living afar or overseas, were equally involved during and after the calamity. It is not uncommon for Filipinos overseas to send back money regularly to their families in addition to when there is a personal or community disaster. Three of the women had families overseas who sent some financial help after they heard of the flood. Churches also offered food, shelter and, more importantly, hope to affected people. Barangay officials also helped to find safety, then assisted in ensuring that everyone was fed and not hungry.

Kawang-gawa (charity)

All the women respondents received some form of assistance during the floods. There were many sources of aid, and generous donations came from international donors such as the Red Cross, Oxfam, and bilateral aid agencies, among others. NGOs, churches, television media organizations, and private Filipino citizens all over the Philippines and around the world were generous in their support (Dassanayake 2013).[3] Most of the assistance was coordinated through the local government units (LGUs) at the provincial, municipal, then village levels. All the villages where participants resided had some aid distribution, with food being the most common. Rice rations were distributed between one and four times a week by local government staff. People were grateful for all these *biyaya*—or blessings—from all the donors who were strangers or people with influence and power.

The duration of formal aid ranged from a one-off donation to up to three months. Private assistance of relatives went even beyond three months. Three of the women respondents reported only a one-time food pack assistance, but the majority received help from anywhere between two weeks and three to four months. In the evacuation areas and some villages, there was a constant stream of food donations for three to four months after the flood. Despite the government's food aid and other forms of assistance, those in evacuation centers wanted to move back and rebuild their lives as quickly as possible. Those who were in their homes were also getting anxious to return to work and earn a living to augment their rations, which were not usually enough for the bigger households. They also wanted to buy a variety of food and not be dependent on the food pack contents.

Long-term coping mechanisms and empowerment

In the 18–24 months after the flooding, most of the women respondents conceded that it was a struggle to recover from the disaster. Many were using several coping mechanisms, trying different tactics to earn income, but found there was very minimal support for long-term recovery. Women have been portrayed by the media on many occasions as helpless and vulnerable people affected by disasters. In some ways, this is partly true. However, these media portrayals are very different from what the women experienced and how they negotiated the challenges of food insecurity in their households. In their homes, they managed life's daily struggles to ensure their family members remain healthy. Sometimes, this attitude puts women themselves at great disadvantage health and nutrition-wise (FNRI 2010).[4] Their social networks have ensured that their families coped in these calamities whether it is the next-door neighbor, an extended family member, the church, or their association with the local village officials.

In the Ketsana/Parma disaster situation, women engaged actively in income-generating jobs after the flood. This was done to earn some money or augment the income of their partners/husbands who had lost their jobs and livelihood potential. Two women respondents had a stable fixed income. Most of the households lost their husband's income. Most of the women respondents had to think of alternative ways of earning. Many of them discovered their earning potential by taking up domestic work for other people who were willing to pay. Three women respondents became collectors engaged in a small-town lottery while two others found long-term small business opportunities by putting up an eatery and food vending home business after some capital fund assistance from relatives.

In general, it seems most women gained a sense of additional empowerment after the flood. Most decided they needed to earn an income to support their families. To enable this to happen, in some instances, their husbands took on carer roles that were traditionally ascribed to the mothers. In other villages, it was a collective kind of empowerment that evolved from the flood experience. Through the assistance of the local government agriculture and nutrition office, households became driven to ensure livelihood and food security by vegetable planting in their backyards. This was a similar experience in the post-flood sustainability projects in more urbanized villages supported by the NGOs (Rom and Hill 2011).

Bahala na

The tendency to leave things to fate and chance is captured by the *bahala na* (leave it up to God) mentality of Filipinos. This attitude has been interpreted from a Western perspective as a fatalistic attitude and unquestioning acceptance. However, this attitude has been argued and debated by Filipino academics to indicate not resignation, but transcendence, to the 'let's do it' attitude fueled by courage and hope that all will be well (Enriquez 1976).

One of the women respondents said, "*Bahala na*!" when she and her husband were faced with the decision to evacuate due to the fast-rising floodwaters. They had to get out with all their children in tow and bravely evacuated amidst the floodwaters even if they risked it all for long-term safety. It is an adaptive coping mechanism subscribed to when faced with challenging situations.

Disaster preparedness

There were many lessons learned during and after the flood. After having experienced the effects of a disaster, all women respondents agreed that their households must prepare as much as they can when there is a warning of a possible cyclone, listen to the news and became more aware of local village advisories. They also put aside some clothes in a plastic bag, stored their belongings in a higher storage part of the house, and stocked rice, canned goods, and noodles, to last at least a day. However, many admitted they had been unable to maintain food stocks as they had experienced food insecurity.

Conclusion

The lived experiences of the respondent-mothers demonstrated the impact that the floods had on local conditions, especially about food security. Households experienced food insecurity at various levels. Access to food was difficult for some time due to the lack of supplies and the loss or lack of income. In some cases, families changed food preferences and the way food was allocated to household members. While there were no deaths or serious injuries, households experienced initial hunger, some illness, and anxiety, which was influenced to a significant extent by the socio-economic status of the household. The EWEs destroyed some housing and disrupted access to health services.

Various coping strategies were used during and after the floods to ensure households, particularly children, had access to food supplies. In many circumstances during and after the disaster, women acted positively and used their capacity to assert and try to change their dire state. This caring role is ultimately translated to the special care and coping strategies described above. Challenges in the performance of this social role highlight the need for acknowledgment and support for women in crises, to ensure that better nutrition and health outcomes are achieved for children and households more generally.

Notes

1 According to the International Panel IPCC Special Report on Extremes (IPCC 2012), "An extreme (weather or climate) event is generally defined as the occurrence of a value of a weather or climate variable above (or below) a threshold value near the upper (or lower) ends ('tails') of the range of observed values of the variable."

2 http://www.unicef.org/progressforchildren/2006n4/malnutritiondefinition.html [Accessed November 8, 2014].

3 Examples of Filipino generosity are many in 'normal' or disaster conditions. Every time there is disaster in the Philippines, Filipinos abroad mobilize their groups and networks to support the survivors.

4 See FNRI (2010) report on mothers' malnourishment and food insecurity.

References

Abarquez, I. and Z. Murshed (2005) *Community-based Disaster Risk Management: Field Practitioners' Handbook*, Bangkok: Asian Disaster Preparedness Center.

Babu, S. C. and G. B. Mthindi (1995) "Developing decentralized capacity for disaster prevention: Lessons from food security and nutrition monitoring in Malawi," *Disasters* 19(2): 127–139.

Bagchi, K., A. Musani, L. Tomeh, and A. Taha (2004) "Nutrition in humanitarian crises," *Eastern Mediterranean Health Journal* 10(6): 747–753.

Bankoff, G. (2004) "In the eye of the storm: The social construction of the forces of nature and the climatic and seismic construction of God in the Philippines," *Journal of Southeast Asian Studies* 35(1): 91–111.

Bankoff, G. (2007) "Living with risk: Coping with disasters hazard as a frequent life experience in the Philippines," *Education about Asia* 12(2): 26–29.

Bevaola, K. and Q. Alam (2012) "Local wisdom-based disaster recovery model in Indonesia," *Disaster Prevention and Management* 21(3): 351–369.

Campbell-Lendrum, D. and R. Woodruff (2007) "Climate change: Quantifying the health impact at national and local levels," in A. Prüss-Üstün and C. Corvalán (eds) *Environmental Burden of Disease Series, No. 14*, Geneva: WHO Public Health and the Environment.

Dassanayake, D. (2013) "Generous Britons raise incredible £23 million in 48 hours for Philippines typhoon appeal," http://www.express.co.uk/news/world/443039/Generous-Britons-raise-incredible-23-million-in 48-hours-for-Philippines-typhoon-appeal [Accessed November 14, 2014].

Datinguinoo, V. M. (1999) "The female famine investigative reports on Filipino women in the 1990s," in C. C. A. Balgos (ed) *Her Stories*, Quezon City: Philippine Center of Investigative Journalism, pp 9–14.

Davis, A. P. (1996) "Targeting the vulnerable in emergency situations: Who is vulnerable?" *The Lancet* 348(9031): 868–871.

Delica, Z. G. (1993) "Citizenry-based disaster preparedness in the Philippines," *Disasters* 17(3): 239–247.

Ebi, K. L. and J. K. Schmier (2005) "A stitch in time: Improving public health early warning systems for extreme weather events," *Epidemiologic Reviews* 27(1): 115.

Enarson, E. and L. Meyreles (2004) "International perspectives on gender and disaster: Differences and possibilitiesm," *International Journal of Sociology and Social Policy* 24(10–11): 49–93.

Enarson, E. and B. H. Morrow (1998) *The Gendered Terrain of Disaster: Through Women's Eyes*, Westport, CT: Praeger.

Enriquez, V. (1976) "Sikolohiyang Pilipino: Perskpektibo at Direksiyon," in R. Pe-Pua (ed) *Sikolohiyang Pilipino: Teorya, Metodo, at Gamit*, Quezon City: University of the Philippines Press, pp 5–21.

Ezzati, M., A. D. Lopez, A. Rodgers, S. Vander Hoorn, and C. J. L. Murray (2002) "Selected major risk factors and global and regional burden of disease," *The Lancet* 360(9343): 1347– 1360.

FNRI (Food and Nutrition Research Institute) (2010) *Philippine Nutrition Facts and Figures 2008*, Taguig City: Philippine Food and Nutrition Research Institute.

Fothergill, A. (1998) "The neglect of gender in disaster work: an overview of the literature," in E. Enarson and B. H. Morrow (eds) *The Gendered Terrain of Disasters: Through Women's Eyes*, Westport, CT: Praeger, pp 11–25.

Ikeda, K. (1995) "Gender differences in human loss and vulnerability in natural disasters: A case study from Bangladesh," *Indian Journal of Gender Studies* 2(2): 171–193.

IPCC (Intergovernmental Panel on Climate Change) (2007) "Climate change 2007. Synthesis report," Contribution of Working Groups I, II, and III to the Fourth Assessment Report. https://www.ipcc.ch/2007 [Accessed November 1, 2014].

IPCC (Intergovernmental Panel on Climate Change) (2012) *Summary for Policymakers: Managing the Risks of Extreme Weather Events and Disasters to Advance Climate Change Adaptation*, Cambridge, UK, and New York: IPCC.

Islam, M. R. (2011) "Vulnerability and coping strategies of women in disaster: A study on coastal areas of Bangladesh," *The Arts Faculty Journal* 4(June 2010-July 2011): 147–169.

Johns Hopkins and the International Federation of Red Cross and Red Crescent Societies. (2008) "Chapter 9: Food security and nutrition in emergencies," https://reliefweb.int/report/world/johns-hopkins-and-red-cross-red-crescent-public-health-guide-emergencies-second-edition [Accessed December 5, 2022].

Kim, N. (2012) "How much more exposed are the poor to natural disasters? Global and regional measurement," *Disasters* 36(2): 195–211.

Landa-Jocano, F. (2001) *Anthropology of a Filipino People V: Filipino Worldview Ethnography of Local Knowledge*, Quezon City: PUNLAD Research House Inc.

Levac, J., D. Toal-Sullivan, and T. O'Sullivan (2012) "Household emergency preparedness: A literature review," *Journal of Community Health* 37(3): 725–733.

Malilay, J. (1997) "Floods," in E. K. Noji (ed) *The Public Health Consequences of Disasters*, Oxford, UK: Oxford University Press, pp 287–301.

McMichael, A. J., J. W. Powles, C. D. Butler, and R. Uauy (2007) "Food, livestock production, energy, climate change, and health," *Lancet* 370(9594): 1253–1263.

Maxwell, D., B. Watkins, R. Wheeler, and G. Collins (2003) "The coping strategies index: A tool for rapidly measuring food security and the impact of food aid programmes in emergencies," paper presented at the FAO International Workshop on Food Security in Complex Emergencies: Building Policy Frameworks to Address Longer-term Programming Challenges, Tivoli, 23–25 September.

Nelson, G. C., M. W. Rosegrant, J. Koo, R. Robertson, T. Sulser, T. Zhu, C. Ringler, S. Msangi, A. Palazzo, M. Batka, M. Magalhaes, R. Valmonte-Santos, M. Ewing, and D. Lee (2009) *Climate Change Impact on Agriculture and Costs of Adaptation*, Washington, DC: International Food Policy Research Institute.

Noji, E. K. (2005) "Public health in the aftermath of disasters," *British Medical Journal* 330(7504): 1379.

O'Brien, G., P. O'Keefe, J. Rose, and B. Wisner (2006) "Climate change and disaster management," *Disasters* 30(1): 64–80.

Open University UK. (2011) "Nutrition module: 8. Household food security," http://www.open.edu/openlearnworks/mod/oucontent/view.php?id=321%3f [Accessed May 10, 2014].

Palmer, T. N. and J. Raisanen (2002) "Quantifying the risk of extreme seasonal precipitation events in a changing climate," *Nature* 415: 512–514.

Parry, M. L. and C. Rosenzweig (1994) "Potential impact of climate change and world food supply," *Nature* 367: 134–138.

Parry, M. L., C. Rosenzweig, and M. Livermore (2005) "Climate change, global food supply and risk of hunger," *Philosophical Transactions of the Royal Society of London. Series B. Biological Sciences* 360(1463): 2125–2138.

Rashid, S. F. (2002) "Gender and floods in Bangladesh," http://www.bracresearch.org/reports/gender_floods.pdf [Accessed February 26, 2013].

Rom, P. J. and A. Hill (2011) "Opportunities from Ondoy: From calamity to social enterprise," Resource Management in Asia Pacific Seminar, February 24, Coombs Building, Australian National University.

Sauerborn, R. and K. L. Ebi (2012) "Climate change and natural disasters: Integrating science and practice to protect health," *Global Health Action* 5: 1–7.

Schmidhuber, J. and F. N. Tubiello (2007) "Global food security under climate change," *Proceedings of the National Academy of Sciences* 104(50): 19703–19708.

United Nations Children's Fund (UNICEF). (1990) *Strategy for Improved Nutrition of Children and Women in Developing Countries*, New York: UNICEF.

United Nations Children's Fund (UNICEF). (1998) *The State of the World's Children 1998*, Oxford and New York: Oxford University Press.

Webb, P. (2009) "Malnutrition in emergencies: The framing of nutrition concerns in the humanitarian appeals process, 1992 to 2009," *Food and Nutrition Bulletin* 30(4): 379–389.

White, P. (2005) "War and food security in Eritrea and Ethiopia, 1998–2000," *Disasters* 29: S92–S113.

World Bank. (2011) "Philippines – typhoons Ondoy and Pepeng: Post-disaster needs assessment – main report," World Bank, https://openknowledge.worldbank.org/handle/10986/2776 [Accessed June 30, 2014].

Young, H. (1999) "Public nutrition in emergencies: An overview of debates, dilemmas and decision-making," *Disasters* 23(4): 277–291.

Young, H., A. Taylor, S. A. Way, and J. Leaning (2004) "Linking rights and standards: The process of developing 'rights-based' minimum standards on food security, nutrition and food aid," *Disasters* 28(2): 142–159.

5

Children and Disaster Risk Reduction: Building Resilience from Education, Local Government Units, and Communities

Roxanna Balbido Epe

Introduction

We are living in a world that is continually challenged in addressing disaster risks, coping with the impact of disasters, and building the resilience of the population, especially among the vulnerable groups and communities. The risk of disaster is increasing at the global level, and the world is becoming more hazardous to live in (Bradshaw 2013, 15–16). The Philippines ranks third as most at risk in the *WorldRiskReports* (Jeschonnek et al 2016, 2018; Jeschonnek, Kirch, and Mucke 2017), and has alleviated its risk index to the ninth rank in 2019. The Disaster Risk Reduction in the Philippines Status Report 2019 reveals data related to a continuum of threats that are also compounded with the country's young demographic profile. Children represent 38% of the Philippine population (CRC Coalition 2020, 4) and 31.4% of them are poor (Philippine Statistics Authority 2021).[1] Thus, there is a need to reduce disasters risks and vulnerability, management of disaster risks, and build resilience through an integrated and child-centered approach.

Securing the welfare of the people and focusing on the vulnerable groups of the population have been a priority toward sustainable development. Among the vulnerable groups, the children are "particularly vulnerable [...] at greater risks [...] and most affected by disasters" (Küppers et al 2018, 26). Angelika Böhling and Piere Thielbörger, in their foreword to the *WorldRiskReport 2018*, emphasize the crucial need for action that requires "comprehensive and participatory concept to protect children in fragile situations and strengthen their rights [...] as the only way to create the foundations for coming generations to develop their life perspectives, particularly in high-risk countries" (Jeschonnek et al 2018, 3). Furthermore, the increasing frequency and evolving challenges of disasters impact resulted in a paradigm shift of the global and national disaster risk reduction and resilience (DRRR) frameworks from a reactive approach to an overarching, inclusive, and proactive approach.[2] Thus, a comprehensive and integrated child-centered DRR, managing disaster risk and vulnerability, and building resilience have been integrated into the global development agenda—the Millennium Development Goals (MDGs) and the Sustainable Development Goals (SDGs). These priorities are translated into local adaptation based on global DRRR instruments like the Hyogo Framework for Action (HFA) and the Sendai Framework for Disaster Risk Reduction and other child-centered DRRR instruments[3] and frameworks promoted by the international community as bases for local action. The Philippines enacted the Philippine Disaster Risk Reduction and Management (DRRM) Act (Republic Act 10121 of 2010) with the National DRRM Framework (NDRRMF) and the National DRRM Plan (NDRRMP) (NDRRMC n.d.) for "safer, adaptive and disaster-resilient Filipino communities towards sustainable development" (NDRRMP 2011, 6). The Department of Education (DepEd) espoused the Comprehensive DRRM in Basic Education Framework[4] that is aligned with the global and national DRRR instruments and the global/regional School Safety instruments.[5]

This chapter is based on an independent and follow-up study on integrated child-centered DRRM (IC-CDRRM) programs and activities of DepEd in 25 schools, the LGUs of Maribojoc and Loon, and 26 select communities that were the project implementing partners of the School-based Disaster Risk Reduction and Management in the Philippines.[6] The information and data in the discussions were collected using qualitative research methods through key informant interviews (KIIs)[7] and focus group discussions (FGDs)[8] conducted in the summer of 2018 with the key stakeholders from the education sector, the LGUs and communities of Maribojoc and Loon. Having the positionality of being an insider-outsider in qualitative research is an advantage in sharing an objective and in-depth analysis in examining the implementation and scaling up of the integrated child-centered DRRM and resilience building (IC-CCDRRMRB) interventions and initiatives during the Bohol earthquake post-disaster response. The analysis and discussions are based on the conceptual framework of an IC-CDRRMRB that demonstrates the convergence and synergies of the education sector through DepEd with its schools, the LGUs through the Municipal DRRM Councils (MDRRMCs), and the communities through the Barangay DRRM Committees (BDRRMCs).

In this chapter, the first section discusses the need for an integrated child-centered DRRM and resilience building. The second section describes the adoption of the IC-CDRRMRB and sustaining the School-based DRR in the Philippines project interventions. The third section shows the synergies of the education sector, LGUs, and communities in adopting the IC-CDRRMRB including child participation. The fourth section presents scaling up to sustain the IC-CDRRMRB. The conclusion underscores the much-needed IC-CDRRMRB and its requisites for safer and resilient schools and communities and "disaster risk proofing development" (Bradshaw 2013, 39–40) toward sustainable development.

Integrated Child-centered DRRM and Resilience Building (IC-CDRRMRB)

While children are particularly vulnerable, at higher risk, and most affected by crises or disasters, their needs, voices, and representation are most often not given enough attention in the development processes of different social institutions. Children are among the "most socially excluded and economically insecure in any society or community are least able to access or control resources needed" in and post-disasters or crises (Enarson 2000, viii). These relative deprivations of children are exacerbated by the contentious issue of an old-school perception that adults know better or what is best for children while the latter are young and incapable of being stakeholders in development processes. However, the United Nations Convention on the Rights of the Child (UNCRC 1989) identifies children as independent rights holders, and the state and social institutions have to uphold, protect, and fulfill the rights of children, development of their full potentials, security, and well-being in all contexts. Around "one of every four children worldwide lives in an area affected by disasters" (Jeschonnek et al 2018, 7; Küppers et al 2018, 26). In the Philippines, about four in every ten Filipinos are children (CRC Coalition 2020, 4). Children now and in the future will be challenged with more and more severe catastrophes. They respond to disasters or crises differently compared to adults.

Children's vulnerability to disasters is due to several factors. Not only of their young age, "children are physically inferior to adults and not as fast or strong" (Küppers et al 2018, 26–27). Children's psychological reactions are different to those of adults, and "their coping strategies are not as developed as those of adults. During and after disasters, they may not be in a position to understand and come to terms with what has happened" (Küppers et al 2018, 28). The *WorldRiskReport 2018* identifies the physical[9] and psychological[10] factors of children's vulnerability that make them most at risk of disasters and their impact (Küppers et al 2018, 63). They are also more vulnerable to being less protected during and post-disasters.

Interrelated to children's vulnerability is the higher risks for children than adults of disasters/crises impact, because of their inferior capacities and "often not as protected by the law" (Jeschonnek et al 2018, 7) or lack of legal and institutional support and protection on the violations of their rights. Their physical, psychological, and emotional fragility or limitations have made them "more quickly overburdened and lose control more quickly [...] and insufficient care or mental traumas can massively and, in the worst case, irreversibly inhibit a child's development" (Küppers et al 2018 26–28). They are at "greater risk of being traumatized or exposed to toxic stress, which affects a child's mental health and development" (Küppers et al 2018, 28). They bear the brunt of losing or suffering disruption in their education during and post-disasters due to the damaged school infrastructures and teaching materials, or most often schools are utilized as evacuation centers. Thus, it would be challenging or impossible for economically disadvantaged children to return to school (UNDRR 2019, 9).

Vulnerability and risks to hazards and disasters/crises can be mitigated by educating, empowering, and involving the community members that should start at a young age. Thus, the need to integrate DRRM and resilience building in the education sector where children or the young population spend most of their productive time in schools learning and developing their potentials. Addressing disaster risks, vulnerability, and resilience of the population and social institutions to prevent or reduce the adverse impact and costs of disasters necessitates an IC-CDRRMRB. The schools, communities, and LGUs including relevant government agencies, non-government organizations (NGOs), and the private/business sector are the social institutions and key stakeholders in building safer and resilient schools and communities.

In this chapter, the *integrated child-centered DRRM and resilience-building* refers to the comprehensive, inclusive, child-centered, participatory, and collaborative efforts of partners such as the education sector, the government entities and their partnerships with NGOs and the private/business sector, and the communities with the active participation of children in, and putting them at the core of, the development processes, programs, and services, to reduce disaster risks including climate change-related risks, save lives, and make safer and resilient schools and communities. Representation of children in DRRM bodies[11] is notably important while respecting children's rights, giving attention and appropriate response to their needs, vulnerabilities, and risks; and recognizing their capacities and potentials as valuable partners in DRRMRB. It conforms with the Philippines policy and development agenda and NDRRMF, and the DepEd's policy and development agenda that promote access, quality, and governance with the three pillars of the Comprehensive School Safety (CSS) and the Comprehensive DRRM in Basic Education Framework.

Adopting IC-CDRRMRB and sustaining the School-based DRR in the Philippines Project Interventions

This section presents how the education sector, the municipal LGUs, and communities of Maribojoc and Loon have implemented the IC-CDRRMRB and sustained the School-based DRR in the Philippines project interventions.

Integrated Child-centred DRRM and Resilience-building in the Education Sector

All the School-based DRR in the Philippines project interventions have built the capacity and enhanced the capabilities of partners in the education sector in adopting an IC-CDRRMRB. All the DepEd research participants have emphasized that officials, district supervisors, schools' heads, teachers, and learners have applied all the knowledge and experiences they gained from the said project, like the DRRM and Psychological First Aid (PFA) training/seminars, Hazard, Vulnerability and Capacity Assessments (HVCAs), DRRM/Contingency planning workshops, advocacy, information, education and communication (IEC) activities, and other relevant interventions including child rights, child participation, and education in emergencies (EiE) through a more inclusive, participatory, and integrated approach. They have followed and complied with all the DepEd directives. They have implemented the IC-CDRRMRB in all schools and have sustained all project interventions at the division and district levels. Even non-pilot schools of the said project also adopted the same approach while the School DRRM manual was still in the development-production stage.

All DepEd respondents highlight their compliance to all the requirements indicated in national directives and instruments.[12] The national DepEd Order No. 37, series of 2015 on the Comprehensive DRRM in Basic Education Framework guides schools in assessing, planning, and implementing their prevention and mitigation, preparedness, response, and recovery and rehabilitation activities. It has institutionalized the child-centered DRRM and resilience building in basic education, to ensure all schools have a conducive and safe learning environment (Pillar 1, Safe Learning Facilities). Under CSS Pillar 2 (School Disaster Management), they conducted risks assessments with the child-led school watching teams, hazard mapping, joint HVCA and DRRM/Contingency Planning exercises with the community. The School DRRM and Contingency Plans are integrated into the Barangay/Community DRRM and Contingency Plans, which are all integrated into the Municipal DRRM and Contingency Plans. They have viable School DRRM Teams composed of teachers, children and parents that are updated annually. Their schools have the School DRR information boards,[13] the early warning system in place, hazard maps, evacuation information, DRRR key messages and signages that are posted in a conspicuous area of their campuses. Every classroom has a DRR corner for IEC materials and key messages, hazard map and evacuation information. They conduct IEC activities[14] and quarterly joint school-community–LGU multi-hazard emergency drills. They fully support child-led initiatives/ activities, children's representation in all structures, and their participation in development processes.

The research participants mentioned that the focus of CSS Pillar 3 (DRR in Education) is on mainstreaming the age-appropriate DRRR, climate change education (CCE), and EiE in the K-12 curriculum. The integration is daily in all subjects in the elementary and it is a separate subject for grades 11 and 12. It also includes capability building interventions like the orientation, training of trainers, and roll-out training for teachers and administrators on the implementation of the Comprehensive DRRM in Basic Education including the PFA training for teachers, guidance counselors, and school heads with the development and utilization of multi-media instructional/teaching and learning materials (TLMs), lesson exemplars, the School DRRM Manuals (1 and 2), PFA, and other relevant modules.

All aforecited interventions are incorporated in the three-year School Improvement Plan (SIP) and Annual Investment Plan (AIP) including allocation of resources that are integrated into the education development plans at the district, division, and regional levels. The DepEd teachers and administrators have stressed the safety and protection of all learners on and off campus and proactive participation of children in their school DRRMRB activities and development processes. The Public Schools District Supervisors are responsible for establishing strong partnerships and coordinating with the different partners in the communities, LGUs, national government agencies (NGAs),[15] NGOs, and the private sector in their area of jurisdiction. They ensure timely information dissemination on weather conditions and other impending emergencies to different schools' heads and barangays/ communities-chairpersons. They regularly monitor, assess, and ensure that all schools have followed the guidelines and proper implementation of the School DRRM and resilience-building activities, and have management oversight. Activities and implementation of IC-CDRRMRB conform to the vision of the education sector that all are "safe and resilient to risks and impacts of disasters and emergencies" (DepEd 2020, 9). The education sector's partnerships with the communities and the LGUs have facilitated their implementation of an IC-CDRRMRB.

The education sector with its government schools have invested and are on track in adopting the IC-CDRRMRB and sustaining their activities in collaboration with the communities and the LGUs. While the policy and development agenda of the education sector has been well supported with the national and local directives, DRRR frameworks, and resources to implement DRRMRB interventions through a comprehensive and integrated child-centered approach, there is a need for augmentation of investments and partnerships to sustain IC-CDRRMRB efforts considering that resources are always limited not only in the study sites in Bohol but elsewhere in the Philippines. This underscores the continuing challenges and resource requirements with the growing number of hazards, more frequent occurrence of disasters, and worsening severity of disaster risks and their impact among children—who are the most vulnerable.

Integrated Child-centred DRRM and Resilience-building in LGUs

The LGUs are duty-bound to provide protection and development interventions for young people. Thus, the Municipal LGUs of Maribojoc and Loon have adopted an IC-CDRRMRB in their local policy and development agenda. Like the education sector, the Municipal LGUs have learned a lot and sustained all the School-based DRR in the Philippines project interventions. They have enhanced their capabilities and strengthened their capacities on IC-CDRRMRB. They adhere to international and national DRRR instruments and policy agenda and ensure interventions are in place and viable.

All the LGU key informants say that children are represented in the MDRRMC. Their voices are heard and incorporated in the local development processes of their policy and development agenda. The LGU of Maribojoc has enacted their Child-Centered DRRM Code.[16] Loon LGU has implemented two executive orders supporting the adoption of child-centered DRRM and institutionalizing the representation of children in the barangays/communities DRRMCs and MDRRMCs.[17] They ensure that children participate in the implementation of their DRRM programs and activities, such as joint LGU-school-community multi-hazard emergency drills, IEC activities, and child-leaders participation in HVCA, DRRM and Contingency Planning, and MDRRMC meetings and reporting. They have allocated resources and continue to invest in children, DRRM, and resilience building. The LGUs have involved all communities, the education and other sectors (such as business, religious, and people's organizations – groups of individuals in a community or locally established grass-roots associations with common interests and share the same needs, concerns, goals or purpose, and benefits, such as the associations of parents and teachers, senior citizens, persons with disabilities, farmers, fisherfolks, women's groups, youth organizations, and among others), and NGAs in adopting IC-CDRRMRB.

The LGUs implementation of IC-CDRRMRB is a tangible outcome of recognizing child rights and potentials, child participation, and the much-needed attention on children's vulnerabilities and greater risks to disasters. This demonstrates a recognition that disasters are "development risks" and "linked to development processes," thus "disaster risks to be integrated into development or 'disaster risk proofing' development" (Bradshaw 2013, xi, 21, 39–40) through an integrated and child-centered approach for safer and resilient schools and communities. While there are significant improvements of LGUs that adopted IC-CDRRMRB with policy commitment and investment on children, but sustaining these activities, however, are subject to their financial considerations. Priorities of local chief executives and politicians would matter and affect the sustainability of LGUs IC-CDRRMRB. Thus, it is necessary to sensitize local politicians on child rights governance and IC-CDRRMRB.

ICDRRM and resilience-building in communities

The communities or Barangay DRRM Committees participants of the two FGDs highlight the support and sustained the interventions of the School-based DRR in the Philippines project that has improved their DRRM capacity and resilience building capabilities, such as multi-hazard emergency drills, early warning system (EWS) signages, child-leaders participation, IEC activities, HVCA, DRRM and contingency planning, community profiling with gender and vulnerability disaggregated database, and cluster approach during emergency response. They realized the importance of child participation, considered the needs and voices of children and other vulnerable groups in their communities, and include them in their planning and implementation of their communities' DRRMRB programs and services. They mention that:

> "children have their concerns and ideas. They have identified the hazards in their school and the community. They also raise their vulnerabilities and risks and the desire to be proactive participants in addressing them. They can make hazard maps and have their own plans too. All these are new realizations for us adults and leaders in the communities. Thus, we made a shift in our approach to involve the children, listen to their voices, and tap their capacities to protect them, build their resilience, and facilitate their development." (Maribojoc and Loon Barangay DRRM Committees FGDs participants 2018)

The community leaders and BDRRMC members are responsible for protecting and fulfilling the rights and welfare of their constituents. They tend to prioritize children and other vulnerable groups, especially during emergencies or disasters. They also observed more proactive women's participation in community DRRMRB, especially in training, seminars, and orientation sessions. Their experience of working closely with the education sector and the LGUs in improving their DRRMRB efforts through an integrated and child-centered approach manifests their realization of the need for inclusion, participation, and empowerment of children in "disaster preparedness systems [...] compiling information on risks for and with children" (Jeschonnek et al 2018, 7). The need for adequate attention on children's voices and involvement in community development processes and disaster risk proofing community development in shaping resilient communities is, therefore, acknowledged in this way. Improvements would also require similar attention on the augmentation of budget and expanding collaborations for resource mobilization to ensure sustainability of their efforts and adoption of IC-CCDRRMRB. It is a reality check that budgetary appropriations are most often inadequate to meet the actual requirements for growing populations that would be affected by more disasters and the much-needed interventions in all phases of DRRM and resilience building.

Synergies and tripartite partnership of the education sector, LGUs, and communities

This section describes the collaborations of the education sector and community, the education sector and the LGUs, the communities and the LGUs, and the participation of children in DRRM and resilience building.

Education sector and community partnership

The DepEd schools have established a strong collaboration and complementation with the communities or barangays. They are always represented including the child-leaders in the BDRRMCs. The schools and the communities are mutually supportive and proactive in their DRRM and resilience-building activities such as hazard mapping, joint school, and community HVCA, DRRM and Contingency Planning, multi-hazard emergency drills every quarter, EWS signages, IEC activities and meetings, *Brigada Eskwela*, and rescue operations. They regularly coordinate and share relevant information, advisories, and updates of activities through oral or written communication, text brigade, or social media. Open communication, reporting, monitoring, and providing feedback are observed, complementing disaster prevention and mitigation, preparedness, response, and rehabilitation and recovery measures at the local level.

The communities through the barangays councils provide daily security support to the schools through the *tanods*.[18] Some barangay councils provide funding for the construction of school toilets while others donate materials for school water, sanitation, and hygiene (WASH) facilities. Communities also allocate budgets for schools annually, depending on their capacity, to augment the schools' financial requirements for their maintenance or activities. They help establish schools' EWS and are responsible for the indigenous EWS consisting of the *kuratong* (traditional bamboo device) and *budjong* (traditional tool using big shell). The communities' hotlines or emergency contact information is always posted in schools. Maintaining or augmenting support for schools is also at stake vis-à-vis the priorities of community officials and their budget appropriations. That is why it is crucial to mainstream IC-CDRRMRB in community or local governance that requires policy support at the national and different levels of LGUs.

Education sector and LGUs partnership

The DepEd at the division, districts and schools have strengthened their partnership with the LGUs and the NGAs that are supportive of school DRRMRB. They support the LGUs policy and development agenda related to safer and resilient schools and communities. The education sector is represented by the Schools District Supervisor, District DRR Coordinator, and the District Children's Group representative(s) in the LGUs DRRM Councils where they regularly share their reports/updates and emerging concerns during regular or special meetings.

The LGUs provide technical, manpower, and financial support, information, and relevant resources on DRRMRB to the education sector. Budgets for schools' projects or activities, child-led initiatives, material assistance such as fire extinguishers, and construction materials are also provided. LGUs apportion certain DRRM funds for schools annually. They act as advisers, coordinators, and links between the education sector and other government DRRM partners at different levels of government, NGOs, and the private sector in seeking support and accessing resources to enhance and sustain their IC-CDRRMRB.

The education sector and the LGUs are closely collaborating in solid waste management; adopting information management and communication protocol; coordination, monitoring, and reporting on emergency preparedness and updates on potential hazards like the onset of typhoons, disaster response, and recovery efforts. The LGUs' hotlines or emergency contact information are disseminated or posted in schools. Overall, the education sector and the LGUs have complemented their roles and responsibilities to ensure safer and more resilient schools and communities. All these significant improvements on proactive partnerships and complementation of the education sector and the LGUs must be sustained or further enhanced. However, the political will and priorities of LGUs' executives and officials, as well as the leadership of the education sector, are critical factors in sustaining and further strengthening their collaborations in promoting IC-CDRRMRB. Likewise, it is important to secure policy commitments at all levels of LGUs governance (barangay, municipal/city, provincial) to ensure that relevant structures, development processes, resource allocation, and mobilization are in place, regardless of who is in power or authority.

Communities and LGUs partnership

The communities and the municipal LGUs have established mutual full support and cooperation in adopting an integrated child-centered DRRM and resilience building. Their strengthened partnerships facilitate inclusion, participation, open communication, reporting, monitoring, providing feedback, and resource complementation. They participate and provide inputs during HVCAs and DRRM/Contingency planning, synchronize interventions and timelines as well identify areas of complementation based on the integration of communities' DRRM/Contingency plans into those of the LGUs. The DRRM information management and communication protocol also ensure that the EWS is in place and functional in the area.

The communities are well represented in the MDRRMCs through their barangay chairpersons, BDRRMCs focal persons, and community child/youth leaders. The communities also tap into the technical support of the LGUs in their implementation of IC-CDRRMRB to include HVCA, hazard and evacuation mapping using the GPS, DRRM/Contingency planning, IEC activities, and joint multi-hazard emergency drills. Relevant records or reports are submitted to provide timely feedback to their respective LGUs. Access to the LGUs annual budgets and DRRM funds ensure that their projects or activities promote safer and resilient communities. Resource assistance to communities for disaster preparedness ensures that LGUs are responsive to their needs. Some of these resources include the provision of handheld radios, disaster emergency kits, hazard maps, training/seminars, and IEC materials. LGUs hotlines/emergency contact information are always disseminated/posted in the communities.

Bankoff et al (2015, 14) highlight the importance of recognizing "people's own priorities, perceptions, belief systems and actions, and involving the people affected and their interests in the planning and implementation of DRR" would propel convergence and synergies of various partners to adopt and sustain IC-CDRRMRB. Sustaining and further strengthening the partnerships can be critical at the intersections of the politics of LGUs and community officials. Thus, securing policy commitment and enactment thereof are crucial requisites in this context.

Child participation in DRRM and resilience-building

The IC-CDRRMRB partners in the education sector, LGUs, and communities have fully supported and encouraged children's participation in their respective development processes and child-led initiatives and activities. Data suggest that children are active and enthusiastic in the implementation of IC-CDRRMRB in schools and communities. They are involved in local DRRM bodies.[19] Their voices are heard and their inputs are considered in the schools, communities, and LGUs. Boys and girls equally participate in promoting safe and resilient schools and communities. The education sector, communities, and LGUs are following the mainstreaming of gender equality and inclusion in their programming and servicing.

In schools

All children are actively involved in all DRRM and resilience-building activities as mandated by DepEd. They support and complement their teachers, communities, and LGUs in facilitating their DRRMRB activities. All research participants have rated children's participation as excellent (10/10 scale). They say that "children really love and enjoy their involvement and serious in doing DRRMRB-related activities. They performed voluntarily."

Children have organized their Children's DRR Groups and elect their officers, to lead and facilitate their child-led initiatives/activities, and to represent their voices in DRRM bodies. They confederate at the district level[20] and conduct activities in schools and districts meetings. Children formulate their plans that are integrated into their respective Schools' DRRM and Contingency Plans. Their participation is guided by the Schools' DRR focal persons and the Schools District DRR Coordinators. They attend the school DRRM Team meetings and demonstrate their initiatives in taking charge of a committee or certain tasks, such as child-led school watching, hazard and evacuation mapping, EWS, first aid group, and running errands. Child-leaders also facilitate orientation sessions for their peers and successors to ensure the sustainability of their DRR groups and child-led initiatives.

The Child-led School Watching Team, represented by all grade levels except kindergarten, is responsible for identifying the hazards, assessing, monitoring, and reporting the risks, and evacuation mapping. The team presents their output to the school DRR focal person, and they prepare their school hazard map which is updated and presented at every School Planning Team meeting for the updating of the SIP, AIP, and Annual Procurement Plan. Child leaders are involved in schools' HVCA, DRRM, and Contingency planning.

All research participants say that "children have loved and sustained their child-led initiatives/projects and activities."[21] They implement the child-led Brigada Kontra Dengue (Brigade against Dengue) and participate in the Brigada Eskwela (School Brigade) and tree or mangrove planting. They also spearhead the WASH in Schools activities and cleanliness of their campus. Their free time in schools is devoted to child-led initiatives, which are supported by the Parents and Teachers Association (PTA) and their communities. They monitor, document, provide feedback, and report their activities to their peers and DRRM bodies. They share photo documentation, narrative reports, key messages, signages, and their stories. They have cross-learning visits, adopt, and replicate the good practices and child-led initiatives in other schools.

Child-leaders have participated in DRRR-related training and workshops for children to build their capabilities in DRRM and resilience building. They apply their learnings from the different capability-building interventions in their child-led initiatives and other school DRRMRB activities such as information campaigns, hazard mapping, and transferring knowledge to students in lower grades and PTA members. All children, including those with special needs, actively participate in the joint school–community–LGU multi-hazard emergency drills conducted every quarter. Child leaders assist during the emergency drills and act as class marshals.

Research participants in the education sector emphasize that *they are happy with the behavioral change of their learners*. Aside from their increased knowledge on DRRR, they develop and have enhanced their communication, facilitation, assessment, planning, monitoring, and reporting skills. They become their partners in implementing an IC-CDRRMRB in schools. Children *take their roles and responsibilities seriously* in DRRMRB *with passion and initiative*. School kids actively engage in DRRR discussions that are beyond their teachers' expectations.

In communities

Children's voices are represented through their schools' child-leaders and youth leaders in communities. They can attend community DRRMC meetings or activities during weekends with the consent of their parents. Children participate in community development initiatives. Community interview informants mention that "children are more active and interested than adults. Their concerns and views are beyond our thoughts that we often are unaware of or have not given enough attention." Vulnerability and risks to hazards and disasters can be mitigated by educating, empowering, and involving the community members that should start at a young age. Hore et al (2018, 1) mention that "an increasing body of research demonstrates that children can, and do, play an active role in reducing disaster risks, and the benefits of involving children in disaster risk reduction not just to the children themselves but to families, schools and wider communities."

In Maribojoc and Loon LGUs

Children participate in local DRRM Councils,[22] share their voices, and report the progress of their Children's DRR Groups and child-led initiatives during MDRRMC meetings. They are involved in the LGU DRRM planning and activities such as joint multi-hazard emergency drills, waste segregation, IEC, and hazard mapping. Children cooperate and are consulted by DRRM partners of LGUs. The LGUs key informants mention that "children are aware of disaster preparedness and mitigation. They are listening, observant, sometimes co-facilitate in DRRR IEC activities, and some are fire brigade volunteers at their grade level. They also share their knowledge to their families."

Narratives on child participation demonstrate that children are capable partners in disaster risk management. They are active participants in building safe and resilient schools and communities. Mucke (2018, 9) highlights the need to prioritize the involvement of children and "giving them a say" in disaster management. They "should be actively involved in disaster preparedness" (Jeschonnek et al 2018, 7). Furthermore, studies show that "children play a major role in recovery work [...]. Recognizing children's perspectives and capacities is a vital part of the process of building community resilience" (Mort et al 2016, 4). A better understanding of their vulnerability and risks of disasters and their impact "can inform more effective policy, enhance resilience and reduce the impact of future emergencies" (Mort et al 2016, 4).

While children are very enthusiastic as proactive partners in IC-CDRRMRB and show the drive to sustain their initiatives and involvement, the education sector, communities, and LGUs must ensure their full support to maintain the momentum of their participation or even scaling up their initiatives. Children need resources, technical support, and the guidance of adult partners, as well as facilitation of linkages and mobilization for cross-learning visits, and sustaining their participation and initiatives. Likewise, investing more in children is equally important, especially in government schools that cater to most of the economically challenged schoolchildren with greater vulnerability and risks to the impact of and recovery from disasters; otherwise, the sustainability of their initiatives or activities would be compromised.

Scaling up IC-DRRM and resilience building

Within the humanitarian and development discourse, DRRM and resilience building are at the core of their priorities toward sustainable development.

People's understanding of hazards, vulnerabilities, risks, resilience, their experiences of disasters, and "cultural production of risk, and their responses to it" (Bankoff et al 2015, 1) shape how various stakeholders and affected populations implement more meaningful interventions that promote child-centered, participatory, inclusive, and collaborative approaches. This section shows how the education sector, the LGUs and communities have scaled up to sustain their implementation of an integrated child-centered DRRM and resilience building in Bohol, Philippines.

Scaling up an IC-CDRRMRB in the education sector

The education sector in Bohol has rolled out provincial adoption in implementing an IC-CDRRMRB and replicating all the interventions and initiatives of the School-based DRR in the Philippines pilot schools. The DepEd has provided relevant training, IEC interventions, and planning workshops for all school administrators, DRR coordinators or focal persons, all teachers, and schoolchildren. Scaling up is also reinforced with the provision of multimedia TLMs, advisories, and relevant resource-links on DRRR, Climate Change Education, and Education in Emergencies through their official website. The IEC materials from Save the Children, an international organization, are reproduced by the schools for each class using their maintenance and other operating expenses (MOOE) allocation. Each teacher has taken extra effort in accessing other print and online materials and being creative in their comprehensive integration of DRRR in their classes. Children's rights, voices, and participation are central to their comprehensive and integrated approach in schools DRRM and resilience-building efforts. Thus, children are fully supported and empowered as partners in DRRM. Scaling up to sustain an IC-CDRRMRB is crucial in fulfilling the thrust of the education sector on "building a culture of safety, reducing risks and ensuring learning continuity" (DepEd 2020, 8). As the DepEd Secretary Leonor Magtolis Briones states, "education cannot wait" (DepEd 2020, 3). Thus "children should be reintegrated into a functional public school system as quickly as possible for them to gain access to society" (Jeschonnek et al 2018, 7).

The provincial rollout of DepEd in adopting IC-CDRRMRB is assured in government schools. However, there is a concern if the private schools are similarly adopting the IC-CDRRMRB or it may suggest that there is a gap in harmonizing the DRRM and resilience-building efforts of the government and private schools. While most of the economically disadvantaged children are in government schools, it does not ensure that those well-off children in the private schools are less vulnerable to hazards or the impact of disasters. This is a concern that the education sector with the government entities and communities should look into and explore ways to include the private schools and harmonize their DRRM and resilience-building efforts with the government schools, the communities, and the LGUs. Thus, scaling up of the education sector should also ensure that the private schools also adopt a similar approach to IC-CDRRMRB. Complementation of resources would be advantageous to bridge the gaps on resource allocations, knowledge gains, and awareness levels on DRRR, CCE, and EiE of all school children. Scaling up to sustain and further improve IC-CDRRMRB in the education sector is also dependent on the availability of budget appropriations that would trickle down to every school's annual investment plan and MOOE allocation.

Scaling up an IC-CDRRMRB in the LGUs

The LGUs of Maribojoc and Loon continue to scale up through more comprehensive and improved DRRMRB programs and services. They have strengthened their MDRRMCs through proactive partnerships with the communities, the education sector, and other stakeholders such as NGOs, private, and business groups. Children and other vulnerable groups including persons with disabilities, senior citizens, and pregnant women are represented in the MDRRMCs and they are involved in LGUs' development processes. They support communities' and schools' viable DRRM structures with more capable responders. The improved capacities and capabilities of LGUs Emergency Response Teams (ERTs) provide support for strengthening the communities and schools ERTs with more skills enhancement interventions, equipment, and ready supplies for emergency response and recovery. Established framework agreements with local suppliers ensure availability and timely delivery of supplies and emergency goods during relief operations. LGUs ensure full implementation of their DRRM, Contingency and Annual Investment Plans, proper utilization of their DRRM Funds, and enhanced children's protection and participation.

LGUs with the NGO(s) provide leadership and first aid response training for children and youth. They produce IEC materials and provide them to all the communities and schools. They continue to closely monitor, assess, and update their hazard and evacuation mapping, EWS, information boards, signages, and key messages. The DRRM and Contingency Plans of all communities and schools are integrated with synchronized interventions and schedules. Close coordination, information management, communication protocol, monitoring, and assessment within the LGUs and with the communities have been strengthened thus far.

The LGUs scaling up to sustain their implementation of IC-CDRRMRB is in the right direction to further enhance their capacities in disaster risk proofing their local development for safe and resilient schools and communities. Putting children's rights and voices central to their efforts, acknowledging the advantages of children's participation, and investing more in DRRM and on children reverberate the "recommendations for integrating child-centered approaches into DRR efforts" (Hore et al 2018, 1) and resilience building. Scaling up or further enhancing the LGUs' adoption of IC-CDRRMRB for safer and resilient schools and communities are subject to their yearly internal revenue allotment (IRA), priorities, and utilization of the mandatory 5% of the IRA for DRRM and resilience building. The LGUs need to be resourceful to access other means of support to augment their capacity because even without a major disaster, the resource requirements for disaster prevention and mitigation, and preparedness are huge. This is really far more than their existing budgets could afford, especially for those areas with more people that have been trying to recover and rehabilitate from a major disaster like in the study sites in Bohol.

Scaling up IC-CDRRMRBs in the communities

The communities have strengthened their respective DRRMCs with more inclusive and proactive memberships and skilled emergency responders. They continue to intensify their IEC activities and joint community–school–LGU multi-hazard emergency drills, to increase the awareness and enhance the capabilities of local people. Children's voices are heard and are supported in their community participation and child-led initiatives. Every household is required to have a Contingency Plan and emergency kit or 'go bag.' DRRM and Contingency Plans, hazard and evacuation mapping, information boards, signages, key messages, and EWS are updated. Relevant DRRR IEC materials from government organizations and NGOs are reproduced and guarantee resource complementation to the education sector.

Communities represent the basic level of local governance and are at the forefront in achieving the country's vision of "safer, adaptive and disaster-resilient Filipino communities toward sustainable development" (NDRRMC-NDRRMF 2011, 17). During this study, close coordination and partnerships within their communities and with their schools and LGUs were enhanced. Their efforts in scaling up to sustain an IC-DRRM and resilience building are fundamental and much needed in that context. The communities as the basic level of governance share similar concerns with the LGUs in scaling up and sustaining their efforts for safer schools and communities.

Conclusion

An integrated child-centered approach reinforces disaster risk-proofing development, DRRM, and resilience building. The education sector, the communities, and the LGUs have recognized the need to give sufficient attention to children's rights, voices, participation, and empowerment;

and as their valuable partners in advancing an integrated child-centered DRRM and resilience building. Their support of project interventions provides a better understanding of disaster risks and vulnerabilities, how they affected children, their limitations, and their ability to prepare and respond to disasters. Thus, these insights "will facilitate more meaningful DRR interventions" (Bankoff et al 2015, 1), and facilitate the adoption of an IC-DRRM and resilience building through tripartite partnerships of the education sector, communities, and the LGUs. This also demonstrates good governance in investing more in DRRMRB and on children, empowerment, and involvement of children and communities in local development processes, securing policy commitment, and scaling up efforts for sustainability.

The convergence and synergies of the education sector, communities, and LGUs demonstrate strengthened partnerships with a better understanding of their roles, shared responsibilities, and accountabilities in disaster risk-proofing their development initiatives, and in promoting adaptive, safer, and more resilient schools and communities. The integration of the schools' DRRM and Contingency Plans into those of the communities that are all consolidated in the LGUs' DRRM and Contingency Plans affected by the Bohol earthquake disaster indicate evidence-based, needs-responsive, and IC-DRRM and resilience-building programs and services, maximizing resources and coordinating effective implementation. Scaling up efforts to sustain the implementation of an IC-CDRRMRB are essential to achieve positive behavior change on disaster prevention and mitigation, preparedness, response, and recovery and rehabilitation.

In all contexts of scaling up in the education sector, communities, and LGUs, to sustain and further improve IC-CDRRMRB for safer and resilient schools and communities it is inevitable that there is always a policy gap, no matter how exemplary are the policy development agenda, proper full implementation, and mobilization of resources. The bottom line is that resources are always limited to meet the actual requirements vis-à-vis the emerging challenges and needs in all phases of DRRM and resilience building with more hazards, vulnerabilities, and risks identified, and the increasing threats, frequency of occurrence, and costs of the impact of disasters. While nothing is perfect, it is crucial to ensure judicious utilization of funds and expand resource mobilization to access support from civil society organizations and private/business groups. These moves would complement and augment the existing allocations that primarily rely on the fiscal appropriation of the national and local governments.

Children have unique capacities and potentials that if properly educated and empowered with an enabling environment can make a difference in DRRR. Children tend to be more open to new knowledge and guidance; they are also eager to learn and explore their curiosities, especially those that matter to or affect them. Those involved in an integrated child-centered DRRM and resilience building in schools and communities in Bohol are more capable now of protecting themselves from, responding to, and coping with the risks and impact of disasters. Enhancing their development, security, and survival potentials would translate to their families, schools, and communities' safety and resilience. Hence, it is imperative to adopt, scale up, and sustain IC-DRRM and resilience building for adaptive, safer, and resilient schools and communities toward sustainable development.

Importantly, the requisites of IC-DRRM and resilience-building include sensitivity, adequate attention and appropriate response to the needs and voices of children; proactive partnerships between the education sector, communities, and LGUs through good governance and child-centered development processes; securing policy commitment from the national government, LGUs, and communities to ensure effective implementation, scaling up, and sustainability of IC-CDRRMB; and mainstreaming of child rights governance at all levels. Empowered children with the knowledge and skills in disaster risk reduction and resilience could make a difference in schools and communities.

Notes

1 Children's population in the Philippines is pegged at 38% (38.38 million) based on the 2015 Population Census, and that the "Philippine population surpassed the 100 million mark (results from the 2015 Census of Population)" (CRC Coalition 2020, 4).

2 The paradigm shift of the global and national DRRR frameworks from a reactive approach that focuses on response interventions to an overarching, inclusive, and proactive approach with four thematic areas: prevention and mitigation, preparedness, response, and recovery and rehabilitation.

3 Other international child-centered DRRR instruments are the following: United Nations Convention on the Rights of the Child in 1989 (UNCRC) which defines "the child as an independent bearer of rights," and it "also applies without restrictions in emergency situations" (Jeschonnek et al 2018, 7); Comprehensive School Safety Framework—a global framework in support of The Global Alliance for Disaster Risk Reduction and Resilience in the Education Sector and The Worldwide Initiative for Safe Schools] (UNDRR 2017); and child-centered DRR adopted by Save the Children and other like-minded organizations working for the welfare, protection, and development of children.

4 The Comprehensive DRRM in Basic Education Framework adopted by the Department of Education based on Department Order (DO) 37, series of 2015 is based on the global, regional (ASEAN) and national Comprehensive School Safety frameworks and international and national DRRM and resilience-building instruments. It strives to: 1) protect learners and education workers from death, injury, and harm in schools; 2) plan for educational continuity in the face of expected hazards and threats; 3) safeguard education sector investments; and 4) strengthen risk reduction and resilience through education (DepEd n.d., 8). It is the guide for all entities in the education sector in "assessing, planning, and implementing their specific prevention and mitigation, preparedness, response and recovery and rehabilitation interventions" through the three pillars of the Comprehensive School Safety [Pillar 1—Safe Learning Facilities; Pillar 2—School Disaster Management; and Pillar 3—Disaster Risk Reduction in Education integrated into schools curricula

that are scaled up to include Climate Change Education (CCE) and EiE] to strengthen resilience in Basic Education (DepEd n.d.).

5 The School Safety instruments include the Comprehensive School Safety and the ASEAN Common Framework for School Safety. The Comprehensive School Safety is the global framework "to advance the goals of the Worldwide Initiative for Safe Schools and the Global Alliance for Disaster Risk Reduction and Resilience in the Education Sector and to promote school safety as a priority area of post-2015 frameworks for sustainable development, risk reduction and resilience" (UNISDR and Global Alliance for Disaster Risk Reduction & Resilience in the Education Sector 2017, 1). The ASEAN Common Framework for School Safety is the regional adaptation of the Comprehensive School Safety. It expounds on the "Operationalisation of the global Comprehensive School Safety Framework for the ASEAN context, and in concurrence with the global efforts for integrating disaster risk reduction in the education sector" (ASEAN Safe Schools Initiative 2016, n.p.).

6 The School-based DRRM in the Philippines is a three-year post-disaster development project (2014–2016) supported by the Prudence Foundation and Save the Children that involved the collaboration of the DepEd, the LGUs, and the communities of Maribojoc, Loon, and Bohol province, as well as relevant national government agencies and NGOs. The goal of the project was to help schools and communities to be better prepared for disasters through safe and enabling learning environment, saving lives, and increasing resilience of project partners (target beneficiaries) in the two pilot municipalities (Save the Children 2016, 5). It provided capacity building and relevant interventions that increased the awareness and strengthened the capacities of regional partners of government agencies, local government units, community leaders, the private sector, teachers, and students on DRR and climate change adaptation (CCA). It enhanced the links and partnerships of the education sector through the Department of Education and its schools, the municipal and provincial government units, the communities, and the private sector in adopting an IC-DRRM and resilience building.

7 The key informant interviews involved 12 participants (six men and six women), who were among the project implementing partners of the School-based Disaster Risk Reduction and Management in the Philippines. Eight key informants (four men and four women) were the DepEd schools' district supervisors, division and district DRR coordinators, select principals of pilot schools in Maribojoc and Loon, and the head of governance and operations of DepEd – Bohol Division office. Four KII participants (two men and two women) were key officials like the local chief executive or the Loon mayor, the municipal DRRM officers of Maribojoc and Loon LGUs, and the municipal planning and development coordinator of Maribojoc LGU. They have provided substantive information using the structured KII instruments.

8 Four FGDs were conducted in this independent study with a total of 23 participants (12 men and 11 women). There were two FGDs with the DepEd and the communities in each of the municipalities of Maribojoc and Loon. The DepEd FGDs involved 15 participants (five men and ten women). They were the heads of schools such as the principals and designated teachers in charge and head teacher as well as a district DRR coordinator. The FGDs with the communities were participated by the officials and DRRM focal persons (seven men and one woman), who were members of the Barangays DRRM Committees.

9 The *WorldRiskReport 2018* identifies the physical factors of children's vulnerability, namely: "small stature; not so strong physically; thin skin, are cold, and sweat very quickly; have heat strokes more quickly; store less water; dehydrate quickly; frequent contact with the ground and the outdoors through playing on the ground can promote the spread of pathogens; low level of vaccination; fast breathing; more susceptible

to respiratory diseases; move more slowly and unsteadily; swim poorly or not at all; require a balanced and nourishing diet; more quickly affected by malnutrition and undernourishment; and higher risk of lasting physical impairments than among adults" (Küppers et al 2018, 63).

10 The psychological factors of children's vulnerability as cited in the *WorldRiskReport 2018* include "high risk of overtaxing; high risk of traumatization; stress inhibits development; more unstable emotionally; and easier to manipulate" (Küppers et al 2018, 63).

11 DRRM bodies refers to the School DRRM Teams, Barangay/Community DRRM Committees, and Municipal/City/Provincial DRRM Councils.

12 The Department of Education issued relevant directives that support the Comprehensive School Safety Pillar 2 or School Disaster Management implementation. Prior to the issuance of DepEd Order No. 37, series of 2015 on August 12, which is the Comprehensive DRRM in Basic Education Framework, the Department of Education issued DO No. 50 series of 2011 on July 11 for the Creation of the DRRM Office as the focal and coordinative unit for DRRM-related activities. The directive ensures the institutionalization of the education sector's culture of safety at all levels, systematize protection of education investments and continuous quality education services delivery. The Department of Education also issued DO 21, series of 2015 on June 1 to ensure consistent and coherent operations in promoting DRRM and resilience building. It defines the DRRM Coordination and Information Management Protocol including the roles and responsibilities of the DepEd region, schools division and schools including the DRRM coordinators which are guided by the provisions of the NDRRM Plan 2011–2028 and RA No. 9155 or Governance of Basic Education Act of 2011. The DepEd DRRM Office was scaled up to DRRM Service and created permanent positions for DRRM coordinators at the national, regional and division offices, to ensure DRRM policies and programs are fully integrated at all levels with the issuance of Department Memorandum 112, series of 2015.

13 The School DRR Information Board shows the basic profile information, the DRRM structure, emergency/hotline details (school, community, LGU, PNP, BFP), early warning system guide, and GPS coordinates of the school.

14 IEC activities include DRRR orientation, seminars, for a including dissemination of DRR and resilience (DRRR) key messages during flag ceremonies, and meetings.

15 The different partners in the LGUs include the LGUs DRRM Councils (Barangay, Municipal and Provincial levels), Municipal Health Office (MHO), Municipal Social Welfare and Development Office (MSWDO), Municipal Agriculture Office, Provincial Health Office (PHO), Provincial Social Welfare and Development Office (PSWDO), Provincial Agriculture Office (PAO). Partners in the national government agencies are the Department of Health (DOH), Department of Social Welfare and Development (DSWD), Bureau of Fire Protection (BFP), Philippine National Police (PNP), Department of Interior and Local Government (DILG), Department of Science and Technology (DOST), Philippine Institute of Volcanology and Seismology (PHIVOLCS), Philippine Atmospheric, Geophysical and Astronomical Services Administration (PAGASA), and Department of Agriculture (DA), among others.

16 The Child-Centered DRRM Code of Maribojoc LGU was enacted on March 14, 2016.

17 Loon LGU has implemented Executive Orders No. 02 (Directing Municipal and Barangay DRRM Offices, Councils and Committees in the Municipality of Loon to adopt child-centered approach in their DRRM plans) and No. 03 (Institutionalizing the membership of young people and children's representatives in Barangay and Municipal DRRM Councils in the Municipality of Loon), series of 2016 as part of their policy commitment for the adoption of an integrated child-centered DRRM and resilience building.

18 The *tanods* are the community or barangay security team on duty every day in schools before and during flag ceremonies, school activities, and dismissal of classes.

19 Children are involved in local DRRM bodies like the Schools DRRM Teams with DepEd Bohol Division Memo No. 232, series of 2016 ("Institutionalizing Children and Youth Participation in the School and Local Disaster Risk Reduction and Management Structures"), the Barangays/Communities DRRM Committees and Municipal LGUs DRRM Councils, and their Children's DRR Groups.

20 All grade levels are represented by their child-leaders in the Children's DRR Groups and confederate at the district level (District Children's DRR Group/Club) represented by child-leaders of all schools.

21 School children's child-led initiatives/projects are their *Gulayan sa Paaralan* (vegetable gardens in school), school DRR kiosk/corner/laboratory, waste management, recycling of waste materials into handicrafts, and promoting DRR key messages and emergency preparedness information campaigns to peers, parents, and community members.

22 The Districts Children's DRR Groups leaders are the children's representatives in the LGU DRRM Council (LDRRMC) with the guidance of the Schools District DRR coordinators.

References

ASEAN Safe Schools Initiative. (2016) "Rollout manual for operationalization of ASEAN common framework for comprehensive school safety". https://www.preventionweb.net/files/51261_publicationaseanrolloutmanual.pdf [Accessed September 30, 2021].

Bankoff, G., T. Cannon, F. Krüger, and E. L. F. Schipper (2015) "Introduction: Exploring the links between cultures and disasters," in F. Krüger, G. Bankoff, T. Cannon, B. Orlowski, and E. L. F. Schipper (eds) *Cultures and Disasters: Understanding Cultural Framings in Disaster Risk Reduction*,., 1–16. London: Routledge.

Bradshaw, S. (2013) *Gender, Development and Disasters*, Cheltenham, UK: Edward Elgar.

CRC Coalition. (2020) "Still in the sidelines: Children's rights in the Philippines," (The Monitoring Report of the Civil Society Coalition on the Convention on the Rights of the Child (CRC Coalition) on the Implementation of the United Nations Convention on the Rights of the Child (UNCRC) 2009–2019). https://www.csc-crc.org/wp-content/uploads/2020/11/CRC-COALITION_FINAL-NGO-ALTERNATIVE-REPORT.pdf [Accessed December 6, 2021].

DepEd (Department of Education). (n.d.) "The comprehensive DRRM in basic education framework," *School Disaster Risk Reduction and Management Manual Booklet 1*, 8–22. http://depedbohol.org/v2/wp-content/uploads/2012/06/DRRM-Manual_Booklet-1_Final-1.pdf [Accessed September 29, 2021].

DepEd (Department of Education). (2020) "Strengthening resilience in basic education," https://www.deped.gov.ph/wp-content/uploads/2020/11/09_TLM_DRRMS-Brochure_20190830.pdf [Accessed September 30, 2021].

Enarson, E. (2000) "Gender and natural disasters". InFocus Programme on Crisis Response and Reconstruction, Working Paper 1, Recovery and Reconstruction Department, Geneva, Switzerland: International Labour Organization Publications https://www.ilo.org/wcmsp5/groups/public/---ed_emp/---emp_ent/--ifp_crisis/documents/publication/wcms_116391.pdf [Accessed April 8, 2020].

Hore, K., J. C. Gaillard, D. Johnston, and K. Ronan (2018) "Child-centred risk reduction research-into-action brief: Child-centred disaster risk reduction," Save the Children and Risk Frontiers, with support from MFAT (NZ), C&A Foundation and C&A. https://www.preventionweb.net/files/61522_childcentreddrrr2abriefeng2018.pdf [Accessed August 10, 2020].

Jeschonnek, L., L. Kirch, and P. Mucke (eds) (2017) *WorldRiskReport Analysis and Prospects 2017*, Berlin, Germany: Bündnis Entwicklung Hilft. https://reliefweb.int/sites/reliefweb.int/files/resources/WRR_2017_E2.pdf [Accessed September 17, 2021].

Jeschonnek, L., P. Mucke, J. Walter, and L. Kirch (eds) (2016) *WorldRiskReport 2016*, Berlin and Bonn, Germany: Bündnis Entwicklung Hilft (Alliance Development Works) and United Nations University – Institute for Environment and Human Security (UNU-EHS). https://reliefweb.int/sites/reliefweb.int/files/resources/WorldRiskReport2016_small.pdf [Accessed September 17, 2021].

Jeschonnek, L., P. Mucke, J. Walter, and L. Kirch (eds) (2018) *WorldRiskReport 2018*, Berlin and Bochum, Germany: Bündnis Entwicklung Hilft and Ruhr University Bochum – Institute for International Law of Peace and Armed Conflict (IFHV) https://reliefweb.int/sites/reliefweb.int/files/resources/WorldRiskReport-2018.pdf [Accessed September 17, 2021].

Küppers, B., F. Mischo, T. Pazdzierny, and F. Strube (2018) "Most disaster victims are children," in L. K., P. Mucke, and J. Walter (eds) *WorldRiskReport 2018*, Berlin and Bochum, Germany: Bündnis Entwicklung Hilft and Ruhr University Bochum – Institute for International Law of Peace and Armed Conflict (IFHV), pp 26–34 https://reliefweb.int/sites/reliefweb.int/files/resources/WorldRiskReport-2018.pdf. [Accessed September 17, 2021].

Mort, M., M. Walker, A. L. Williams, A. Bingley, and V. Howells (2016) "Children, young people and flooding: Recovery and resilience," Lancaster University. https://eprints.lancs.ac.uk/id/eprint/82478/4/Children_Young_People_and_Flooding_Report_Final_singles.pdf [Accessed August 12, 2020).

Mucke, P. (2018) "The situation of children in disasters," in L. Kirch, P. Mucke, and J. Walter (eds) *WorldRiskReport 2018*. Berlin and Bochum, Germany: Bündnis Entwicklung Hilft and Ruhr University Bochum – Institute for International Law of Peace and Armed Conflict (IFHV), pp 9–14 https://reliefweb.int/sites/reliefweb.int/files/resources/WorldRiskReport-2018.pdf [Accessed 17 September 2021).

National Disaster Risk Reduction and Management Council (NDRRMC). (n.d.) "Implementing rules and regulations of Republic Act No. 10121," https://ndrrmc.gov.ph/2-uncategorised/1547-implementing-rules-and-regulation-of-republic-act-no-10121 [Accessed December 5, 2021].

NDRRMC-NDRRMF (National Disk Risk Reduction and Management Framework. (2011) "The Philippine national disaster risk reduction and management framework," https://www.adrc.asia/documents/dm_information/Philippines_NDRRM_Framework.pdf [Accessed December 5, 2021].

National Disaster Risk Reduction and Management Plan (NDRRMP). (2011) "The national disaster risk reduction and management plan 2011 to 2028," https://www.dilg.gov.ph/PDF_File/reports_resources/DILG-Resources-2012116-420ac59e31.pdf [Accessed December 5, 2021].

Philippine Statistics Authority. (2021) "Children in the Philippines," https://psa.gov.ph/statistics/children [Accessed December 7, 2021].

Save the Children. (2016) "School-based disaster risk reduction in the Philippines: Endline evaluation report," Save the Children with the Support of Prudence Foundation.

United Nations Convention on the Rights of the Child (UNCRC). (1989) "United Nations Convention on the Rights of the Child," https://www.ohchr.org/Documents/ProfessionalInterest/crc.pdf [Accessed July 22, 2020].

UNDRR (United Nations Office for Disaster Risk Reduction). (2017.) "Comprehensive school safety framework: A global framework in support of The Global Alliance for Disaster Risk Reduction and Resilience in the Education Sector and The Worldwide Initiative for Safe Schools," https://www.preventionweb.net/files/51335_cssbooklet2017updated.pdf [Accessed September 24, 2021].

UNDRR (United Nations Office for Disaster Risk Reduction). (2019) "Disaster risk reduction in the Philippines: Status report 2019," Bangkok, Thailand: United Nations Office for Disaster Risk Reduction (UNDRR), Regional Office for Asia and the Pacific https://www.unisdr.org/files/68265_682308philippinesdrmstatusreport.pdf [Accessed August 30, 2021].

UNISDR and Global Alliance for Disaster Risk Reduction & Resilience in the Education Sector. (2017) "Comprehensive school safety," https://www.preventionweb.net/publication/comprehensive-school-safety-0 [Accessed September 24, 2021].

PART II

Haiyan Aftermath

6

From Disaster Woes to Economic Windfall: The Case of San Juan Women of Sta. Rita, Samar

Anita G. Cular

Introduction

I have long admired individuals who engage in entrepreneurial activities, and it always prompts me to ask them how they started. During one of our family members' birthday celebrations, we went to visit the San Juan by the Bay Floating Restaurant for a gathering. It was then that I discovered initial information about their ecotourism activity and that this was a post-Haiyan livelihood. This first visit was followed by several more as I gathered data through interviews from the members of the San Juan Women's Association.

The incredible impact of super typhoon Haiyan caught the attention of the world, and it is common knowledge that it brought massive damage to Eastern Visayas (EV). Based on the National Disaster Risk Reduction and Management Council report (NDRRMC 2013), there were 3,424,593 families and a total of 16,078,181 individuals, 12,139 barangays, and 44 provinces affected by super typhoon Haiyan (NDRRMC 2013, 2). In terms of the damages brought on dwellings, there were 550,928 totally damaged and 589,404 partially damaged houses, or a total of 1,140,332 dwellings (NDRRMC 2013, 4). Damages to infrastructure (roads, bridges, ports, water systems, power supply, school buildings, health facilities, drainage systems, telecommunications, government buildings, and so on) were estimated at PHP9,584,596,305.69 (NDRRMC 2013, 6), while those of the social services sector (education, health, and housing) amounted to PHP42,981,516,687.33 (NDRRMC 2013, 7). Lastly, in the production sector (agriculture, fisheries, mining and quarrying, trade and services, and tourism), the damages were estimated at PHP24,431,165,763.52 (NDRRMC 2013, 8). As for loss of life, there were 6,300 deaths attributed to drowning from the typhoon storm surge and flooding; 93.68% of which, or 5,902 were from EV. There were 1,062 missing persons, 94.63% or 1,005 of whom were from EV, and 28,688 were injured, 91.28% or 26,186 from EV (NDRRMC 2013, 3). The number of casualties and missing and injured individuals shows the enormity of how EV suffered from super typhoon Haiyan.

Indeed, damages to life, personal and real estate properties, and infrastructures, as well as the production and social services sector were tremendous. Super typhoon Haiyan delivered an external shock to households that distorted their normal ways of life; earnings from livelihoods either in farming or fishing were affected and in small and medium businesses (in goods and services, labor in the construction sector, etc.) were halted. Consequently, modest provision of basic needs like food, shelter, and clothing became the crux of these people's survival.

These negative impacts of typhoon Haiyan were commonly felt by those who survived. For some survivors, their immediate survival and short-term strategy was migration to nearby cities. This exodus of residents from Leyte and Samar was evident in the multitude of people queuing in the airport terminal of Tacloban City and at bus or van stations when travel services became available. However, not everyone migrated. Many chose to stay and subsisted via available relief goods given to them. As reconstruction and rehabilitation efforts were being made by the government from 2014 to 2016, local non-government organizations (NGOs), international non-government organizations (INGOs), and assistance from private foundations also proliferated in the region. With the outpouring of this assistance from international and local donors, these resources were tapped by individuals and households who worked their way toward economic recovery and a chance to improve their economic well-being. Unknowingly, the efforts of these individuals of San Juan are contributing to the attainment of Sustainable Development Goals (SDGs) that each country aspires to achieve by 2030.

This chapter is an ethnographic study about the residents of San Juan, Sta. Rita, Samar, who survived super typhoon Haiyan. As a qualitative approach, data were collected through a series of visits, interactions, and interviews among the research participants, mostly the women of Sta. Rita. They were interviewed in their respective workplaces in July 2018, March 2019, and September 2019; at the onset of the study until its completion, respectively. To maintain the privacy of the research participants, pseudonyms were used in the presentation of their accounts. Their narratives describe how they have evolved into resilient micro-entrepreneurs and now operate an ecotourism activity that is contributing productively to the local economy of San Juan.

According to Cohen (2018, 1), when a person is confronted with a tragedy, natural disasters, health concerns, and other challenges, resilience means how well a person can adapt to these events. Being able to recover easily and swiftly when compared to others is an indication of good resilience. Walsh (2011, 149–161) emphasizes that a family's resilience is "the family's ability to withstand and rebound from disruptive life challenges, strengthened and more resourceful." Walsh (2011) presents a framework that includes *belief*, *structure*, and *communication*, which refer to shared belief and finding meaning in adversity, cultivating hope and positive outlook; family connectedness and mutual support; and open and clear, tolerant and encouraging communication.

In the Philippines, resilience is aptly represented through an analogy of the bamboo plant, which is known to bend but not break. The bamboo is a symbol of humility, to go with the natural flow, and this attitude is considered important in dealing with the everyday vagaries of life. It also symbolizes the power to spring back after experiencing adversity (e.g., Haiyan), as well as growth and renewal. Everyone has a growth potential; "a continuous growth that is steady and incremental will be remarkable most especially when it is compared with what or where one used to be" (Reynolds 2011, 2).

To borrow the following quotes by author Yasmin Mogahed, "Resilience is very different from being numb. Resilience means you experience, you feel, you fail, you hurt. You fall. But, you keep going" (Mogahed 2018). Mary Holloway, a resilience coach, says that "Resilience is knowing that you are the only one that has the power and the responsibility to pick yourself up" (Holloway 2018). From an economic perspective, Hallegatte states:

> [...] to be microeconomic resilient depends on the ability of households to recover from or adjust to the negative impacts of external disturbance, which depends on the distribution of losses, on households' vulnerability, such as their pre-disaster income and ability to smooth shocks over time with savings, borrowings and insurance, and on the social protection system, or the mechanisms for sharing risks across the population. (Hallegatte 2014, 1)

Indeed, based on observations and results from this study, the women participants and their respective families and community epitomize strength of character and resilience when confronted with disaster and vulnerabilities.

Transition from desperation to self-empowerment: Women of San Juan, Sta. Rita, Samar

Sta. Rita is a municipality on the island of Samar in EV. It is considered a third-class municipality and had a population of 41,591 in 2015 (PSA

Figure 6.1: Map, Sta. Rita

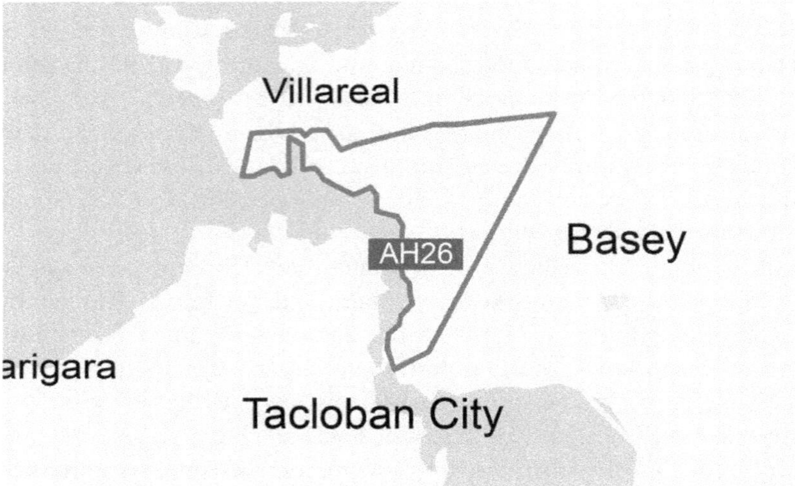

Note: Sta. Rita is the Samar municipality nearest to Tacloban City. Across the San Juanico Bridge is Sta. Rita, where Barangay San Juan belongs.

Source: Map data, Google Maps (2019).

2015). It is connected to Tacloban City on the island of Leyte by San Juanico Bridge. As one exits from Tacloban City going to the Samar Islands, the very first Samar town encountered is Sta. Rita. Thus, Sta. Rita is the nearest town to Tacloban City, where typhoon Haiyan wrought its worst fury (see Figure 6.1).

San Juan, on other hand, is one of the 38 barangays of Sta. Rita with a population of 938 in the 2015 census (PSA 2015). The residents are dependent on farming, fishing, and construction work/labor. It is a small village near to San Juanico Bridge that links Samar and the Leyte islands.

San Juan Women's Association: Forerunner of the Floating Restaurant and Boardwalk by the Bay

In a study by Lopez (2016, 2) in the Philippines, the post-disaster impact on women is exacerbated due "to food shortage, shelter, compromised health services, access to potable water, sanitation and hygiene facilities, sexual and reproductive issues." However, Lopez emphasized that women take the lead in post-disaster care where "women have shown strong leadership" when it comes to "sustaining recovery and reconstruction by identifying new sources of income and taking initiatives in demanding improvements administered by local official" (Lopez 2016, 3).

The San Juan Women's Association (SJWA) was established in January 2013 with 39 members. Before typhoon Haiyan, the association's members were engaged in mat weaving and cooking. Their woven mats were peddled in Tacloban City, and their cooked foods were sold in their barangay. Then, on November 8, 2013, typhoon Haiyan struck. As a result, the organization became inactive. The raw materials for mat weaving were no longer available. However, an opportunity to revive the association emerged via the Barangay Chairperson, a woman who actively searched for assistance when she felt that her barangay was being left out by the local government of Sta. Rita. While other barangays were given livelihood assistance, none was given to San Juan. She articulated this concern in a gathering of barangay chairpersons of Sta. Rita when a contact person of the Bantay Kalikasan-Sagip Kapamilya ABS-CBN Foundation in Samar took interest in San Juan and noted that a tourism project would be a great idea, given its proximity to San Juanico Bridge.

Based on my interview with the Barangay Chairperson, as early as February 2015, the tourism officer of the provincial government doubted the potential of the San Juan Bay Project when they surveyed the area. Contrary to the assessment of the tourism officer, the representative of ABS-CBN Foundation was convinced that the area had potential as a tourist destination based on a visual inspection. The representative brought the Chairman of ABS-CBN Foundation, the late Gina Lopez,[1] to the area to meet the Barangay Chairperson. According to the Barangay Chairperson, Lopez questioned "why it takes an outsider to see the potential of San Juan by the Bay project, and such is not seen by a local of Samar." She further challenged the Barangay Chairperson and said that "there will come a time when there will be no poor residents in San Juan." The ABS-CBN Foundation, in its effort to contribute to the communities' disaster recovery efforts, required that the initial fund be given to an organized group or a community association. Hence, the SJWA was the only beneficiary association. This donation was legitimized through a resolution made by the Barangay Council.

The San Juan Floating Restaurant and Boardwalk by the Bay was set up when the ABS-CBN Foundation gave the initial materials worth PHP1.68 million for the construction of the boardwalk and the initial two (big and medium) cottages for the floating restaurant. This financial donation gave the impetus for the project. The labor input was supplied by the residents of the barangay through volunteer work as well as free food for the volunteers, while the barangay officials contributed some of their personal funds. The construction started in May 2015. To sustain the construction for almost a year, the barangay used their General Fund amounting to PHP120,000.00. From February 17, 2016 until October 2016, the San Juan Floating Restaurant operated on a limited basis due to its ongoing construction. Full operation commenced in November

2016. Since then, the SJWA, through the San Juan Floating Restaurant and Boardwalk by the Bay, has been open to the public daily. Aside from the funds donated by the ABS-CBN Foundation, it also gave five kayaks to the SJWA. The Department of Tourism – Regional Office 8 (DOT-8) gave three more kayaks, a television set, a freezer, a gas range, and uniforms (T-shirts) to the members of the association. Except for the kayaks, the rest of DOT-8 assistance was used in the association's canteen and catering services.

Field interviews revealed that the local government unit (LGU) of Samar initially doubted the potential of the project,[2] but ultimately supported the association in the endeavor. Through Governor Sharee Ann Tan (2010–2019), Samar LGU allocated a budget of approximately PHP7million in 2017 for the construction of a barangay road, a concrete boardwalk, two big cottages, two medium cottages, a bigger kitchen, and a comfort room as part of the Samar ecotourism program.

Since its initial launch, the association has operated the San Juan Floating Restaurant and Boardwalk by the Bay. Cognizant of the fact that the operation of the ecotourism activity requires the assistance of men, the association's active 28 members are no longer exclusively women, but a mixture of 19 (remaining women from the original count of 39 original members) and 9 men members who share and are bounded by a common aspiration for a better life.

In July 2018, the association became a cooperative. It is now known as San Juan by the Bay Service Cooperative with a new set of officers and members of the Board of Directors. The transition from an association to cooperative spells a positive development of the ecotourism activity. As a cooperative, it is under the accreditation and oversight supervision of the Cooperative Development Authority (CDA), a government agency that regulates and assists cooperatives. The CDA monitors the operation of cooperatives involving human and financial management and provides opportunities for training for its members and officers. Members are guided by cooperative values like self-help, self-responsibility, democracy, equality, equity, solidarity, honesty, openness, social responsibility, and caring for others. These values are emphasized during seminars for members of the cooperative.

San Juan is a small barangay with a population of fewer than a thousand. The residents' aspirations and dreams of a better quality of life are embraced by the cooperative. The San Juan by the Bay Service Cooperative, through their Floating Restaurant and Boardwalk by the Bay, has a grand mission to help the community through the provision of suitable livelihood. The cooperative now involves men in different aspects of their operation although women comprise the majority; they seemingly work together to pursue their aspirations in life.

The San Juan Floating Restaurant and Boardwalk by the Bay

The San Juan Floating Restaurant and Boardwalk by the Bay is an ecotourism activity. As a post-disaster livelihood program, it satisfies the two-pronged objective of ecotourism: providing livelihood and protecting the environment. The San Juan Bay is an important area of work by the cooperative by planting and maintaining mangrove trees. One of the benefits of mangroves in coastal barangays like San Juan is the protection they offer from storm surges. Thus, people are encouraged to plant and protect them.

Cular (2015, 3) suggests that ecotourism is a sustainable form of tourism within natural and cultural heritage areas. It also means protection and management of natural resources and the preservation of culture, indigenous knowledge, and practices. Ecotourism also provides economic benefits to host communities with trickle-down effects. Furthermore, it means that there is community-based participation in environmental education and the use of ethical practices. The thrust of ecotourism is not only to conserve these natural tourist spots but also to generate various economic activities that will improve the lives of the people in the community. These include investment and development of small enterprises, environmental education, conservation awareness, and empowerment of the local communities. Sustainable ecotourism activities provide income support to families and consequently improve their living conditions.

The San Juan cooperative is protecting the Bay area and at the same time generating a livelihood as it offers recreation activities such as kayaking, strolling along the boardwalk amid red mangroves, promoting local food in their floating restaurant, and swimming in designated swim pens. All of these activities charge a fee. Local and foreign tourists are attracted to the scenic view of the San Juanico Bridge fronting the floating restaurant. Tourists are said to be in commune with nature and the smell of fresh air makes recreation with family and friends more enticing and pleasurable.

Socio-economic background of the members

The socio-economic background of the members of the cooperative is common with the predicament of the poverty-stricken segment of the population in the country. The members of the association are either marginal fishers or farmers or hired laborers in construction projects in nearby Tacloban City. Some women provide laundry services in Tacloban, while some sell goods in nearby cockpit areas. Their average monthly income is below PHP10,000.00 or approximately US$192.30. Five members are 4Ps beneficiaries of the government.[3] Except for a widow, all are married and have children, with eight children being the highest number. Most of them are elementary educated. The housewives, when free from their household chores, help their spouses in either farming or fishing activities. A way out from depressing economic conditions is through micro-entrepreneurship, which was made possible by external assistance. Chandy (2013, 3) states that "most of the world's poor, micro-entrepreneurship is the only way to eke out a living. Urban poor households and rural farmers in developing countries who are self-employed are micro entrepreneurs." People engage in micro-entrepreneurship because it provides a way of helping themselves out of poverty, gain income and improve their lives.

Entrepreneurship experience of the San Juan women

Hind and Driss (2017) note three typologies of women entrepreneurs. The first type refers to women entrepreneurs who come from wealthy backgrounds with financial capacity and possess professional skills or appropriate training. The second type includes women entrepreneurs who have experienced personal difficulties and are coping. Their concern is independence and self-fulfillment. Generally, these women have a certain level of knowledge and sufficient training, enabling them to do more business. The third type of woman entrepreneur is economically active, but has very limited know-how and training. Their involvement is not a choice but is borne out of necessity as a response to economic breakdown brought about by widowhood, divorce, and other factors. The low income generated by them is used to satisfy their basic needs.

The women entrepreneurs of San Juan belong to the third category. Their entrepreneurship was a response to an economic breakdown due to the disaster brought by typhoon Haiyan. With a low level of education, they were not prepared nor trained to be entrepreneurs. The motivation to ease their tight financial circumstances was paramount, but the non-monetary and non-economic benefits they derived from their entrepreneurship were also important in terms of personal outlook and behavior.

This section presents the experiences of the San Juan women as micro-entrepreneurs. These experiences demonstrate the overall growth of women as individuals and their transformation as productive members of society.

Economic empowerment

Before typhoon Haiyan, the preoccupation of women participants in San Juan was varied. Some were engaged in mat weaving, and they occasionally peddled their finished products in Tacloban City. Some women were engaged in cooking local food for sale in their community. Some of them were helping their husbands on the farm and in fishing activities. Others accepted laundry and sold goods (food and non-food items) at cockfighting events on Sundays. However, these activities did not preoccupy them on a weekly or monthly basis. Thus, their income derived from these activities was intermittent and not sustainable. During their free time, they stayed in their respective homes usually doing unpaid household chores and tending to the needs of their family. This all changed after typhoon Haiyan.

The preoccupations of women changed when the San Juan by the Bay Floating Restaurant was opened. Women summarized the changes in their lives with the following quotes:

> "We can decide individually. We feel liberated because we have our own income." (Mana, 46 years old)

> "We are helping financially our families too and it is no longer difficult to obtain money when we need it most. We help in household maintenance. I was able to sustain the college education of my daughter." (Thurn, 44 years old)

It is common knowledge that women's involvement in traditional agriculture is unpaid work, especially if they are working in their farms or rice fields. With their engagement in the ecotourism activity, the women workers now receive a monthly income. A cook earns PHP7,000.00 (US$132.76), a sweeper PHP2,000.00 (US$37.93), and gatekeepers and assistants in the canteen PHP4,000.00 (US$75.866). These women have a regular monthly income in post-disaster Haiyan, which is very different compared to their unstable pre-Haiyan income. With a permanent source of income from the cooperative, they can engage in short-term borrowings when the need arises. They have become financially independent. Women's valued economic contribution in their households and the community, and their relative cash liquidity are indicators of the women's economic empowerment. They claimed that they are empowered this time: *"We feel liberated because we have our own income,"* according to Mana, 46 years old.

Entrepreneurship

As housewives, women in San Juan during their pre-Haiyan days were spent either in unpaid farm work or housework. This was very common and they had plenty of spare time. Their usual interactions were with their neighbors. San Juan is a rural area that is commonly characterized by residents' lack of money and lack of employment opportunities. Its landscape and perceptions of people have changed with the opening of ecotourism activities in the area. The women said:

> "We feel like professionals." (Lea, 40 years old)

> "We feel like managers; we are busy." (Mana, 46 years old)

Who would think that super typhoon Haiyan would pave a way for these women to collectively act like entrepreneurs? These women are elementary educated and not trained to run a business. But despite their low educational background, they can perform the tasks of receptionists, a coordinator, a secretary who receives calls and reservations, and sometimes they act as tour guides. These are tasks expected of a well-rounded professional entrepreneur. Indeed, it gave them a feeling of being in charge of their cooperative's business operations. They are busy six days a week from 7:00 in the morning until the guests have left, usually at 4:00 in the afternoon. According to an interviewee:

> "We have this new feeling of being in charge." (Lala, 38 years old)

Noting that their daily schedules in handling the information or reception area of their business make their day more exciting and busier, a gatekeeper said:

> "Every day we take a bath early because, at 7:00 in the morning, we are ready to report to our post." (Lea, 44 years old)

Personal transformation

Since domestic activities were the preoccupations of rural poor women in San Juan, their outward appearance is not considered important. They wore old and shabby clothing. Sometimes they would dress themselves with casual wear when they went to the nearby city and the town center. Personal grooming then was not a priority in their daily routine. Putting on lipstick was not a common practice among the women in San Juan before the Haiyan disaster. However, all these things changed greatly when their entrepreneurial activity started. They said:

> "Every day I put on lipstick unlike before. We are able to wear presentable clothing." (Lala, 38 years old)

> "We are time conscious now. We feel we are beautiful. We feel glamorous." (Lea, 40 years old)

> "We have appreciated it. We become more disciplined." (Thorn, 44 years old)

Their personal transformations were not only manifested physically, that is, how they changed their personal appearance through good grooming—like taking a bath early every day, having casual wear if not in their uniform T-shirts, wearing their identification cards like employees in the formal sector, putting on lipstick and powder—but their social and emotional outlooks have improved as well. These feelings were notably nonexistent in their pre-Haiyan days. Another positive outcome is the value of discipline that they have developed in their work. Being appreciated by guests brings about feelings of happiness for the recipient of the compliment. These women were no different. Being appreciated by someone due to work well done creates a lot of positivity.

Learning social graces

The opportunity of women in San Juan to learn was made possible through their ecotourism activity. Their usual conversations before were limited to the members of their family and their neighbors. Now, however, their circle of social encounters has expanded as their social spaces widened. These comments were conveyed:

> "We say good morning, good afternoon to our guests. We become exposed to various guests with different backgrounds." (Lala, 38 years old)

> "We have attended seminars and trainings for business skills, and we have learned a lot." (Mana, 46 years old)

Dealing and conversing every day with different guests of varied backgrounds and personalities was a challenging job for them, coupled with the fact that the English language is not their mother tongue. A receptionist said:

> "Sometimes we speak English and we give directions to our guests." (Lala, 38 years old)

Nevertheless, they were able to do it successfully. Every encounter broadens their horizons, and bits and pieces of information they gathered from their guests provide them with new ways of thinking. Furthermore, social graces are important social skills that these women have been able to nurture as they continue to accomplish their daily tasks. Added to their knowledge is the informal education from training and seminars related to their work. These include basic training in handling guests and how to run a business.

Employment in ecotourism activities after super typhoon Haiyan was made possible through the concerted efforts initially by ABS-CBN Foundation, and then later the regional and local governments. These partnerships and financial assistance are seen in the improved physical structure of the boardwalk, the floating restaurant, and other amenities. The ecotourism activities have ushered in opportunities for women workers to gain experiences that will last a lifetime. Personal growth shared during interviews include the feeling of empowerment, the feeling of being in charge, and of being financially independent, but most of all their feelings of being happy:

> "We are happy and light-hearted." (Lea, 40 years old)

In the report of Meguro and Osaki (n.d.) on the empowerment of women in post-disaster reconstruction for the APEC Policy Partnership on Women and the Economy (PPWE), they suggest that self-esteem and confidence gained from the interactions with other women are part of their capacity building which are needed to sustain their entrepreneurial activity. These were observably gained by the women in Barangay San Juan. These women are the front-liners; they receive the guests they meet and talk with a lot of people from various backgrounds. Women became multi-skilled, fueled by their desire for a better life for themselves and their families.

The social impact of the ecotourism activities on the women as members of their association is in stark contrast to their pre-Haiyan daily activities. When working on the farms, they were exposed to inclement weather. When not weaving mats or cooking, they were just confined in their homes waiting for the day to end. With no source of permanent income, life was difficult. Women claimed they were not confident and independent then. They were not conscious of their looks or appearance; they looked dowdy. Although super typhoon Haiyan was the worst disaster to happen in their lives, it ironically ushered in an opportunity for them to improve their economic status through the social responsibility of organizations like ABS-CBN and the role and assistance of the local and national governments to change the lives of the Haiyan survivors for the better.

Entrepreneurship: San Juan Floating Restaurant and Boardwalk by the Bay

The income generated by the association comes from the entrance fees paid the tourists, rental of the cottages, and income from their canteen and catering services. The entrance fee for 5 to 14 year-olds is PHP10 (US$0.21) per person, while for those 15 years old and above is PHP20 (US$0.42) per person. Table 6.1 exhibits the annual visitors and the revenue generated from 2016 to 2018.

Table 6.1: Number of visitors and revenue generated, 2016–2018

Year	Average no. of tourists	Income in PHPs	Equivalent in US$ based on the prevailing exchange rate for the relevant year
2016	700	PHP14,000.00	49.8130/$ or US$281.05
2017	1,300	PHP26,000.00	49.9230/$ or US$520
2018	1,350	PHP27,000.00	52.7240/$ or US$512.100

Source of PHP price per dollar (exchange rates): Bangko Sentral ng Pilipinas. Online Statistical Database, https://www.bsp.gov.ph

The facilities offered by the Cooperative are the following: one large cottage, two medium cottages, and one small cottage with the corresponding rental fee of PHP3,500.00, PHP1,500.00, and PHP1,000.00 respectively. The Cooperative also operates a canteen where pre-ordered food is available and catering services are also offered. Assuming that the rental of cottages is realized every weekend, their total monthly revenue amounts to PHP60,000.00 (US$1,001.47) per month. Income from the canteen operation is estimated at PHP32,000.00 (US$606.42) per month.

The projected income for all the cottages (large, medium, and small) every year is estimated at PHP720,000.00 (US$13,656.02), while their canteen or catering service amounts to PHP390,000.00 (US$7,390). The total proceeds from all their revenue-generating undertakings (entrance fee, cottage rental, canteen/catering service, rentals of tables, and kayaks), are estimated to run to PHP1million. The San Juan women from an economically disadvantaged background, who have been working together in their entrepreneurial activities as a cooperative, are aiming toward becoming millionaires someday. The peak days of their operation are on weekends. Reservation is necessary if one wishes to rent the cottages, otherwise, the tourist will end up renting a table in their canteen.

The members of the women's association are indeed practicing micro-entrepreneurs. They have employed strategies to increase their income by harnessing local resources. For instance, the seed capital used for their canteen operation was an interest-free loan from the Barangay Chairperson, who is a member of the association, in the amount of PHP13,000.00 (US$246.56). From the income of their canteen operations, they were able to repay all the seed capital to the Barangay Chairperson. As they continue their profitable operation, their accumulated profits were used as their seed capital to venture into catering services, which started in May 2018.

Another indicator of entrepreneurship is the ability of the women to flow their earned income back into their operation to expand their business. One tangible proof was the use of the income from their operation to build their information kiosk at the entrance, the parking area, two additional cottages (one small and one medium), and a floating comfort room. These additional physical structures are an indication of their thriving micro-entrepreneurship with ongoing expansion.

Operation of the San Juan Floating Restaurant and Boardwalk by the Bay

In 2019, there were 28 members of the cooperative with women representing the majority (19), taking turns in managing their daily operations. In their first year of operation, all those who were working in the ecotourism services were purely volunteers because they did not receive any honorarium. They aimed to keep their business rolling.

The flow of the operations of the San Juan Floating Restaurant and Boardwalk by the Bay commences at the information kiosks. As guests arrived at the site, they are welcomed by the women on duty, where guests are asked to sign in the logbook for record purposes and the determination of the entrance fee. The registration and payment slip are brought by the guest to the floating restaurant, where the cashier is stationed and payment is accepted there together with whatever food was ordered and consumed. Then the guests travel down the boardwalk (see Figures 6.2 and 6.3) until they reach the floating restaurant by the bay. An alternative for the elderly or disabled guests who are not able to tread on the boardwalk is to use the floating cottage as a big boat for an extra fee. Upon reaching the destination, they can opt to rent a floating cottage or simply use a table with chairs on the floating canteen. Guests can also pre-order their food from the catering service or they can order them upon arrival at the site. Aside from these options, guest may rent kayaks while others simply enjoy swimming in the bay. The women at the entrance also coordinate with the other members of the association who are managing the floating restaurants and the men in charge of the floating cottage for transport purposes. They inform them through cellphone calls on how many guests have arrived, how many are renting cottages, and other details.

Ecotourism activities are translated into revenue by the Cooperative and income to their members. Following the initial voluntary work of the members in their first year of operation, they decided to give a meager honorarium to its members in their second year of operation. An allocation of PHP5,000.00 or US$100.15 (based on the 2017 exchange rate) for an honorarium was allocated, and this was divided among those members who rendered service or work for the month. The following year, in June 2018, their honorarium was increased to PHP6,000.00 or US$113.80 (2018 exchange rate). The women members who were assigned at the entrance

Figure 6.2: Original boardwalk

Note: The original boardwalk was made of bamboo. Photo by A. G. Cular, July 2018.

each received PHP60.00 (US$1.07) per day. Then, in November 2018, their honorarium was increased from PHP60.00 to PHP133.00 (US$2.52) per day.

Male members, on the other hand, were earning more income because each was paid PHP300.00 (US$6.00) per day for construction or repair jobs, such as fixing the bamboo boardwalk. In addition, when the floating cottage was used for transporting the guests to and from the entrance to the site, they were paid PHP500.00 (US$10.01), and this was divided among the men who were involved in the conveyance work. Members who were

Figure 6.3: Concrete boardwalk

Note: The boardwalk is 400m in length, 200m of which is concrete.
Photo by A. G. Cular, September 2019.

in charge of the canteen and the catering service (the cashier/cook and the dishwasher/helper) received initially an honorarium equivalent to 10% of the income of the canteen and for every catering service. This time, their monthly income is pegged at PHP7,000.00 or (US$132.76) for the cashier or cook, and PHP4,000.00 or (US$75.86) for the dishwasher/helper. The type of work remains gendered with men performing physical labor as carpenters and boat transport.

Economic benefits

The cooperative is projected to earn PHP1million from its ecotourism entrepreneurial activities. According to the members, the PHP133.00 (US$2.522) received per day by the gatekeepers as their honorarium is small and not even enough to pay for refreshments in a café shop or eating out in a fast food restaurant in the city. But from the members' point of view, PHP4,000.00 (US$75.866) income per month is already a windfall; otherwise, their households would be dependent on their spouses' daily income from farming, fishing, and construction. For the women, having a meager but sustained income spells a big difference. They are no longer reliant on their husbands' incomes. They can help their family in financing their expenditures, and they do not worry much about where to obtain money when they need it most. The PHP4,000.00 or US$75.866 monthly income is comparable with the salary of household help, or the salary of a laundrywoman, or a salesperson in a department or grocery store. However, there is an existing policy of the cooperative of 'no work, no pay', which seems to be a common practice in commercial establishments in the city, too.

The San Juan women's involvement in micro-entrepreneurship is similar to the study of Franck (2012, 73) in Penang, Malaysia, where she found that "women's micro entrepreneurship can be motivated by a wide range of factors including: to earn an income; interest in doing business; increased flexibility and autonomy; possibility to combine with family obligations; and renegotiating spatial practices." This study indicates that necessity and choice are the factors behind the motive to enter into entrepreneurship. Thus, the motive of San Juan Women to earn income to augment the family income, as well as to have their own financial flexibility and independence relate well with this study. The sustained operation of the ecotourism services paves the way for poverty alleviation and a brighter future for the members.

Prospects of the cooperative: The Floating Restaurant and Boardwalk by the Bay

The future operation of the cooperative and their business seems promising. According to the members, they will see to it that it will be sustained in providing ecotourism service to the public and sustained income to the members. As they say, "*uugupan namon*" or they will be behind to fully support it and manage it well. They claimed it as their very own. They have already worked for a year voluntarily, and now, even with a meager honorarium, they are motivated to work harder (see Figure 6.4).

Given the fact that the assteociation received additional assistance from other entities such as the local government of Samar and the Department of Tourism, this synergy between the government, the private sector, and

Figure 6.4: The gatekeepers

Note: The gatekeepers (from left to right): Lorna, Nelia, and Nora. At the back is Feding, the sweeper at the entrance and its environs. Photo by Carlito Cular, September 2019.

the stakeholders of San Juan signals a positive development and expansion of the ecotourism activities. At any given time, ecotourism projects can simultaneously accommodate more guests. An improvement in the road network provides better transportation to the Bay area and, therefore, encourages more tourists. Consequently, this would mean more income for the cooperative. As of my last visit on September 18, 2019, the construction of the concrete boardwalk was ongoing and was 50% completed.

Lopez (2016, 3) indicates that "access and control of productive resources and results of wealth-creating careers are important factors for the independence of both women and men." This finds relevance in the study where the members of the San Juan by the Bay Service Cooperative do have access to and control of the ecotourism activities in the Bay area. They are committed to continually improving and sustaining their livelihood by working together with other support agencies and organizations. In due course, their socio-economic status is projected to be better than their pre-Haiyan conditions.

San Juan Women and SDGs

Gender equality is one of the 17 SDGs to transform our world by 2030. This goal may not be clear to the women in San Juan but they are contributing impressively to the achievement of some other SDG goals like no poverty, zero hunger, decent work opportunities, and economic growth at the community level. Aside from fostering women's empowerment, the SJWA, and later the cooperative, is working toward a sustainable future where poverty is reduced through local partnerships and the well-being of the people in rural communities is improved. Their motivation to engage in entrepreneurial activity is rooted in their desire to minimize the ills of poverty and the security of food for the family. Gender equality enables equal access to opportunities and decent work is enjoyed by the women as they work together with male members of the cooperative in San Juan.

Concerning SDG 'Life Below Water,' the women are contributing to the management and protection of marine and coastal resources. Their ecotourism activity necessitates the maintenance of mangrove plantations surrounding the San Juan Bay area. The mangrove plantation protects the shorelines and serves as a natural barrier from storm surges in case of typhoons. Super typhoon Haiyan in 2013 was a clear reminder of the danger of death from storm surges that women and their families do not wish to experience again.

Conclusion

The post-Haiyan socio-economic conditions of the 19 women members of the San Juan by the Bay Service Cooperative show marked improvement compared to their pre-Haiyan situation. This improvement is demonstrated in their overall well-being. Being happy and content with their achievements, no matter how inconsequential they may appear to others, is a personal triumph for them. Being able to socialize with other individuals has given them a window for self-improvement and the realization that they can engage in other productive activities aside from farming, fishing, weaving, cooking, or doing laundry; that of being micro-entrepreneurs.

Super typhoon Haiyan left marks of destruction and trauma, yet women survivors chose to rehabilitate their communities. Economic resilience is the only alternative that marginalized women in farming and fishing industries look forward to. In this case, external assistance from a private foundation, as well as from the local and national governments, were vital to uplift the worst conditions of households in post-disaster communities. The vicious cycle of poverty and socio-economic problems among the people in San Juan was transformed through timely and relevant interventions. The support given by the ABS-CBN Foundation and the collaboration with other stakeholders, specifically the local government of Samar, paved the way for the entrepreneurial activities of the SJWA, which ushered positive economic and social impact into their lives. The role of women in the community as leaders is much recognized in understanding the socio-economic conditions in the area. The female Barangay Chairperson stood out when she passionately complained about why her community was left out when livelihood assistance was given to other communities. Her complaint and initiative paved the way for acquiring financial support and establishing the Floating Restaurant and Boardwalk by the Bay. A dismal economic situation post-Haiyan has been transformed into sustainable ecotourism by local people.

Indeed, the members of San Juan by the Bay Service Cooperative are a resilient group of individuals whose desire to improve their lives has been successful so far. Their low level of education was not a barrier; many women believe that there is always room for improvement no matter how difficult life is. Their willingness to act, organize, collaborate, and seek external assistance are contributing factors to their success. The most important factor points to having an open mindset and a positive attitude in moving forward to alleviate their socio-economic status. These women transformed themselves as agents of change in their economic, physical, and social outlooks. Their gains have a multiplying effect on their respective families, and toward the community in Barangay San Juan. With sustained local employment opportunities and income generation, Barangay San Juan is projected to have better outcomes than its pre-Haiyan conditions. As the Barangay Chairperson shared in an interview, the challenge is still on and "that gradually no single family in Barangay San Juan will remain poor." At this time, they are headed in the right direction.

Like the resilient bamboo plant, the women in San Juan have participated in a slow but sustained economic windfall that improved their outlook in life compared to their pre-Haiyan conditions. The women and their families have turned the tides of disaster to become resilient micro-entrepreneurs and empowered individuals.

Notes

1 She served the Philippines in her capacity as Secretary of the Department of Environment and Natural Resources.

2 Based on the response and account of the Barangay Chairperson during the interview.

3 4Ps or Pantawid Pamilyang Pilipino Program, also known as conditional cash transfer (CCT), is the anti-poverty program of the government that grants subsidy to the poorest households in the country, to improve their health, nutrition, and education. The conditionalities of the grant are primarily for the parents to send their children to school and to the health care centers on a regular basis.

References

Chandy, R. (2013) "How micro-entrepreneurs could change the world". Deloitte Institute of Innovation and Entrepreneurship, pp 1–11 https://www.london.edu/lbsr/how-microentrepreneurs-could-change-the-world [Accessed November 13, 2018].

Cohen, H. (2018) "What is resilience?" https://psychcentral.com/lib/what-is-resilience/ [Accessed September 15, 2019].

Cular, A. G. (2015) "Ecotourism and value adding: Source of livelihood among the villages in Basey Samar, Eastern Visayas, Philippines," *The Social Sciences* 10(6): 826–831.

Franck, A. K. (2012) "Factors motivating women's informal micro-entrepreneurship experiences from Penang, Malaysia," *International Journal of Gender and Entrepreneurship* 4(1): 65–78. DOI:10.1108/17566261211202981

Google Maps. (2019) Map data. https://www.google.com/maps/place/Santa+Rita,+Samar/@11.4045932,124.8429571,11z/data=!3m1!4b1!4m5!3m4!1s0x330869aff9f7da47:0x1debdaa2c31b9caf!8m2!3d11.4500484!4d125.0167866 [Accessed June 16, 2019].

Hallegatte, S. (2014) "Economic resilience: Definition and measurement," Policy Research Working Paper No. 6852. Washington, DC: The World Bank, https://openknowledge.worldbank.org/handle/10986/18341 [Accessed November 6, 2018].

Hind, B. and F. Driss (2017) "An exploratory study of women entrepreneurship in Morocco," *Journal of Entrepreneurship Research and Practice* (2017): 1–19. DOI:10.5171/2017.869458

Holloway, M. (2018) "66 resilience quotes to help you during tough times," Jay Liew, Liew.@GreatBigMinds, https://www.greatbigminds.com/resilience-quotes [Accessed November 9, 2018].

Lopez, A. A. N. (2016) "Reducing the vulnerability of women in disaster stricken areas," National Defense College of the Philippines (NDCP) Policy Briefs, No. 1, pp 1–6. https://www.ndcp.edu.ph/wp-content/uploads/2021/12/EPB-LOPEZ_Reducing-the-Vulnerability-of-Women-in-Disaster-Stricken-Areas.pdf [Accessed April 30, 2019].

Meguro, Y. and A. Osaki (n.d.) "Good practices on economic empowerment of women in post-disaster reconstruction in Tohoku and the Asia-Pacific," APEC Policy Partnership on Women and the Economy (PPWE), https://www.mofa.go.jp/mofaj/files/000088221.pdf [Accessed September 16, 2019].

Mogahed, Y. (2018) "Top 20 Yasmin Mogahed quotes". https://www.youtube.com/watch?v=SQs0QT9CPpA [Accessed November 2, 2018].

National Disaster Risk Reduction Management Council (NDRRMC). (2013) "Final Report re Effects of Typhoon 'Yolanda' (Haiyan)," https://ndrrmc.gov.ph/attachments/article/1329/FINAL_REPORT_re_Effects_of_Typhoon_YOLANDA_HAIYAN_06-09NOV2013.pdf, [Accessed January 19, 2020].

PSA (Philippines Statistics Authority) (2015) "2015 census", https://psa.gov.ph/content/highlights-philippine population-2015-census [Accessed March 17, 2021).

Reynolds, G. (2011) "Presentation zen be like a bamboo: Lessons from work and life," https://www.presentationzen.com/presentationzen/2011/06/be-like-the-bamboo.html [Accessed September 26, 2019).

Todaro, M. (1997) *Economic Development*, London, UK: Longman.

Walsh, F. (2011) "Family resilience: A collaborative approach in response to stressful life challenges," Resilience and Mental Health. Challenges Across the Lifespan (January): 149–161. DOI:10.1017/CBO9780511994791.012

7

Disaster Events and Role Transitions: Narratives of Filipino Rural Women after Typhoon Haiyan

Rowena S. Guiang and Ervina A. Espina

The Philippines ranks as the third most at-risk country in the world (Radtke et al 2018). It is a disaster-prone country due to its geographical location. On average, the Philippines experiences 20 typhoons annually (Information on Disaster Risk Reduction of the Member Countries n.d.). Typhoons are a commonplace experience in the Philippines. The most frequently hit areas are located in the center, as well as the eastern part of the country. Inhabitants of these areas are inured to these events, therefore, minimal preparations are made.

The eastern part of the country is the first to be hit when typhoons enter the Philippine area of responsibility (PAR). On November 8, 2018, one of the strongest typhoons (typhoon Haiyan) to ever hit land devastated the islands of Samar and Leyte (Region VIII) in the Visayas. It left in its wake thousands of casualties and tons of debris. The extent of the damage wrought by the event disrupted the normal day-to-day functioning of the locals (in this case, the women of San Juan, Sta. Rita, Samar). Access to basic needs such as food, shelter, clothing; sanitation, and locals' sense of safety and security were also disrupted. Aside from the material losses, the social-psychological aspects of people's lives were also affected, which necessitated major adjustments on their part.

One of the vulnerable sectors which was adversely affected by typhoon Haiyan were the women of Barangay (Brgy.) San Juan, Sta. Rita, Samar. The town is located right beside the iconic San Juanico Bridge, which connects the islands of Samar and Leyte. Barangay San Juan is situated in a coastal area which makes it vulnerable to disasters. It is one of the barangays which was adversely affected when Haiyan devastated Region VIII.

This chapter expands the idea that disasters expose women's vulnerabilities as they contend with the after-effects of disaster events—in this case typhoon Haiyan. Women are vulnerable to begin with, but the impact of typhoon Haiyan, which resulted in the loss of livelihood, houses, and material things, among others, calls to attention the effects of these losses to women's role transitions and role expansion when they are forced to contend with the challenges associated with these events. Women experienced role transitions and role expansion as they responded to the effects of typhoon Haiyan as a life event, even as they continued to perform their traditional gender roles. We define life events as situations that affect women's multiple roles—roles before and after the disaster—and the role transitions that took place. We contend that although these role transitions may have already taken place before the disaster, life events like typhoon Haiyan enable the overlapping of women's multiple roles and the assimilation of new roles like becoming the breadwinner, becoming an active community worker, and so on. Such a situation results in the blurring of role boundaries for women, which then impacts the accomplishment of these roles even as they struggle to find balance in these changes.

Despite the struggles accompanying the disaster brought by typhoon Haiyan, the women of Brgy. San Juan, Sta. Rita, found their voices through the help of an external organization. Their recovery was facilitated by the Bantay Kalikasan program implemented by ABS-CBN Lingkod Kapamilya Foundation, a socio-civic foundation whose thrust, among others, is to help survivors rebuild their communities.

The women's stories reflect their struggles and finding their voices in the midst of the challenges they faced because of the impact of typhoon Haiyan. Women's active role and engagement in disaster recovery reflects the UN's agenda for Sustainable Development specifically to achieve gender equality. Toward this end, women must be recognized for their active involvement and contribution to community building. Investing in women's empowerment ensures sustainable development. Intervention strategies must put a premium on mechanisms for the participation and engagement of low-income women. Their capabilities and strengths in disaster and post-disaster contexts must be harnessed to reinforce and maintain shifts in gender roles.

Data for this research was generated through a modified interviewing process described as a 'walking interview.' It is a qualitative tool that facilitates free responses from the experiences of respondents within a specific location. Responses from these interviews were organized using thematic analysis by Braun and Clarke (2006) to produce our insights of women's narrative of their role transitions.

Disaster and women's vulnerabilities

Typhoon Haiyan highlighted the vulnerabilities of the women of Brgy. San Juan, Sta. Rita. They suffered tremendous losses especially since they lived in a community at risk of disaster, located a few meters from the sea. Most of the houses in the area are built of lightweight materials and are unable to withstand powerful winds. Even the mangroves lining the coast offered meager protection from the storm surge that inundated their entire community. Due to the massive impact of the disaster (that is, material losses), survivors experienced disorientation as they struggled with multiple material losses and life disruptions (Dynes 2005; Weiss and Korn 2006). As shared by one participant:

> "Our house was submerged in water. We ran towards the plaza because our house was destroyed. Our clothes, everything was taken away by the strong waves. When we looked at our house, it was gone; destroyed. We did not recover our things." (Lydia, 57)

The losses accompanying the disaster were catastrophic for the women of Brgy. San Juan, Sta. Rita. They lost not only their material possessions (houses, clothes) but also the physical spaces (gardens, rice fields, pig pens, fish pens) for most of their livelihood activities. Their sources of livelihood (fruits, vegetables ready for harvest, livestock, fish) were lost as well. A participant shared:

> "We lost everything after the typhoon. We lost our house here and the other one near San Juanico bridge where we had our fish pens. We lost everything. We also lost all our livestock, we lost our carabao." (Melinda, 52)

These losses delay the recovery of women, especially vulnerable women from trauma (Enarson, Fothergill, and Peek, 2007; Drolet et al 2015). In fact, for these women, it was a struggle to stay alive, especially during the emergency phase of the disaster, since aid and relief were not easily accessible to their community. This situation is not surprising since there was difficulty in accessing some areas like in Brgy. San Juan in Sta. Rita. One participant narrated:

> "It was good that the police stationed at San Juanico Bridge shared the food that they were able to get from people who passed by, they gave us coffee." (Melinda, 52)

Another woman shared:

> "Yes, we received assistance and relief goods, I am not saying that we did not get any, but some relief goods did not get to us." (Edna, 54)

Women living in poverty especially those in Brgy. San Juan, Sta. Rita, have low education attainment, limited mobility, low income, and limited decision-making powers and are in a disadvantaged position compared to other groups, such as men and women living in higher-income urban settings (cf. Horton 2012; Bradshaw 2014; Reyes and Lu 2015). The reality is that disasters leave everyone exposed to wide-scale devastation and post-disaster challenges (Shenk et al 2010; Srivastava 2010), but more so for women who live in poverty. Poverty incidence is high in the Philippines. Data from the Philippines Statistical Authority (PSA 2019) for the first semester of 2018 reveal that Western Samar, where Sta. Rita is located, recorded the second-highest poverty incidence across the whole of Region VIII. This situation reveals that three in every ten families in the area are poor, meaning that their income is below the poverty threshold or the minimum income required to be able to provide basic food and non-food needs (PSA 2019). One woman shared:

> "I remember how difficult my life was. I worked as a servant. I was able to send my children to school until the sixth grade only; since we didn't have coconut trees (for copra) nor farmlands (to augment our income). That's why I engaged in a variety of odd jobs like mat weaving, working as a servant offering to clean someone's garden and farmlands." (Antonia, 70)

Another participant reported:

> "Even before the typhoon, our life was already quite difficult. Because of the kind of work my husband has (fishing and construction work). But God provides so even if we do not earn much we somehow manage to get by. We have to because we cannot find other sources of income." (Lydia, 57)

The women's difficult circumstances before the typhoon were exacerbated by the challenges in the period after the typhoon; they became more disadvantaged. In addition, existing unequal gendered power relations further limit their access to basic health services, deprive them of material resources, cause them to suffer social isolation, and loss of livelihood opportunities, which contributes to the most difficult challenge of the recovery phase (Samuels 2012; Reid 2013). Therefore, both their poverty and traditional roles at home put women in a vulnerable position (Reyes and Lu 2015).

One participant shared the pain of the many difficulties that she and her family experienced as a consequence of the typhoon, saying:

> "One of my biggest problems is money… of finding money to be able to send my child to school… I still have not fully recovered from my problems (i.e. loss of livelihood) from the past few years, primarily the financial problems, things like that." (Edna, 54)

Despite their poverty situation, vulnerabilities, and the challenges encountered post-Haiyan, women participants in Brgy. San Juan still manifest the role of survivors rather than miserable victims of disaster (Jayarathne 2014). This is reflected in their desire to rebuild with their families.

The struggle to survive and rebuild

The women of Brgy. San Juan faced enormous losses, most especially their houses and livelihoods. Given the immediacy of these needs, they had to find the means to address their lack of housing facilities and livelihood for them to survive. However, the difficulty in accessing resources impacted their capacity to rebuild their lives. Rebuilding their houses and their community was made more difficult by the lack of resources, which contributed to a long and painful process of recovery. Low-income rural women generally do not have savings to fall back on during emergencies, and even if they do it is still not enough for their emergency needs. Theirs is a life enmeshed in a cycle of borrowing which is why in times of emergencies, they do not have the resources to address their needs. Thus, the myriad problems faced by women before typhoon Haiyan was compounded by their lack of financial resources in the rebuilding phase, both at the individual household level, as well as the community level. One of the participants recalled their difficult circumstance, saying:

> "It was a slow process. I am grateful things are slowly getting back to what it was before the typhoon. I was also worried about our livelihood… our house was mostly destroyed and had to be repaired. People's properties were destroyed. Everybody had to do repairs. It was like that (after the storm). I borrowed from CARD [a lending company]. I was able to buy things to replace what was lost. Not all at once, slowly, one at a time." (Edna, 54)

Another woman shared her experience in replacing what they lost:

> "One of my children who are in Manila sent us money so that we can buy a cooking pan. That's all that we could manage to buy. Since there was no electricity in Tacloban we went to Calbiga [a city in Samar]

since it was spared from the typhoon. They had electricity there. That is where we went to get the money that my child sent. It was just 700 pesos." (Lydia, 57)

Before the typhoon, house repairs, even minor ones, were not even a priority to the participants given their limited income. After super typhoon Haiyan destroyed their dwellings, house reconstruction and major repairs became their primary concern. Owing to the destruction, they did not have any choice but to construct new houses. A participant shared:

> "We had to replace everything. Our house… we had to replace everything. Before, we just repair things if they are broken. But it won't do this time because it was all so sudden." (Edna, 54)

Rebuilding houses was quite difficult because of the scarcity of materials and exorbitant prices of housing materials if and when they were available. Most of the hardware stores in Tacloban City, the center of commerce, were destroyed. Building materials had to be sourced from other areas further away. Consequently, they were much more expensive.

The loss of livelihood also meant that families did not have the extra funds for reconstruction and that women were also incapacitated in other aspects of their disaster lives. A participant reported:

> "It was exhausting looking for work after the typhoon because you had to find money to replace what you had lost. Yes, we were supported by organizations like the Red Cross, we were getting help from many people. But it was not the same as when we were able to earn and have income. Eventually, however, lending agencies started to open. I was able to borrow from CARD and I was able to start replacing the things that we lost." (Lydia, 57)

Another woman noted:

> "Our house was in disarray. Generating income was slow during that time." (Antonia, 70)

The aftermath of the disaster magnified women's vulnerabilities as they were confronted with the challenging circumstances of losing their livelihood. A participant reported how difficult it was after the typhoon:

> "I can say that those were very difficult times for us. The fruits from our trees were ready to be harvested. I was planning on selling them. But it was all gone. That really added to our difficulties. The bananas,

> the plants, the other fruits trees were gone. ... The loss of livelihood was distressing. On top of that our house was a mess. It had to be repaired since it was destroyed." (Edna, 54)

For women whose sources of livelihood depended on their produce, it was indeed a tremendously challenging time. Fordham (1988) argues that individuals who experience the most disadvantages in their daily lives are likely to be the most vulnerable to the negative impacts of a disaster. This reality is reflected in the lives of the poor women of Brgy. San Juan, Sta. Rita, Samar, whose lives were adversely affected by the disaster. Poverty reflects their vulnerability to the impacts of typhoon Haiyan as a life event that resulted in the myriad challenges that drastically changed their lives.

Disaster as life event

Life events (LEs) are defined as occurrences that compel people to make permanent or temporary adjustments to their lives (Cleland et al 2016). Literature on LEs describes them as something negative associated with stresses, and negative life changes (Lock et al 2012; King and Ogle 2014). However, LEs are not only detrimental to health and well-being but could be beneficial as well (Dorenwend 2006; Cleland et al 2016). This chapter considers disaster as an LE as it underscores the idea of occurrences like typhoon Haiyan which force people to make changes in their lives. These changes facilitate role transitions which may already have unfolded before the disaster event but are made more significant after the disaster. Some women participants, for example, have taken the role of breadwinner as their husbands are not able to provide for the family. One woman shared how she had to take a more central role in providing for the family after the typhoon:

> "My husband and I both work to augment the family income. But he is getting old, he cannot work as a carpenter anymore, so I now have to work for the family (especially after typhoon Haiyan) [...]. It was exhausting then [right after the typhoon] because my husband is old. He is 58, that's why he cannot work anymore." (Edna, 54)

This situation seems to reflect that husbands have taken on a back-seat role while women have stepped up in assuming their roles. The norms, values, and established mores of society limit women from overcoming their secondary state of being relegated to the sidelines (Reid 2013). In this situation, disaster as an LE is an opportunity to enact changes about gender disparities (Drolet et al 2015) highlighting the crucial role of women in assuming more gender roles despite the struggles they face.

Disaster and role transitions

Traditional gender roles differentiate men's role as the provider and women's as the caregiver, such that their lives revolve around their private and domestic duties (Perrone 2009). However, gender stereotypes are contentious since changes are perceived to take place as women and men take on new roles with corresponding differential demands because of LEs such as disasters (Wrigley-Asante 2011; Bosak et al 2017). Typhoon Haiyan provided the opportunity for the women of Brgy. San Juan to take on roles that were formerly male-dominated. As one participant shared:

> "I am taking more responsibilities now, particularly in making sure my child gets an education since my husband is getting older and weaker." (Edna, 54)

Another woman participant shared how she went on about a typical day fulfilling her various roles in the community and at home:

> "[...] I do the laundry as fast as I can, just three pieces, particularly the clothes of my children so that they have clean clothes for school. Then I clean the house. I report to the eco-resort after I take a bath, then go home for a while in the afternoon. I report to work at 7 am and go home when my shift is over." (Lydia 57)

Disaster events produce spaces in which gender relations are reconfigured, for example, when mothers take on hard physical labor (farm work, gathering firewood in the mountains, animal herding) considered to be masculine, while shouldering the burden of domestic feminized roles (Hyndman and de Alwis 2003; Cupples 2007). Such situations reflect the multiplicity of women's gender roles which are often disrupted, and which oftentimes undergo transitions during disaster events most especially in the case of typhoon Haiyan. A participant said:

> "I even went across the bay (to Tacloban City). We went there to look for mangrove seedlings. It was like that back then. My youngest child who was about ten years old during that time helped me. He's 16 now. We had to do that so that we could eat." (Lydia, 57)

This experience is in sharp contrast with their previous domestic responsibilities in the privacy of their homes. After the disaster, women took on more active and public roles, especially in livelihood activities which were mostly performed by men. Their contribution to the family's survival is highlighted as they ventured outside the confines of their home and its associated domestic responsibilities. Their engagements outside the house are now recognized and given value in the community. Women's important contribution is acknowledged not only in the private sphere but also in the public domain (Reyes and Lu 2015; Tariq and Syed 2017). A participant explained:

> "In a way, I'm happy that the storm happened, because good things came out of it. Because this resort would not be here if not for the typhoon… we [our community] did not get noticed before that." (Melinda, 52)

The recognition of women's community engagement is of particular significance especially since community work is considered an extension of women's domestic role. In the Philippines, low-income rural women have always been involved in communal activities like "*pintakasi*" or regular barangay clean-up or those who serve as Barangay Health Workers (BHWs). In addition, women are engaged in voluntary work (i.e., organizing barangay activities).

> "I was active in the barangay clean-up. Especially when my child was younger and studying here (grade school in the barangay) […]. I actively participated then." (Lydia, 57)

Community work is shared work that is usually voluntary. However, these social activities are not easily categorized as public or private but take place both in the public and private domains. What is important to note is that community work enables women to have a social life that bridges their domestic life and productive life (Hansen 1987).

After typhoon Haiyan, women participants, especially mothers, took on new roles, which resulted in role transitions. Generally speaking, these changes may take the form of role expansion or role reversal; from mostly undertaking reproductive roles to more active roles inside and outside of the home while coming to terms with new practices and involvement in their community. Women take on more roles and responsibilities for housework than men regardless of age or life stage (Horne et al 2017). Simply put, women take on more roles and responsibilities before, during, and after a disaster (Reyes and Lu 2015). These changes are reflected in the lives of the women in Brgy. San Juan in Sta. Rita, Samar. A participant shared:

> "Because of the typhoon, we had this eco-resort, there is change. Before I was holding a bolo [knife/small sword]. Now I have a ball pen. I'm not saying that I no longer use a bolo because I still tend

> to my garden at home. I still continue with those responsibilities however, something has changed, I also now use a pen every day." (Melinda, 52)

Despite the challenges accompanying women's role transitions against a backdrop of disaster as an LE, women seemed to experience a *seamless flow* from one role to another (Fothergill 1999). By this we mean that women seemed to cope well with all the changes; they continuously performed their traditional gender roles alongside the assumption of new roles as a result of the impact of typhoon Haiyan. Despite encountering initial difficulties, women have learned to integrate existing roles with that of their new ones. This may have been the result of community members helping one another and receiving help from external sources like non-government organizations (NGOs).

NGOs and women's agency

NGOs are considered as a subgroup of the broader nonprofit sector that provides diverse aid and services geared toward increasing the welfare of people in poor areas, including disaster-affected countries (Werker and Ahmed, 2008; Towe, Acosta, and Chandra 2017). In times of disaster, NGOs are at the forefront in terms of response and recovery, depending on the various phases of disaster. They collaborate with governments, the private sector, civil society, academia, and communities at risk (Lassa 2018), such as those in coastal areas vulnerable to disasters. After typhoon Haiyan, the presence of international, national, and local humanitarian organizations was felt in the areas affected by the disaster, most especially in Samar and Leyte. Their presence is felt because they are "part of the global disaster risk reduction (DRR) ecosystem as they continue to shape both humanitarian emergencies action and the DRR agenda at different levels where their presence is needed and valued and their contribution is uniquely recognized" (Lassa 2018, 1).

In the Philippines, there are several NGOs as well as socio-civic organizations which assist people. In the context of disaster, this entails providing relief assistance, and in the emergency phase, providing relief goods. In the recovery phase, they assist in the rehabilitation of material structures. As a case in point, a community in Samar that was devastated by typhoon Haiyan received support from a local organization. The ABS-CBN Lingkod Kapamilya Foundation through the Bantay Kalikasan program chaired by then Ms. Gina Lopez provided the much-needed support through skills training to develop local tourism in the area. Through this initiative, women in the community were provided with opportunities to exercise making important decisions to achieve favorable outcomes such as financial income, personal hygiene, social networking, and increased self-confidence. As a result, women are now able to express their individual and collective agency in their need to recover from the disaster because of the support provided by external organizations such as the ABS-CBN Lingkod Kapamilya Foundation. The women of Brgy. San Juan banded and worked together to establish the eco-resort. A participant shared:

> "My husband also works there [at the eco-resort]. In the beginning, our group was a women's association. But we cannot really run this place on our own so our husbands helped us. Gina Lopez of ABS-CBN provided us with this livelihood. This is really a big help to us because it's difficult for us to find work. Gina Lopez's foundation helped us greatly. ABS-CBN provided us with livelihood. She really helped us a lot. (Lydia, 57)

Because of coming together as women in the eco-resort project, they gained a sense of confidence and competence which changed the way they perceived themselves and how they conducted themselves. Women can cope despite the initial difficulties in adjusting to the changed situation brought about by a disaster event.

Becoming empowered

Women participants contributed much to the rehabilitation efforts of the community, and they also played a central role in ensuring the survival and well-being of their families. Engagement in volunteer work in the community, specifically the eco-resort established with the support of the ABS-CBN Lingkod Kapamilya Foundation and eventually supported by the local government, offered them an opportunity to develop new skills and broaden their horizons. They also gained new experiences that contributed to their economic recovery and personal growth. One woman participant shared this in terms of the changes that happened to her:

> "My mind is like a dull knife, which has been sharpened." (Edna, 54)

Another woman reported:

> "I rarely venture out of the barangay and do not regularly visit Tacloban, the only people I interact with are people here in the barangay. Right now, there are many guests like you. I meet people like you; celebrities (from Manila) come here too. I am happy. We get a chance to work too since I just used to stay home." (Lydia, 57)

Women in the process gained agency (i.e., individual, collective) that demonstrated empowerment in the course of recovering from their experience from typhoon Haiyan. This is expressed in their economic independence; their capacity to earn, as well as the responsibility of running the eco-resort together with the other women in their community. In comparison to pre-disaster life, women's contribution to disaster recovery efforts reflects the vital role they play as equal partners in development. Women empowerment and gender equality are reflective of the United Nations advocacy for Sustainable Development Goals.

Life after the storm

Disasters are negative LEs, but they can also be considered opportunities for change. These changes may have been brought about because of the situation where women endured and struggled because of their experience from a disaster. They are survivors and not hapless victims of the disaster. While rebuilding was an arduous and grueling endeavor, it was facilitated by the assistance of external organizations like the ABS-CBN Lingkod Kapamilya Foundation, which provided the means for livelihood, and microcredit financing to the community in addition to the local government's assistance. Access to microcredit eased women's financial burden. A participant shared:

> "It [using lending companies] was of great help to me. Others tell me not to borrow money so that you won't have problems with debts. But I argued that if I don't, where will I get a big amount all at once. Most of us here in Barangay San Juan only earn enough for food. There is no extra money for our children's school expenses. We can barely manage to give our children money for their fare. When I was able to borrow PHP10,000 I was able to pay for my child's tuition fee, buy uniform and give transportation allowance. (Lydia, 57)

Women's multiple roles (i.e., reproductive, productive, community) have become interconnected, resulting in easing role transitions. Thus, community and individual recovery were facilitated by women's active participation in the efforts to rebuild their communities with the help, for example, of the ABS-CBN Lingkod Kapamilya Foundation.

Centrality of motherhood

Women in the Philippines are expected to marry and have children; with motherhood associated with femininity. Low-income rural women participants consider motherhood to be very important, and as mothers, they accept that sacrifices in the fulfillment of their duties are necessary. The continued narrative of sacrifice to be a good mother, a culturally valued identity in the Philippines, is an expected traditional gender role of women.

In the context of disaster, protecting children and making sure they survive the aftermath of the disaster is an important duty of mothers. Similar to mothers in most societies, their caretaking role takes precedence during disasters (Fothergill 1999). As such, women participants regard their motherhood role as significantly more important than other roles they undertake. A mother shared:

> "My primary responsibilities are to give to my children and provide for their needs. It's really about taking care of them." (Melinda, 52)

The motherhood experiences of rural low-income women during the recovery period revolved around providing education to their children. This responsibility is already challenging during ordinary times because of limited income and was made more complex by the difficulty of the post-disaster circumstances, the loss of men's breadwinner role, and the expense of college education. Children who were in high school before the disaster are now in college and had to be sent off to the city. Fulfilling the maternal duty of providing for their children's education impacts women's multiplicity of roles which is further observed among low-income women. A participant noted:

> "To be honest, right after the storm, my child begged me, she said, 'Mother, I really want to go to school.' But I was so… I said, 'I can't afford to send you to school, you know that your father cannot find work, so what do I do?' So, I had to find a way for my child. She needs me." (Edna, 54)

Parents embrace the idea that education is a way out of poverty and difficult life circumstances. The child is considered the family's salvation from poverty, which explains why mothers focus on sending their children to school regardless of financial situations. The success of children is equivalent to the success of the family. In addition, children are a source of self-esteem; low-income women have limited sources of esteem, thus, despite the hardships that sending them to school would entail, mothers and their families make sacrifices. This is one of the main struggles of low-income mothers who send their children to school at the expense of their own needs. A participant commented:

> "I am having a hard time right now because I have to send my child to college. I am really having a hard time… nobody is helping me. So I am forced to borrow from lending companies like CARD. It's

> like that. That's where I can borrow a big amount. And I pay it off through my earnings from my work here [the eco-resort]. (Lydia, 57)

While public school education in the Philippines is free with the recent enactment of Republic Act 10931, providing for free college education in state universities and colleges, a college education remains difficult to access for low-income families. School expenses, as well as transportation costs to the city, further strain the already-limited finances of low-income women in rural areas. This is echoed by a participant who said:

> "What was added to the family's expenses is my child's tuition expenses. When my child was still in high school, there was not much expense… I don't know where to get the money to pay for the school expenses of my child who is in college, the expenses for transportation too. I didn't have this problem before because my children were still studying here in the barangay but right now one is studying in Tacloban. So sometimes when I cannot find the money for the fare I tell my child not to go to school." (Lydia, 57)

Motherhood as a source of strength

Despite the challenges of mothering in the aftermath of the disaster, women participants found it to be a source of strength, serving to ground them during difficult and uncertain times. Their children need them; therefore, they must continue living. For them, their sacrifices as mothers are their source of strength. According to a participant:

> "Losses can be endured as long as the family and children are safe, that we are together." (Melinda, 52)

The dream that their child will have a better future served as their motivation to continue making sacrifices after the disaster event. They see the fulfillment of their maternal duty as an essential part of who they are.

> "My problem right now is how to pay for my child's school expenses. I'm not sure if women are all like that. I mean I try to find ways to look for money so that we can pay [for school expenses]." (Edna, 54)

Continuing narratives: Life is sacrifice

The disaster and the difficult circumstances that followed challenged women's ability to fulfill their caretaking role as mothers. The impact of disasters not only resulted in the expansion of women's roles (i.e., breadwinner) resulting in trying to cope and manage their added roles as well as their changed circumstances. Theirs is a narrative of sacrifice as they continuously struggle to adjust to the various role transitions accompanying disaster as an LE. One woman shared her continuing efforts to address the needs of her family:

> "There is not much difference in my life before and after the storm. … The financial difficulties remain but it's not so exhausting anymore. Unlike the previous years. … Things are a little better now since my child has graduated from college and can contribute a little to the family's income." (Edna, 54)

Low-income women's activities and considerable efforts are directed toward ensuring that their family's needs are met. Confronting the material losses as well as simultaneously juggling roles, both old and new, while not particularly easy was taken by women as yet another sacrifice that they deal with. A woman comparing her circumstances before and after the typhoon shared:

> "For me, I'm not really having a difficult time because I am used to hardship. It has always been like that ever since I was a child. So for me, it is sort of the same, [life] entails making sacrifices." (Edna, 54)

Rural low-income women whose lives revolved around traditional gender roles now view their lives as a series of changes. As they struggled to survive, they continued to fulfill their traditional gender roles as well as adopted new ones. Women's experiences reflect the delineation of gender roles in traditional societies where the gender divide is very clear. While the changes were accepted without much difficulty, women's lives continue to be defined by the continued sacrifices they make for their families. The fulfillment of their gender roles demands a lot of self-sacrifices that add to their already strenuous life to ensure the well-being of their family.

Disasters burden women with added responsibilities and roles, yet these changes have afforded them opportunities. Recognition of women's active role in disasters, and in recovery efforts both in the home and in the community is important to highlight their participation as empowered stakeholders in community building. To this end, external support mechanisms need to be in place to sustain the changes and transitions in women's roles.

Conclusion

Disaster impacts the economic, social, and psychological aspects of individuals. Women learn to navigate role transitions such as becoming the breadwinner, decision-maker, and community worker after experiencing typhoon Haiyan. These roles compound their other myriad roles such as mother, wife, daughter, community volunteer, and informal worker. Our walking interviews with women in Brgy. San Juan point to the need to investigate the psychological impact of disasters on women's experiences that are not often recognized in the Philippines. The economic impact and the need for more sources of livelihood especially among rural low-income women intersect with other social dimensions post-disaster. Also, the results of the study call for deeper interconnectedness of disaster as a life event, role transitions, and women's life stages in disaster studies in the Philippines and elsewhere.

References

Bosak, J., A. Eagly, A. Diekman,, and S. Sczesny (2017) "Women and men of the past, present, and future: Evidence of dynamic gender stereotypes in Ghana," *Journal of Cross-Cultural Psychology* 49(1): 115–129. DOI: 10.1177/0022022117738750.

Bradshaw, S. (2014) "Engendering development and disasters". *Disasters* 39(S1): S54–S75 doi:10.1111/disa.12111.

Cleland, C., A. Kearns, C. Tannahill, and A. Ellaway (2016) "The impact of life events on adult physical and mental health and well-being: Longitudinal analysis using the GoWell health and well-being survey," *BioMed Central Research Notes* 9(1): 470.

Cupples, J. (2007) "Gender and hurricane Mitch: Reconstructing subjectivities after disaster," *Disasters* 31(2): 155–175. doi:10.1111/j.0361-3666.2007.01002.x.

Dorenwend, B. (2006) "Inventorying stressful life events as risk factors for psychopathology: Toward resolution of the problem of intra-category variability," *Psychological Bulletin* 132(3): 477–495. DOI: 10.1037/0033-2909.132.3.477.

Drolet, J., L. Dominelli, M. Alston, R. Ersing, G. Mathbor, and H. Wu (2015) "Women rebuilding lives post-disaster: Innovative community practices for building resilience and promoting sustainable development," *Gender & Development* 23(3): 433–448. doi:10.1080/13552074.2015.1096040.

Dynes, R. (2005) "Community social capital as the primary basis for resilience. Preliminary papers. 344, http://udspace.udel.edu/handle/19716/1621 [Accessed November 7, 2021].

Enarson E., A. Fothergill, and L. Peek (2007) "Gender and disaster: Foundations and directions," in H. Rodrıguez, E. L. Quarantelli, and R. R. Dynes (eds) *Handbook of Disaster Research* New York: Springer, pp 130–146.

Fordham, M. (1988) "The intersection of gender and social class in disaster: Balancing resilience and vulnerability," *International Journal of Mass Emergencies and Disasters* 17(1): 15–36.

Fothergill, A. (1999) "Women's role in disasters," *Applied Behavioral Science Review* 7(2): 125–143. DOI:10.1016/S1068-8595(00)80014-8.

Hansen, K. V. (1987) "Feminist conceptions of public and private: A critical analysis," *Berkeley Journal of Sociology* 32: 105–128. http://www.jstor.org/stable/41035361 [Accessed November 7, 2021].

Horne, R., M. Johnson, N. Galambos, and H. Krahn (2017) "Time, money, or gender? Predictors of the division of household labour across life stages," *Sex Roles* 78: 731–743. DOI: 10.1007/s11199-017-0832-1.

Horton, L. (2012) "After the earthquake: Gender inequality and transformation in post-disaster Haiti," *Gender & Development* 20(2): 295–308, doi:10.1080/13552074.2012.693284.

Hyndman, J. and M. de Alwis (2003) "Beyond gender: Towards a feminist analysis of humanitarianism and development in Sri Lanka," *Women's Studies Quarterly* 31(3–4): 212–226.

Information on Disaster Risk Reduction of the Member Countries (n.d.) https://www.adrc.asia/nationinformation.php?NationCode=608&Lang=en [Accessed December 6, 2022].

Jayarathne, S. (2014) "Women's potential in dealing with natural disasters: A case study from Sri Lanka," *Asian Journal of Women's Studies* 20(1): 125–136, https://doi.org/10.1080/12259276.2014.11666175.

King, K. and C. Ogle (2014) "Negative life events vary by neighborhood and mediate the relation between neighborhood context and psychological well-being," *PLOS ONE* 9(4): e93539, https://doi.org/10.1371/journal.pone.0093539.

Lassa, J. (2018) "Roles of non-government organizations in disaster risk reduction," In *Oxford Encyclopedia of Natural Hazard Science*, Oxford: Oxford University Press. DOI: 10.1093/acrefore/9780199389407.013.45

Lock, S., G. J. Rubin, V. Murray, M. B. Rogers, R. Amlôt, and R. Williams (2012) "Secondary stressors and extreme events and disasters: A systematic review of primary research from 2010–2011," *PLOS Currents*: 4. DOI: 0.1371/currents.dis.a9b76fed1b2dd5c5bfcfc13c87a2f24f.

PSA (Philippine Statistics Authority) (2019) "Poverty situation in Eastern Visayas (first semester 2018)," http://rsso08.psa.gov.ph/article/poverty-situation-eastern-visayas-first-semester-2018#sthash.ttPUxFOF.dpbs [Accessed November 7, 2021].

Perrone, K. (2009) "Traditional and nontraditional work and family roles for women and men," *Journal of Career Development* 36(1): 3–7. doi:10.1177/0894845309340787.

Radtke, K., H. Mann, D. Weller, L. Kirch, and R. Prütz (2018) "The world risk index 2018," in *World Risk Report, 2018: Focus: Child Protection and Children's Right*, Berlin: Bündnis Entwicklung Hilft – Gemeinsam für Menschen in Note.V.Chausseestraße 128/12910115.

Reid, M. (2013) "Disasters and social inequalities," *Sociology Compass* 7(11): 984–997.

Reyes, D. D. and J. Leilanie Lu. 2015. "Gender dimension in disaster situations: A case study of flood prone women in Malabon City, Metro Manila," *International Journal of Disaster Risk Reduction,* http://dx.doi.org/10.1016/j.ijdrr.2015.11.001i [Accessed November 7, 2021].

Samuels, A. (2012) "Moving from great love: Gendered mobilities in a post tsunami relocation neighborhood in Aceh, Indonesia," *International Journal of Urban and Geographical Research* 36(4): 742–756.

Shenk, D., J. Mahon, K. Kalaw, B. Ramos, and I. Tufan (2010) "Understanding the disaster experience of older adults by gender: The experience of survivors of the 2007 earthquake in Peru," *Healthcare for Women International* 31(11): 965–980. DOI:10.1080/07399332.2010.502274.

Srivastava, K. (2010) "Disaster: Challenges and perspectives". *Industrial Psychiatry Journal* 19(1): 1–4.

Tariq, M. and J. Syed (2017) "Intersectionality at work: South Asian Muslim women's experiences of employment and leadership in the United Kingdom," *Sex Roles* 77: 510–522, DOI.10.1007/s11199-017-0741-3.

Towe, V., J. Acosta, and A. Chandra (2017) "Towards more nuanced classification of NGOs and their services to Improve integrated planning across disaster phases," *International Journal of Environmental Research and Public Health* 14(11): E1423, DOI: 10.3390/ijerph14111423.

Weiss, T. and D. A. Korn (2006) *Internal Displacement: Conceptualization and Its Consequences*, New York: Routledge.

Werker, E. and F. Ahmed (2008) "What do nongovernmental organizations do?" *Journal of Economic Perspectives* 22(2): 73–92. DOI: 10.1257/jep.22.2.73.

Wrigley-Asante, C. (2011) "Women becoming bosses: Changing gender roles and decision-making in Dangme West district of Ghana," *Ghana Journal of Geography* 3: 60–87.

8

Women's Participation in Small Island Resilience Building in Concepcion, Iloilo, Philippines

Dhino B. Geges

Background

This chapter discusses women's contribution to community resilience through social entrepreneurship (SE) and social enterprises (SEs). The need to support entrepreneurial activities in boosting economic conditions, especially with the participation of women, clearly demonstrates a crucial element in the building back better (BBB) approach. Likewise, a thorough understanding of how women have been supporting local economic development and rehabilitation in disaster-prone communities was also examined. This may contribute relevant insights to the theory and practice of entrepreneurship and gender studies, as it describes the roles of women and the relationships of various actors in disasters. It has value added in the SE literature as this chapter contributes to further strengthening of empirical approaches in the field.

In terms of disasters, communities residing on small islands are more exposed and vulnerable compared to mainland communities. Moreover, the availability and accessibility of institutional support and assistance in these areas are some of the most common issues among people living on the island. Layers of marginalization can be observed from socio-economic issues (e.g., poverty, alternative livelihood), to environmental constraints (e.g., natural hazards). Thus, this study focused on small island resilience building.

Gender in disaster and resilience studies has been gaining interest among scholars from various fields (Fordham, 1998; Nguyen, 2019). In particular, the work of Enarson (1998) and Bradshaw (2013) posit gender in disaster research agendas, while its development in the literature can be explored still further. As disasters generate casualties in terms of lives and properties, it disempowers different groups and sectors in the community. This is especially noticeable among women as one of the vulnerable groups (Enarson 2000). The vulnerability of women amidst disasters has been elevated on different levels of discourse through time. Some have argued that women's vulnerabilities can be positioned in a broader spectrum of defenselessness to having the capacity to bounce back (Islam et al 2017). Beck (2005) has made a valid observation pointing out that women's contributions in disaster recovery are often ignored, and that most of the rebuilding initiatives are male-dominated, focusing much on infrastructure and physical development rather than socio-economic interventions. This situation needs to be reviewed since there were incidences where women were more severely affected than men (Enarson and Chakrabati 2009). As such, it is necessary to promote and enhance female-led recovery and rehabilitation. This is critical as the voices of women, in both formal and informal decision-making processes, are gradually being recognized; although the level of recognition still needs to be advanced (Baliyan 2014).

Women are crucial in poverty-alleviating strategies due to their particular vulnerability to poverty exacerbated by limited access to social services and its effects on their children and their larger community (Fotheringham and Saunders 2014). In the 2008 Philippine Statistics Authority (PSA) data, 3.57 million women belong to the informal economy (PSA 2009), particularly because of their reproductive roles (e.g., intrinsic childbearing and home-keeping roles). Their productive, reproductive, and community responsibilities require their income-generating activities to be flexible to cope with all the roles ascribed to them. Amidst disasters, the same domestic responsibilities—home-keeping and ensuring the maintenance and survival of the family—have enhanced their abilities to cope in such times by engaging in various forms of employment and bringing food to their families (Ofreneo and Hega 2016). This can be viewed as resilience in the women's sector. As defined in Municipal Ordinance No. 2017–18 or an *Ordinance Declaring November 8 of every Year as "Community Resiliency Day" in the Municipality of Concepcion, Province of Iloilo*, resilience is defined as the ability to overcome challenges of all kinds—trauma, tragedy, threats, personal crisis, and bounce back from difficult experiences stronger, wiser, and more personally powerful. Based on this meaning, resilience is viewed as highly psychological.

Fotheringham and Saunders (2014) looked into strategies for empowering women, particularly through social enterprises, which have recently gained interest as a new poverty reduction model. One of their conclusions was that there is still limited research on the poverty-reducing role of social enterprise for women, and proposed mechanisms and an integrative framework to provide a means of synthesizing our current knowledge while providing the basis for future investigations. According to Dacanay (2013), SEs are a social mission-driven and wealth-creating organizations that has double or triple bottom-lines with the poor as the primary stakeholders (SEPPS).

As Bourquist and de Bruin (2019) suggest, gender operates as an agent in the process of organizational strengthening and social transformation. Historically, however, social entrepreneurship has underscored that women are categorized as key players in the field (Hechevarr̂ia et al 2012). Women's leadership is recognized to achieve the successful implementation of SE initiatives. There are also observations that social enterprises in developing countries, like the Philippines, have the potential to drive women's empowerment from the household to the community level (Haugh and Talwar 2016), which may serve as a bridge in solving other social and gender-related issues rooted to its social mission.

Disaster context and resilience building

The frequency and intensity of disasters in the Asia-Pacific region have increased over recent years. In the last decade, around 1,592 disasters were estimated to have hit the region. These have resulted in high incidences of death and economic decline (UNESCAP 2017). The United Nations Economic and Social Commission for Asia and the Pacific report titled *Disasters Without Borders: Regional Resilience for Sustainable Development*, indicated that there were 512 catastrophes and 177,000 deaths accounted for in Southeast Asia from 2005 to 2014.

Disaster incidences are very common in the Philippines, which has often been cited as one of the most disaster-prone countries in Southeast Asia. The Global Climate Risk Index (Kreft et al, 2014) reported that the Philippines, Cambodia, and India were among the countries most severely affected by disasters in 2013. In addition, the United Nations University-Institute for Environment and Human Security (UNU-EHS) listed the Philippines, along with Vanuatu and Tonga, as the three countries with the highest exposure and risk from natural hazards and disasters in 2014. As for countries most exposed to multiple hazards, the World Bank's Natural Disaster Hotspot (2015) ranked the Philippines in the top eight; and in 2018, the ASEAN Coordinating Centre for Humanitarian Assistance on disaster management noted that the Philippines has the highest multi-hazard risk score compared with other countries within the region. The hazard risk score comes from combining the types of hazards, vulnerability, and the coping capacity of the country to disasters.

Natural hazards continue to threaten the social and human development of the region, reinforcing the prevalence of poverty and growing economic and political inequalities. In November 2013 alone, one of the most devastating typhoons ever recorded in the region—typhoon Haiyan—struck the Philippines. Haiyan has brought tremendous economic loss and affected more than 14.1 million people (NDRRMC 2013).

More evidence reveals the role of risks, shocks, and vulnerability in perpetuating poverty after a disaster (Morduch 1999). The social services and livelihood sectors are commonly affected when disaster strikes. Damages caused by disasters consequently make the poor more vulnerable because they lose their source of livelihood and have no means and support to start again. Meanwhile, gender plays an important role in managing the negative effects of disasters. In this situation, resilience is necessary for the social and livelihood sectors which include women. Most developing countries, including the Philippines, depend heavily on small and medium enterprises as the backbone of the economy. Unfortunately, more than half of these are situated in high-risk areas.

Many of the countries in the region initiate various actions to become more disaster resilient. However, mainstreaming resilience at individual level is no longer a feasible option. Rather, it should be a communal imperative that is best positioned through a collective action (Akhtar 2016, as cited in UNESCAP 2016). This has prompted a call among all countries in the region to integrate local disaster risk reduction and resilience as part of their sustainable development plans and programs. One of the significant responses to this call was the adoption of the Comprehensive Agreement on Disaster Management and Emergency Response (CADMER) by the ASEAN in 2005. The CADMER serves as a regional framework for "cooperation, coordination, technical assistance and resource mobilization in all aspects of disaster management" (CADMER Work Programme, cited in Petz 2014, 10).

Some of the initiatives and experiences in building community resilience are outlined in the Hyogo and Sendai frameworks (UNISDR 2015) and succinctly summarized in the principles of BBB (James Lee Witt Associates 2005; Clinton 2006). These principles are recognized as meaningful in the post-disaster context, as they provide a solid framework for recent humanitarian efforts during conflict and disaster. One of the strategies to strengthen resilience that fully supports the BBB principle is through social entrepreneurship. Social entrepreneurship continues to acquire global recognition as it has established a solid reputation in delivering innovative and sustainable strategies, resulting in social value creation. It has had various impacts (Doherty et al 2009; Galera and Borzaga 2009) that directly solve societal issues (Estrin et al 2013), particularly global poverty and social services delivery Mair, and Seelos, 2007; (Grimes, 2010; Newbert, 2014).

The participation of women in social enterprise management is highly recognized by scholars and development workers in the field. More than workers, they act as leaders of these enterprises. Unfortunately, this idea has not been fully explored in the context of resilience building, thus this raises questions such as What could be the role of women in building resilience through social entrepreneurship? and What are the factors that could support and restrain women's participation? Thus, this chapter provides a case study of women in Concepcion, Iloilo.

Disaster context of the study site

Geographical and socio-demographic profile of Concepcion

The Municipality of Concepcion is one of the eleven municipalities[1] comprising the fifth district of Iloilo Province. It is situated 107 kilometers northeast of Iloilo City and is accessible to all types of land transportation. As a coastal town, Concepcion is bounded in the north by the Municipality of San Dionisio and the Visayan Sea, on the west by the Municipalities of Sara and Ajuy, on the south by the Municipality of Ajuy, and the east by the Visayan Sea. It is a third-class municipality composed of 25 barangays—11 islands and 14 mainland barangays. The study focuses on 11 island villages. The municipality has a total population of 45,396 with 10,142 households. It has a total land area of 9,327.45 hectares and total municipal territorial water of 96,481.56 hectares (Municipal Profile of Concepcion 2016).

Concepcion is one of the extremely devastated areas after typhoon Haiyan made its fifth landfall specifically at Pande Azucar island (NDRRMC 2013). Aside from the poor economic situation, Concepcion is categorized as a high-risk municipality because of its geographical location. Based on its Disaster Risk Reduction and Management (DRRM) Plan, it is prone to multiple hazards that include typhoon, squall, flood, drought, and storm surge.

For the data collection method, the study utilized participant observation, key informant interviews (KIIs), and focus group discussions (FGDs). Related secondary data including document reviews deepen the research findings and analysis. Both quantitative and qualitative methods of data collection were used through a case study approach. Qualitative methods are effective when little is known about a phenomenon and when the aim is to derive ideas from the data and link it to theoretical underpinnings (Langley 1999). Data collection through KIIs was gathered in November 2016 while the data from the Barangay Disaster Risk Reduction Management (BDRRM) FGD session were collected in October 2018. This study adopted a pragmatic

Figure 8.1: Map of Concepcion, Iloilo Province, Philippines

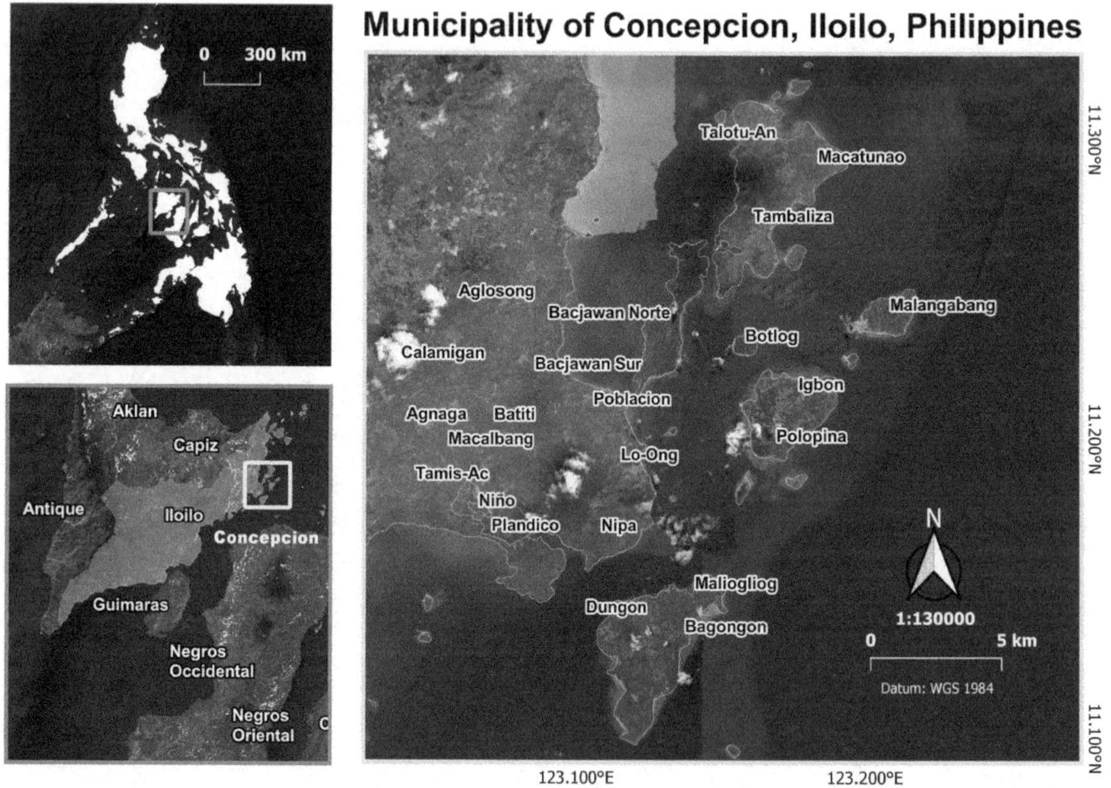

Source: https://gadm.org/maps/PHL/iloilo/concepcion.html

worldview and pluralistic approach as the most appropriate, allowing the researcher to design a framework based on women's lived experiences toward strengthening small island resilience.

Typhoon Haiyan and small island communities

Based on the assessment of the Municipality of Concepcion in the province of Iloilo, the coastal town has all types of natural hazards. All villages are exposed to hydrometeorological hazards, which include storms with storm surges, landslides, and flooding. The area has a high risk of disaster incidence because of its high exposure and vulnerability to these hazards. In November 2013, super typhoon Haiyan made its fifth landfall in Panay island, which resulted in severe damage, especially among the coastal municipalities like Concepcion. Table 8.1 presents the list of damages of typhoon Haiyan.

In terms of disaster management, a meager 5% of the Internal Revenue Allotment (IRA), the annual general fund source of Concepcion Local Government Unit is allocated as the Local Disaster Risk Reduction and Management Fund (LDRRMF), which is mandated by RA 10121 or the Philippine Disaster Risk Reduction and Management Act of 2010. For a municipality with an average of PHP30million as IRA, its LDRRMF is approximately PHP1.5 million. Of the 5% lump sum allocation, 30% is allocated as Quick Response Fund (QRF) and 70% for pre-disaster activities. However, in the case of typhoon Haiyan, the funds were immediately depleted due to the severity of the disaster, which required additional outside funding support. Thus, the support of the national government, private sector, and international donors is needed to respond to the immediate needs of the municipality and to plan for its long-term rehabilitation and recovery using the BBB principle. As a third-class municipality, Concepcion is exploring other means of support.

Table 8.1: List of damages of super typhoon Haiyan in Concepcion, Iloilo

Damages	Frequency
Number of Casualties	23
Number of Persons Injured	1,730
Totally Damaged Houses	6,403
Partially Damaged Houses	2,423
Number of Persons Affected	41,181
Estimated Cost of Damage	PHP717,633,000.00

Source: Municipal Disaster Risk Reduction Management Office, Concepcion, Iloilo.

After Haiyan, the restoration of residents' livelihood activities especially among island barangays of the municipality became one of the main priorities. As a strategy, livelihood and enterprise development projects were carried out through capacitating and organizing people in the community to ensure its sustainability. The following sections present selected organizations and key actors which played critical roles in the rehabilitation process.

Typhoon Haiyan and actions by men and women

Despite the rising concern for the inclusion of gender perspectives in disaster plans, programs, and policies, it is still marginalized and not fully actualized as a core agenda. Ramalho (2019, 12) notes that gender mainstreaming in the broader landscape of DRRM is still framed within the "biologically deterministic stereotypes, ignoring the multiple subjectivities inherent to personal experiences of insecurity, deprivation and calamity, and failing to confront issues of power and powerlessness underpinning vulnerability which is equally critical to the dynamics and outcomes of participatory processes." There is a need to address gender-specific challenges in disaster preparedness (Reyes and Lu 2017). Based on the FGD with the BDRRMC members in Concepcion, the actions and preparations taken by men and women from the island communities before, during, and after typhoon Haiyan were noted.

The division of labor in preparation before the typhoon was based on traditional societal gender roles in the Philippines. The men handled strenuous work: securing the houses, moving the boats to higher places, and the cutting of tree branches. The women, on the other hand, prepared their food and secured their belongings as well as basic items such as food, water, and clothing to bring to the evacuation center.

When the typhoon hit Concepcion, residents were in a panic since they were not fully aware of what to do; they were afraid of the strong winds, heavy rain, and large waves. Accordingly, the roles of the men and women remained the same. The men were outside and helped other families to evacuate using their pump boats while the women took care of the children and elderly, and ensured that they were safe.

After the typhoon had passed, all 24 villages were left with severe damages, most especially in the 11 island villages. Houses were destroyed, some even unrecognizable to the owners. The men continued the physical work in fixing damaged houses and boats and cleaning the streets, while the women and children were tasked to clean the inside of the houses. Some children, however, stayed in the health centers because they had fevers. Some of the elderly did the laundry while others were left in evacuation centers since their houses were destroyed.

Community-based women's organization in social entrepreneurship

The Federation of Kalipunan ng Liping Pilipina (KALIPI) Women's Association in Concepcion, Iloilo Inc. (KWACII) is a women's community-based organization, which was established in April 2005. In 2016, it had a total of 3,268 members across the 25 villages of the municipality (KWACII 2016). The association is registered with the Securities and Exchange Commission (SEC). Their member-women's groups per village are encouraged to apply for other government accreditation and recognition such as the Department of Labor and Employment (DOLE) workers' association and local government units (LGUs) partner civil society organization. This is essential since most of the funding or donor organizations required proof of registration or accreditation as a legitimate community organization.

The KWACII envisions having a God-loving, empowered, proactive, and unified citizenry in a secured and developed community aimed toward sustainable and balanced governance (KWACII 2016). Its mission is to promote equality, integrity, and economic upliftment of women in their organizational functions and deliver appropriate services based on the needs of members. This organization partners with the LGUs and key stakeholders such as people's organizations (POs), non-governmental organizations (NGOs), and government agencies. Microfinance and micro-lending services were the major services of the association that were part of recovery and rehabilitation after typhoon Haiyan. They are also regularly encouraged by the local government of Concepcion to showcase products during the annual tourism fair hosted by the provincial government of Iloilo.

Community members of KWACII have mentioned the establishment of the production center as one of their main contributions that benefit everyone (Figure 8.2). The construction of this community production center was a project achieved with the organization's efforts. Through the bottom-up budgeting program (BUB) of the Department of Interior and Local Government (DILG), the organization, in consultation with the Concepcion LGU, submitted a project proposal for funding. The building serves as its office and venue for educational training and workshops. Other organizations within the municipality may use it for livelihood-related activities free of charge. Alongside this experience was the conscious effort of the organization to promote socially inclusive projects. One of the officers said that:

> *"Tinanggal ang womens production center sa project proposal ginawa na lang na community production center. Baka kasi isipin nila na exclusive lang ito for women. Gender equality din kasi ang pinopromote namin."* (Secretary, KWAICII)

Figure 8.2: Community production center at Concepcion, Iloilo

> ["We took out 'women' in the title of the project proposal and made it into 'community production center' instead of 'women's production center.' We don't want people to think that it is just exclusive for women and because we are also promoting gender equality.]

Project activities

In 2014, the association implemented a microfinancing project in support of the recovery of fishing as a livelihood. It was funded by the Philippine Business for Social Progress (PBSP), a corporate-led social development foundation. The association received a total amount of PHP1,375,000.00 for this project, which benefited 375 women across three groups in Concepcion (PBSP 2015). PBSP selected the KWACII as the only partner beneficiary in Concepcion based on the following criteria: its registration with the SEC or DOLE, service track record of the association, engagement in the fishing industry, experience in microfinancing, and the willingness to implement the project.

Part of the microfinancing project funded by the PBSP was the provision of training activities for them to effectively implement their selected business enterprises. According to the project status report in 2015, 130 women members participated in basic entrepreneurship training (PBSP 2015). Some 35 members attended the gender and development/women's health training, ten in the organizational development/leadership training, and four were trained on financial management including topics like recording and bookkeeping. Educational opportunities like this training helped improve the capacities of members in organizational and project management.

The SEs in the Philippines have gained strong support from civil society. The long history of participation and collaboration of NGOs was instrumental in the emergence of SEs in the country. Ballesteros and Llanto (2017, 1) noted that the Philippines "has a social and cultural environment that is conducive to the framework of social entrepreneurship given that it has had a widespread focus on bottom-of-the-pyramid issues." Laylo (2018, 7) further explains the emergence of SEs in the country citing the enabling factors including "governance, socially inclusive and economic approach, financial services, infrastructure development, and entrepreneurial culture," which have shaped the conceptualization and nourishment of SEs.

Despite the popular adoption of SE, there was little attention on their impact in the post-disaster context. Studies in this area have primarily focused on the definitional angle of SE (Mair et al 2006; Dacin et al 2011). No clear studies document SE as an agent that can be used to strengthen resilience at the local level. While much has been mentioned about its success when it comes to poverty reduction, discussions on how poverty becomes an aggravating factor in disasters are yet to be explored.

Sengupta and Sahay (2017) note the dearth of published studies that explored SE as framed and practiced within the Asia-Pacific region. Thus, scholars from multidisciplinary areas need to examine SE and its practice in countries where the number of studies is relatively low. More importantly, it behooves scholars to investigate the institutional conditions that support the proliferation of SEs (Doherty et al 2009). In Southeast Asia, the majority of the SEs are considered small social ventures. These SEs offer solutions to unemployment, social services delivery, and environmental protection (Santos et al 2009). They are also shaped by historical, political, cultural, and economic forces, resulting in the various conceptual frameworks for SE. However, SE in the region remains unexplored (Dacanay 2009; Santos et al 2009), which makes this case study in Concepcion significant.

Multilayered perspectives of women's vulnerability

The various factors which enable and hinder women's participation in social enterprise and livelihood activities are mirrored on the multiple layers of their marginalization. Their marginalization is rooted in the different facets of their vulnerabilities, including but not limited to the following:

Women's socio-economic vulnerability

The existing poverty situation in island villages directly affects the participation of women members of KWACII. Concepcion ranks as one of the poorest municipalities in the province of Iloilo. This is magnified by 45.26% poverty incidence in 2015 (Municipal Profile of Concepcion 2016). Based on its economic profile, most of the locals have a monthly average income below the poverty line set by the Philippine National Economic Development Authority (NEDA). Previous reports have shown that the municipality has been experiencing high poverty incidence since 2000, where 87% of the households stayed within the poverty threshold. Because poverty is multidimensional, it has implications in education, health, politics, culture, social aspect, and well-being of the population.

Poverty calls for women to diversify their livelihood skills while still attending to household duties. They do this through assistance from the Go Negosyo Center of the Municipal LGU. Opportunities were explored to offer new skills that are still relevant in the fishing and farming sectors. The possible increase in income gives them enabling support to participate in organizational activities. For example, the financial capacity of a family or household dictates the decision of members to engage in social and public activities (e.g., women's month parade), which are usually held at the *poblacion* (the central part of the municipality where commercial activities take place) or mainland. The limited funds from the gender and development (GAD) budget of the barangay are always not enough for their GAD-related activities expenses. Consequently, they are obliged to use their personal money to augment their public expenses.

Women's geographical vulnerability

The proximity of the island villages poses another limiting factor in women's participation since most of the capacity-building interventions are conducted outside their respective villages. Most of the women leaders have to pay for their transportation to go to the mainland or the city just to attend meetings or capability-building activities like training. People who reside on the island villages are dependent on the limited daily trips of commercial passenger boats. It would be an asset to have their own motorized pump boat, especially in cases of emergency. This kind of ownership also pushes women to develop their boat-driving skills and not to depend on their husbands to do this.

Women's environmental vulnerability

The high exposure to natural hazards like storm surges, landslides, and typhoons constantly affects the operations of women's livelihood activities. These prevent them from improving their socio-economic status, because of the extent and nature of damages they experience. Based on the participants' feedback, people residing on the islands always face the danger of squalls while traveling the sea.

Women's financial capacity

The involvement of women's organizations in community-managed savings and credit activities brought a multitude of benefits, particularly in improving financial literacy and financial resiliency. Financial resiliency is observed as the process of enhancing financial resources or assets needed to bounce back if a disaster happens (Boin and Van Eeten 2013), and it is a critical element in rebuilding and recovery activities. There were also stories where women influenced their husbands and children to save money. These activities have earned high regard among women members since everything is managed by the members themselves. In addition, the access to credit was very simple and community-centered compared to credit issued by commercial banks.

Women's kinship and family support

Women leaders have strong relationships with their families and relatives. This value is rooted in women as they lead and manage the organization and encounter several issues and challenges. They also confirmed the changing behavior of their husbands, from being antagonistic to rendering their services, to being supportive of them to effectively perform their responsibilities. One of the women leaders from the island village has shared her experience where her husband extended full support as the president of the association. She revealed how her husband has provided transportation services using his motorcycle every time she needs to attend training and workshops. This instance upholds the value of women leaders to a lifelong learning perspective where education and learning opportunities transcend the boundaries of learning institutions. Participation in capacity-building interventions with the full support of family members seems to be a vehicle in empowering women as leaders. These findings coincide with other studies engaged in promoting gender and disaster risk reduction in the Philippines (Allen 2006; Gaillard 2015). According to Ramalho (2019), the involvement of women in various public spaces tends to push for a redistribution of household functions which can be considered as another facet of resilience-building.

The concept of SEs with the poor as primary stakeholders (Dacanay 2013) is bounded by a common mission of upholding their own financial, social, and ecological well-being. They have been categorized as: farmers, fisherfolk, workers in the informal sector, persons with disability, Indigenous peoples, women, senior citizens, victims of calamities and disasters, youth and students, children, and informal settlers (Dacanay 2012). These are the sectors covered by the Social Reform Act and Poverty Alleviation of the Philippines or Republic Act 8425.

Women's roles and contributions in promoting resilient community

Some studies have shown certain practices and beliefs are discriminatory in times of disaster (Martin and Holtsberg 2014). These practices are rooted in women's vulnerabilities, especially in disaster contexts (Lacson 2015). For example, women's needs are unmet as disasters adversely affect several institutions that provide social services. Ofreneo and Hega (2016) found in the case of Buklod Tao in San Mateo Rizal and the Pambansang Tagapag-ugnay ng Manggagawa sa Bahay (PATAMABA) chapter in Sta. Barbara, Iloilo, that women can demonstrate collective action in creating the means to help them adapt against climate change and natural hazards.

The Philippines took the initiative to have a resolution for equal gender participation to promote and advocate ecologically sound activities during the 55th session of the UN Commission on the Status of Women (UN CSW). The CSW is a global intergovernmental body established to promote gender equality and women's empowerment. The role of women in the entrepreneurial field is an important element in pursuing a gender-inclusive and gender-responsive community (Luchsinger 2015). Women's attributes are also recognized in contributing success despite the patriarchal orientation of communities (Shinnar et al 2012; Urbano et al 2014). Women's organizations perform roles in implementing SE and livelihood activities. Their roles are shaped by gender and power, which are mirrored in the mounting power of women (Kabeer 2008).

The roles of women's organizations are considered essential in the community development landscape as well as in the areas of community resilience building. The following are the identified roles of KWACII as the only women-based organization in Concepcion. It also showcases the ways in which these roles contribute to resilience building:

As a catalyst in improving the socio-economic condition of women members

The KWACII women participants articulated that they have a significant interest in improving their economic productivity to support their families. One participant shared her observation in the shifts of roles in the family:

> "*Yung mga asawa dati, naghihintay lang ng sahod ng asawa. Ngayon naghahanap na din ng paraan para may pagkakitaan.*"
>
> ["The wife usually just waits for the salary of their husband, but now they (the wives) are also actively searching for income-generating activities."] (Dona Bayot, 42 years old)

Women are significantly part of the fishing industry, being the main source of livelihood in the community. They develop skills in the areas of fish drying, marketing, and selling. Women are also given opportunities to further diversify their livelihood skills in response to the call for resilience building. For instance, women in Taloto-an village were allowed to operate a community-based bakery through the support of the Daughters of Charity Foundation. They have experienced issues related to management, but the spirit to keep on moving persisted. They show their accountability and willingness to explore various solutions to the challenges they encounter. Similarly, women members of Tambaliza, Salvacion, Maliog-liog, and Botlog in Concepcion are implementing a cost-shared approach to augment their existing livelihood activities. While the financial support is given individually, the organization still upholds the value of collective responsibility. Indeed, the engagement of women's group in various local economic activities is valuable not just to their families, but also to their communities.

The organization magnified a particular element of solidarity when they underscored the notion of *malasakit* (deep concern), best described as empathy for others. Accordingly, they were able to sustain the flow of their program activities despite the challenges that they have experienced because of *malasakit*. The certain push for *pagmamalasakit* (empathizing for others) has been vital also to stimulate members' participation. Cielito Habito, in his article in the Philippine Daily Inquirer (2016), pointed out that the concept appears to have no direct English translation. The concept of care, concern, stewardship, compassion, and empathy has been suggested by Habito (2016) to be similar in meaning, but does not quite capture the full essence of the word *malasakit*. An approximate definition might be 'to care for something [or someone] like it was one's own.' The author added that *Malasakit sa Kapwa* is altruism extended to someone as if that person were one's self or own kin. It extends to inanimate objects or property, and, in that context, means handling or using something with care because it is not ours, and we understand and empathize with how the owner would feel if the object were damaged or misused.

As a platform in addressing women's issues and advocating women's rights

Establishing a good reputation and getting recognition as an organized women's group has gradually led to providing spaces for women's organizations in island communities to be organized and responsive. The nature of gender inclusion was observed in community gatherings and community projects where contributions and advocacies of women were highly encouraged. This scenario allows the women leaders to continuously work hand-in-hand with members and maintain a positive perception from the community.

Women members have conveyed that they experience a deep sense of unity and strong cooperation in their organization. This has been evident in their regular activities, such as but not limited to, the *Women's Month* celebration that covers various program activities depending on the monthly theme. Members from far-flung areas attend this celebration despite their struggle to reach the mainland, which is the usual venue for the program. Their presence in this kind of activity provides them with a sense of recognition and a platform to further articulate their respective rights and obligations to incentives, assistance, and representation as a community organization.

Women leaders are said to be at the forefront of moving the organization to attain its goals. Spaces are created to hear the voices of the members and confront issues affecting their capacities towards organizational performance. Members are also inspired to do better and continue performing their shared responsibility of promoting women's rights and welfare. In 2017, the leader of KWACCII became a recipient of the Ilongga award. The award is given to outstanding women leaders from various fields and sectors in the province. Their leader received this award because of her dedication to work as a community volunteer under the Kapit-Bisig Laban sa Kahirapan-Comprehensive and Integrated Delivery of Social Services (KALAHI CIDSS) program for numerous years and serving as a role model in pursuing women empowerment in rural areas.

As a platform in improving capacities to manage livelihood projects and organizational functions

Expressions of gender inclusivity and gender sensitivity are practiced by the KWACII participation in construction-related activities during the post-disaster rehabilitation efforts. Gender sensitivity is framed as the inclusion of women in both formal and informal decision-making activities. The process allows a conscious effort to consider the needs and strengths of women as part of the community. Women's voices are recognized in the planning and implementation of local economic development projects, which have contributed to the BBB objectives of the area. One of the members reported:

> "*Noong una hindi ako marunong maglansang, ngayon marunong na ako, dati hindi ako marunong magkumpuni ng bahay, dependent ako sa aking asawa, ngayon marunong na kasi nagtatrabaho ang asawa ko. Nakikita na ang kayang gawin ng lalaki ay kayang gawin din ng babae. Nakikita yung role pag dating sa shared responsibility.*"
>
> ["I didn't know how to nail things, but I learned how to do it. I used to depend on my husband, but now I had to learn how to do it since he has a job already. I can see here that women can also do what

> men can do and that there is shared responsibility."] (Lina Causing, 60 years old)

It was observed in the community that people became unordered because of the conflicts that emerged during the relief distributions. Community residents tend to compare the kind of assistance that they are getting from different donors about their social capital. This has resulted in a more proactive engagement of women's organizations to mediate and revisit possible strategies for making sound decisions that could directly address the said issue. The leadership role by women was recognized as the strength of the executive committee in terms of organizational management and transparency, especially concerning the finance of the organization. One member expressed:

> "*Kung anu ang problema at kaperahan ng samahan ay naishare din sa buong myembro para alam nila ang nangyayari.*"
>
> ["The problem and finances of the association are shared with all members so that they are aware of what is going on."] (Dancel Dominguez, 36 years old)

Women leaders and members have actively participated and represented the women sector in the BDRRMC. The multi-sectoral representation is strongly encouraged to ensure community participation and sustainability of the DRR initiatives. This strengthens the initiative of the community to maintain inclusivity and address marginalization in times of disaster preparedness and response. Women take the proactive role in the conduct of needs assessment, especially after typhoon Haiyan, which resulted in extreme damage to the fishing sector. Likewise, given that KWACII is a recognized women's organization, it is manageable for BDRRMC to easily request their representative.

As a bridge in building networks and linkages

The women's organization is instrumental in building connections and strengthening linkages with numerous actors and stakeholders in the phase of disaster recovery and enhancing livelihood and entrepreneurial activities. This becomes an important asset to the community to continuously address vulnerability and collectively improve capacities in project management, resource mobilization, and income-generating activities. Women have shown their eagerness to maintain a good reputation in their community, which results in strategic partnership building with LGUs as duty bearers as well as with NGOs.

As an instrument in improving organizational development and performance

Members have reported that there is a stronger sense of participation in organizational development-related activities, such as their monthly meetings. Some members of KWACII said:

> "*Kapag nagpapatawag ka ng meeting, parang ang hirap hirap nilang abutin. Ngayon naman ay nauuna pa sila sayo pag sinabing may meeting.*"
>
> ["It used to be difficult to reach members when meetings were called. Now, members look forward to meetings."] (Lina Causing, 60 years old)
>
> "*Ayaw ng mga asawa nila dati na dumalo ang kababaihan ng meeting. Ang sabi ng asawa, "Anung makukuha mo dyan?" Pero ngayon, ang asawa na ang nagtutulak.*"
>
> ["Their husbands didn't allow their wives to join the meetings asking, "What good will they get from that?" This time, the husbands push their wives to join the meetings."] (Diona Villanueva, 38 years old)

To overcome the limitations of budget and financial support for women empowerment training and seminars, the organization learned to institutionalize regulation policies, such as setting criteria for membership due to the big number of interested potential members.

Changes experienced by the members of the organization were also discussed during the FGD in Polopina Village, one of the island barangays in Concepcion. After typhoon Haiyan hit the municipality, there were significant and positive improvements in the organization in terms of leadership, membership, livelihood, and income. The direction and management of the organization were strengthened by having firm leadership. Women leaders were clear about their organizational vision, mission, goals, and objectives (VMGOs). There is an increased rate of membership, which means that typhoon Haiyan might have raised the consciousness of the people about advocacy and the importance of collective mobilization.

The livelihood component of the organization has helped the members to cope with the impacts of typhoon Haiyan. However, the organization still lacks different forms of resources and capital. SE has been a significant source of income for most of the members and provides alternative sources of livelihood. Furthermore, the members learn techniques and strategies in business operation. A target market was also identified and people's capacities for the community-based SE were developed, but training or capacity-building interventions should be sustainable.

Women members spearheaded the improvement of their capacities through attending educational training workshops and learning engagements. These are mostly facilitated by the NGO sector as part of their development projects. Mentorship and coaching are also established among officers and active members of KWAICII. This also paved the way for enhancing their livelihood skills to improve their economic gains.

Conclusions

Women's participation in managing livelihood and entrepreneurial activities contributed to resilience building in island communities in Concepcion, Iloilo. The different roles performed by the women's organization have resulted in improved status of their livelihood and the resilience status of the community where they belong. Data show that after typhoon Haiyan hit the municipality, there were significant and positive improvements in the leadership capacity, membership reach, and livelihood opportunities of the women's organization. Women's involvement in disaster preparedness can be accounted for in fulfilling their functions as subcommittee members in the BDRRMC. Participation of women in diverse community-based engagements especially in social enterprise management is exemplified as an extension of their reproductive, productive, and community roles.

Results also show that while women's leadership drives most of the SE, running them has also given them meaningful participation and a strengthened sense of ownership. Their participation is notable in livelihood activities. Many of them are involved in other parts of the fishing value chain, most especially in fish vending, processing, and other services in the sector. The support of NGOs through financial grants and provisions of capacity-building interventions were instrumental in establishing these SEs.

The role of KWAICII in spearheading CoMSCA is seen as one of the effective ways of sustaining SEs. Their attempt to harmonize the economic, social, and environmental aims (i.e., triple bottom-line principles) can be observed even though the SEs are still new. They were able to promote women's welfare and empowerment and attempted to ensure financial sustainability while practicing ecologically oriented activities. The priority, therefore, is to meet the needs of the poor family members.

The livelihood projects of the KWAICII organization have helped the members to cope with the impacts of typhoon Haiyan. The enterprise has been a significant source of income for most of the members, providing alternative sources of livelihood. Furthermore, the members were able to learn techniques and strategies in business. A target market has been identified, and people's capacities for the community-based SE have been developed. However, the organization still lacks different forms of resources (e.g., financial and human) as well as social capital to sustain the provision of training and other capacity-building interventions.

Gender inclusivity, enhanced participation, effective communication, and leadership management are identified as other critical factors to sustain the women's (KWACII members') livelihood activities. The support of NGOs and the LGU as duty bearers have also contributed to their enhanced capacity, especially toward disaster risk reduction and natural resource management. Resilience in island communities can be achieved once the issue of vulnerability, exposure, and risks are adequately addressed. Providing more spaces for women and other vulnerable sectors (e.g., children and elderly), for example, through women leading and championing SE initiatives, can contribute to a huge impact in attaining community resilience. Mainstreaming gender and inclusion in small island resilience-building efforts can be enhanced by supporting SEs which facilitate women empowerment.

Note

1 The Philippines is divided into political boundaries, namely: region, province, municipality, and barangay. A municipality or town is an LGU consisting of a group of villages or barangay.

References

Allen, K. A. (2006) "Community-based disaster preparedness and climate adaptation: Local capacity-building in the Philippines," *Disasters* 30(1): 81–101.

Baliyan, K. (2014) "Participation of women in household decision making: A case study of Muzaffarnagar district, Uttar Pradesh," *Bhartiya Krishi Anusandhan Patrika* 29: 159–161.

Ballesteros, M. M. and G. M. Llanto (2017) "Strengthening social enterprises for inclusive growth: Philippines," Discussion Paper Series No. 2017–04.

Beck, T. (2005) "Learning lessons from disaster recovery: The case of Bangladesh," Disaster Risk Management Working Paper Series 11. Hazard Management Unit (HMU), The World Bank. https://www.preventionweb.net/files/1590_recoveryBangladesh.pdf [Accessed November 30, 2017].

Boin, A. and M. J. Van Eeten (2013) "The resilient organization," *Public Management Review* 15(3): 429–445. https://doi.org/10.1080/14719037.2013.769856

Borquist, B. R. and A. de Bruin (2019) "Values and women-led social entrepreneurship," International Journal of Gender and Entrepreneurship 11(2): 146–165. DOI: doi.org/10.1108/IJGE-08-2018-0093.

Bradshaw, S. (2013) *Gender, Development and Disasters*, Cheltenham, UK: Edward Elgar.

Clinton, W. J. (2006) "Lessons learned from tsunami recovery: Key propositions for building back better," Office of the UN Secretary-General's Special Envoy for Tsunami Recovery, New York: UN, https://reliefweb.int/report/thailand/lessons-learned-tsunami-recovery-key-propositions-building-back-better [Accessed November 30, 2017].

Dacanay, M. L. (2009) "Social entrepreneurship: An Asian perspective," in J. A. Robinson, J. Mair, and K. Hockerts (eds) *International Perspectives on Social Entrepreneurship*, Basingstoke: Palgrave Macmillan, pp 163–182.

Dacanay, M. L. (2012) "Social enterprises and the poor: Enhancing social entrepreneurship and stakeholder theory," PhD Dissertation, Copenhagen Business School, Denmark.

Dacanay, M. L. (2013) *Social Enterprise and the Poor: Transforming Wealth*, Quezon City, Philippines: Institute for Social Entrepreneurship in Asia.

Dacin, M. T., P. A. Dacin, and P. Tracey (2011) "Social entrepreneurship: A critique and future directions," *Organization Science* 22(5): 1203–1213.

Doherty, B., G. Foster, C. Mason, J. Meehan, K. Meehan, N. Rotheroe, and M. Royce (2009) *Management for Social Enterprise*, London: Sage.

Enarson, E. (2000) "Gender and natural disaster," Crisis Response and Reconstruction Working Paper No. 1. International Labor Organization. https://www.ilo.org/employment/Whatwedo/Publications/WCMS_116391/lang--en/index.htm [Accessed November 30, 2017].

Enarson, E. P. and P. G. D. Chakrabarti (2009) *Women, Gender and Disaster: Global Issues and Initiatives*, Los Angeles, CA: Sage.

Enarson, E. and B. Morrow (eds). (1998) *The Gendered Terrain of Disaster: Through Women's Eyes*, Westport, CT: Greenwood.

Estrin, S., T. Mickiewicz, and U. Stephan (2013) "Entrepreneurship, social capital, and institutions: Social and commercial entrepreneurship across nations," *Entrepreneurship Theory and Practice* 37(3): 479–504.

Fordham, M. (1998) "Making women visible in disasters: Problematising the private domain," *Disasters* 22: 126–143.

Fotheringham, S. and C. Saunders (2014) "Social enterprise as poverty reducing strategy for women," *Social Enterprise Journal* 10(3): 176–199. DOI: https://doi.org/10.1108/sej-06-2013-0028.

Gaillard, J. C. (2015) *People's Response to Disasters in the Philippines: Vulnerability, Capacities, and Resilience*, New York: Palgrave Macmillan.

Galera, G. and C. Borzaga (2009) "Social enterprise: An international overview of its conceptual evolution and legal implementation," *Social Enterprise Journal* 5(3): 210–228.

Grimes, M. (2010) "Strategic sensemaking within funding relationships: The effects of performance measurement on organizational identity in the social sector," *Entrepreneurship Theory and Practice* 34(4): 763–783.

Habito, C. (2016) "Malasakit," *Philippine Daily Inquirer*, August 19. http://opinion.inquirer.net/96610/malasakit [Accessed November 30, 2017].

Haugh, H. M. and A. Talwar (2016) "Linking social entrepreneurship and social change: The mediating role of empowerment," *Journal of Business Ethics* 133(4): 643–658.

Hechevarría, D. M., A. Ingram, R. Justo, and S. Terjesen (2012) "Are women more likely to pursue social and environmental entrepreneurship," in K. D. Hughes and J. E. Jennings eds *Global Women's Entrepreneurship Research: Diverse Settings, Questions and Approaches*, Cheltenhan, UK: Edward Elgar, pp 135–151.

Islam, M. R., V. Ingham, J. Hicks, and I. Manock (2017) "The changing role of women in resilience, recovery and economic development at the intersection of recurrent disaster: A case study from Sirajgang, Bangladesh," *Journal of Asian and African Studies* 52(1): 50–67. https://doi.org/10.1177/0021909614560244

James Lee Witt Associates. (2005) "Building back better and safer: Private sector summit on post-tsunami reconstruction," James Lee Witt Associates, LLC, Washington, DC. https://www.preventionweb.net/files/2070_VL108700.pdf [Accessed December 15, 2017].

Kabeer, N. (2008) "Paid work, women's empowerment and gender justice: Critical pathways of social change," Pathways Working Paper 3. Brighton, UK: Institute of Development Studies.

Kreft, S., Eckstein, D., Junghans, L., Kerestan, C., and Hagen, U. (2014) "Global climate risk index 2015: Who suffers most from extreme weather events? Weather-related loss events in 2013 and 1994 to 2013," Germanwatch. https://www.germanwatch.org/sites/default/files/publication/10333.pdf [Accessed December 15, 2017].

KWACII Kalipunan ng Liping Pilipina (KALIPI) Women's Association in Concepcion, Iloilo Inc. 2016. Organizational Profile. Concepcion, Iloilo.

Lacson, I. (2015) "How disasters affect women," *Rappler*, March 9. https://www.rappler.com/move-ph/issues/disasters/thought-leaders/86237-women-disasters-role [Accessed January 20, 2021].

Langley, A. (1999) "Strategies for theorizing from process data," *Academy of Management Review* 24(4): 691–710.

Laylo, A. (2018) "Macro-level enabling conditions for the formation of social business enterprises in the Philippines," *Asia Pacific Journal of Innovation and Entrepreneurship* 12(1): 5–13. https://doi.org/10.1108/APJIE-12-2017-0044

Luchsinger, G. (2015) "UN woman 2014–2015 annual report," http://www.unwomen.org/media/headquarters/attachments/sections/library/publications/2017/un-women-annualreport-2016-2017-en.pdf?la=en&vs=0. [Accessed January 20, 2021].

Mair, J., J. Robinson, and K. Hockerts (2006) "Introduction," in J. Mair, J. Robinson and K. Hockerts eds *Social Entrepreneurship*, New York: Palgrave, pp 1–13.

Mair, J. and C. Seelos (2007) "Pro table business models and market creation in the context of deep poverty: A strategic view," *The Academy of Management Perspectives* 21(4): 49–63.

Martin, S. and M. Holtsberg (2014) "Why gender matters in natural disasters," https://www.rappler.com/move-ph/ispeak/76101-gender-matters-natural-disasters [Accessed January 25, 2021].

Morduch, J. (1999) "The microfinance problem," *Journal of Economic Literature* 37(4): 1569–1614.

Municipal Profile of Concepcion. (2016) Local Government Unit of Concepcion. Office of the Municipal Planning and Development, Concepcion.

National Disaster Risk Reduction and Management Council. (2013) "Final report re: Effects of typhoon," https://ndrrmc.gov.ph/attachments/article/1329/FINAL_REPORT_re_Effects_of_Typhoon_YOLANDA_(HAIYAN)_06-09NOV2013.pdf [Accessed December 5, 2016].

Newbert, S. L. (2014) "Building theory in social entrepreneurship," *Journal of Social Entrepreneurship* 5(3): 239–242.

Nguyen, H. T. (2019) "Gendered vulnerabilities in times of natural disasters: Male-to-female violence in the Philippines in the aftermath of super typhoon Haiyan," *Violence against Women* 25(4): 421–440. https://doi.org/10.1177/1077801218790701

Ofreneo, R. and M. Hega (2016) "Women's solidarity economy initiatives to strengthen food security in response to disaster," *Disaster Prevention and Management* 25: 168–182.

Petz, D. (2014) "Strengthening regional and national capacity for disaster risk management: The case of ASEAN," https://www.brookings.edu/wp-content/uploads/2016/06/Strengthening-Regional-and-National-Capacity-for-DRM-Case-of-ASEAN-November-5-2014.pdf [Accessed January 30, 2021].

Philippine Business for Social Progress. (2015) "Typhoon Yolanda relief and recovery (fishing value chain development)," Intramuros, Manila.

PSA (Philippine Statistics Authority). (2009) "2008 informal sector survey," https://psa.gov.ph/content/informal-sector-operators-counted-105-million-results-2008-informal-sector-survey [Accessed January 30, 2021].

Ramalho, J. (2019) "Empowerment in the era of resilience-building: Gendered participation in community-based (disaster) risk management in the Philippines," *International Development Planning Review* 41(2): 129–148. https://doi.org/10.3828/idpr.2018.25

Reyes, D. D. and J. L. Lu (2017) "Gender dimensions and women's vulnerability in disaster situations: A case study of flood-prone areas impacting women in Malabon City, Metro Manila," *Journal of International Women's Studies* 18(4): 69–88.

Santos, J., L. Macatangay, M.A. Capistrano, and C. Burns (2009) "Southeast Asia," in J. A. Kerlin (ed) *Social Enterprise: A Global Comparison*,., Lebanon, NH: Tufts University Press, pp 64–87.

Sengupta, S. and A. Sahay (2017) "Social entrepreneurship research in Asia-Pacific: Perspectives and opportunities," *Social Enterprise Journal* 13(1): 17–37.

Shinnar, R. S., O. Giacomin, and F. Janssen (2012) "Entrepreneurial perceptions and intentions: The role of gender and culture," *Entrepreneurship Theory and Practice* 36(3): 465–493.

UNESCAP (United Nations Economic and Social Commission for Asia and the Pacific). (2016) "Disasters in Asia and the Pacific: 2015 year in Review," Bangkok, http://www.unescap.org/resources/disasters-asia-and-pacific-2015-year-review [Accessed November 30, 2017].

UNESCAP (United Nations Economic and Social Commission for Asia and the Pacific). (2017) "Leave no one behind. Asia-Pacific disaster report 2017". Disaster Resilience for Sustainable Development, https://www.unescap.org/sites/default/files/publications/0_Disaster Report 2017 High res.pdf [Accessed November 30, 2017].

UNISDR (United Nations International Strategy for Disaster Reduction). (2015) "Global assessment report 2015," The Pocket GAR 2015, Geneva: United Nations.

Urbano, D., E. Ferri, and M. Noguera (2014) "Female social entrepreneurship and socio-cultural context: An international analysis," *Revista de estudios empresariales, Segunda Época* 2: 26–40.

World Bank. (2015) *World Development Report 2015*, Washington, DC: World Bank.

9

What Could Have Been? Disaster Impact Analysis of Haiyan in Region VIII

Shiely Anne O. Julian and Francine Blaise M. Loja

Introduction

Typhoon Yolanda, internationally known as Haiyan, is one of the most devastating calamities that the Philippines had suffered in the last 20 years. With sustained winds of 295kph and gusts of 360kph, Haiyan hit the Visayan provinces (Eastern Samar, Leyte, Cebu, and Iloilo) on November 8, 2013, and was classified as a Category 5 hurricane, earning the super typhoon epithet (Mullen 2013). With six total landfalls in the different Visayan islands, Haiyan affected over 16 million individuals and left over 6,300 dead, over 1,000 missing, and 28,688 people injured (NDRRMC 2013, 2–3). Overall, super typhoon Haiyan resulted in estimated total damage amounting to PHP89.6 billion and estimated losses of PHP42.76 billion.[1] The area worst hit by the super typhoon was the Eastern Visayas (Region VIII), particularly Leyte, with over 5 million people affected (31.19% of the total 16 million), 5,902 (93.68%) dead, 1,005 (94.72%) missing, and 26,186 (91.28%) injured. A total of 492,856 (43% of the total) houses were destroyed, of which 248,306 were partially damaged while 244,550 were completely wiped out (NDRRMC 2013, 3–4). In Region VIII alone, PHP4.99 billion was the total amount of damages in infrastructure; PHP18.92 billion in the productive sector;[2] PHP41.89 billion in the social services sector;[3] and PHP2.9 billion cross-sectoral.[4] Total damages for the region reached PHP68.7 billion or almost 72% of the total damages of typhoon Haiyan in the Philippines (NDRRMC 2013, 148). But we argue that these are not the only costs brought about by the typhoon. In this chapter, we present a new approach to identifying and estimating disaster impacts such as that of typhoon Haiyan in Region VIII, focusing on the region's economy.

On November 11, 2013, then-President Benigno S. Aquino III declared a State of National Calamity and sent three C130 aircraft to Tacloban City, Leyte, carrying personnel, relief goods, and a communications van. The Department of Trade and Industry (DTI) declared a price freeze in the affected areas (Official Gazette n.d.) and within the first week after Haiyan's onslaught, the Department of Social Welfare and Development (DSWD) established a Donations Desk at the national and regional levels and prepositioned emergency relief resources worth PHP206.58 million (readily available for augmentation) which consists of 1) PHP66.18 million—standby funds; 2) PHP31.06 million—which translates to 16,115,448 food packs; and 3) PHP109.33 million—other food and non-food items. The Department of Health (DOH) extended PHP1,157,504.95-worth of medical assistance while the National Disaster Risk Reduction Management Council (NDRRMC) put up a 'One-Stop-Shop' at the Bureau of Customs office in NAIA (a centralized area for all services to receive foreign aid) and Mactan Airport for the influx of assistance from the international community (Department of Foreign Affairs 2013). The DTI also held a *Diskwento*[5] Caravan in Ormoc.

Different foreign governments and nongovernmental organizations (NGOs) also helped in the post-disaster recovery efforts of the typhoon. Donations in the form of relief goods and monetary aid came from Australia,[6] Canada,[7] Japan,[8] the United Kingdom,[9] and the United States of America.[10] Intergovernmental organizations such as the World Food Programme (WFP) donated 45,000 sacks of rice, while the United Nations Children's Fund (UNICEF) sent 60 tons of aid in the form of shelters and medicine, along with water purification and sanitation equipment. In addition, the International Labour Organization (ILO) provided livelihood assistance to beneficiaries across communities in Leyte (including Tacloban), Northern Cebu, Bohol, Negros Occidental, and Coron, Palawan (International Labour Organization 2014).

Rehabilitation efforts both from the Philippine government and international organizations continued to pour into the country even months after the disaster. On December 16, 2013, the Philippine government, through the National Economic Development Authority (NEDA), launched the Reconstruction Assistance on Yolanda (RAY) which was a "strategic plan to guide the recovery and reconstruction of the economy, lives, and livelihoods of people and communities in the areas affected by Typhoon Yolanda (Haiyan)" (Official Gazette 2013, n.p.). A year after, the Comprehensive Rehabilitation and Recovery Plan (CRRP), amounting to PHP167.9 billion, absorbed the RAY (Hanley et al, 2014, 18); and as of November 2014, PHP52 billion of funds had already been released to the different national agencies (Official Gazette 2014, n.p.). The top five agencies that received the largest share of the released funds were theDSWD, Department of Education (DepEd), Department of Interior and Local Government (DILG), National Electrification Agency, and the Department of Public Works and Highways (DPWH) (Department of Finance 2015, n.p.). Two years after Haiyan, rehabilitation projects were nearing completion, such as the Emergency Shelter Assistance, distribution of learning kits to Haiyan-affected schools, and rehabilitation of damaged airports, bridges, and roads (NEDA 2015, n.p.). The DSWD received PHP17 billion of the PHP73 billion pledged foreign aid. Still, they were able to disburse 84% of the donations and were used for transition shelters, cash for work, ready-to-eat food items, and medicines, among others (as of September 2015) (Official Gazette 2015). As of 2015, a total of PHP14.82 billion-worth of shelter assistance was given to the affected families from Palawan, Panay, Northern Cebu, Samar, and Leyte municipalities (Official Gazette 2015).[11]

Even after Haiyan, the national government continues to integrate a disaster response fund in the yearly budget—enabling the local government to be resilient and financially able to respond immediately in case another calamity happens. In the General Appropriations Act of 2018 by the Department of Budget and Management (DBM), a total of PHP19.6 billion was appropriated for the NDRRM Fund for the reconstruction, rehabilitation, repair, aid, relief and other works or services including pre-disaster activities. Of the PHP19-billion allocation, PHP7.57 billion was allotted for maintenance and other operating expenses while PHP12 billion was appropriated for capital outlays. In addition to that, a Quick Response Fund amounting to PHP7.6 billion was lodged under the budget of different agencies.[12] Financial assistance from international organizations also continued to pour in and according to the United Nations Office for the Coordination of Humanitarian Affairs (UN OCHA), total funding received for the rescue and rehabilitation of Haiyan-hit areas amounted to US$865.15 million as of October 7, 2016 (FTS-UN OCHA 2016, 3).

This monetary aid and relief play a significant role in post-disaster recovery. Previous studies have already proved that transfer payments are of great help in the recovery from disasters (Coffman and Noy 2012). Even short-term effects of disasters become trivial once financial aid is controlled for (Barone and Mocetti 2014). However, this does not happen with the mere presence of monetary aid. The financial assistance received for the post-Haiyan rehabilitation (and for any disaster, for that matter) must not be whimsically and capriciously spent. These must be coupled with projects and strategies, and backed by studies and accurate data, for it to efficiently work. One way of rationalizing these plans and efforts is having a good basis of the economic losses due to calamities, which could also vary depending on the methods used in assessing these impacts. For one thing, disaster impacts are classified into direct damages[13] and indirect losses[14] (Noy and DuPont IV 2018, 327–328). Between these two, direct economic damages are easier to measure as these can be easily obtained from surveys and first-hand accounts. On the other hand, indirect economic losses require statistical estimation on either macroeconomic[15] or microeconomic[16] variables. Examples of such statistical estimation are the Complete General Equilibrium (CGE) model analysis (Tuano et al 2016), difference-in-differences (D-I-D) approach (Anttila-Hughes and Hsiang 2013) and, two-way fixed effects panel model (Berlemann and Wenzel, 2018), among others.

A relatively new approach called the Synthetic Control Method (SCM) developed by Abadie et al (2010) provides a way to construct a counterfactual. The counterfactual is one or more unit(s) that has or has not been subjected to the intervention of interest (e.g., natural disasters). In this context, the counterfactual is the scenario if Haiyan did not happen. Such counterfactual would then be compared with the unit/s exposed to the intervention, which, in this context, is the actual scenario in which Haiyan occurred. Employing this methodology, we provide an estimate of the full economic impact of typhoon Haiyan on Region VIII—the value of forgone future production due to damaged physical and human capital—which is essential in designing DRRM (disaster risk reduction management) initiatives. We do this by using a cross-region panel data of all the regions in the Philippines inclusive of the years 2002 to 2017, giving us a total of 12 years of pre-intervention data (typhoon Haiyan hit the country in November 2013).[17] In particular, we use the real GRDP per capita, released annually by the Philippine Statistics Authority (PSA),[18] and other economic variables such as population density, inflation, and gross capital formation, among others,[19] to measure the difference in the economic performance of the actual Region VIII, hit by Haiyan, and the synthetic Region VIII, had Haiyan not happened. In the following sections, we present the economic profile of Region VIII pre- and post-Haiyan to be followed by a discussion of our methodology and results. Lastly, we conclude with the importance of these results and how these fit in the whole picture of disaster risks reduction and management of the Philippines.

Region VIII's economic profile before typhoon Haiyan

One of the 18 regions of the Philippines, Eastern Visayas is a region mainly consisting of six provinces, one independent city, and one highly urbanized city—Tacloban City (Department of Agriculture n.d.). With a total land area of 2,156,285 hectares, it accounts for 7.1% of the whole country (Department of Agriculture, n.d.). In 2012, it had a population of 4,236,900 with a real Gross Regional Domestic Product (GRDP)[20] at PHP143,420,096 and real GRDP per capita[21] of PHP33,850 (PSA 2019, n.p.). Consequently, Region VIII accounts for 4.4% of the country's population while contributing 2.3% to the country's Gross Domestic Product (GDP) (PSA 2019, n.p.). With its naturally endowed resources that provided communities with sources of livelihood, Region VIII's location fronting the Pacific Ocean allows the local fishermen to sustain their fishing industry. According to the DTI, the region's "sea and inland waters are rich sources of salt and freshwater fish and other marine products" (DTI n.d.). Similarly, its land is used for agricultural purposes to produce crops such as rice, corn, coconut, and sugarcane. Further, the region is also abundant in mineral resources such as chromite, uranium (in Samar), gold, silver, manganese, magnesium, bronze, nickel, clay, coal, limestone, pyrite, sand, and gravel. Yet with all these natural resources, Region VIII has seen a slow and steady increase in GRDP per capita from 2002 until 2011.

In 2012, however, the region experienced a -6.8% GRDP growth rate, which translated to an 8% decrease in GRDP per capita (PSA 2013, n.p.). This was due to the fire incident in Isabel, Leyte, which caused the shutdown of the Philippine Associated Smelting and Refinery Corporation (PASAR) smelting plant where cathodes, the region's top export commodity, are produced (NEDA VIII 2012, 2). Following that event, Region VIII's exports declined by 24% and cathode production by 55%, which resulted in an impact that was felt both in the short and medium term. The said smelting plant was idle for six months until it resumed operations in July 2012. But a year after resuming operations, it sustained heavy structural damages following typhoon Haiyan, and it approximately took four to six weeks to be able to return to its normal operations (ABS-CBN 2013). Hence, in 2012, the region reported an estimated magnitude of people living below the poverty line at 1,882,934 or 45.2% poverty incidence among the population (PSA 2013, n.p.). The data from the PSA (2013) also show that in 2012, the region's annual per capita poverty threshold[22] was at PHP18,076 while the per capita food threshold[23] was at PHP12,732. For the same year, the region recorded a 27.2 income gap[24] and a 10.2 poverty gap[25] (PSA 2013, n.p.). Moreover, Region VIII had a Gini Coefficient Ratio of 0.4834 (PSA 2017, n.p.), which represents the economic/income inequality among the population of the region—that is, for Region VIII, there is a relatively big income inequality with its value almost reaching 0.5, since the lower the Gini Coefficient is, the less inequality there is.

Typhoon Haiyan and its consequences

Undoubtedly, Haiyan is one of the strongest typhoons ever recorded, not only in Philippine history but also in the world. Ravaging the lives of over 16 million people (in over 3.4 million families), the super typhoon resulted in an estimated total damage of PHP89.5 billion and estimated losses of PHP42.7 billion in the country (NDRRMC 2013, 3–5).[26] In Region VIII alone, which was the worst hit by the typhoon, over 5 million people (over 1 million families) were devastated (NDRRMC 2013, 3). Total direct damages for Region VIII amounted to PHP68.7 billion, which brought the region's economy to a downturn with its GRDP per capita decreasing by 3.5% in 2014,[27] or PHP33,771 from PHP35,002 in 2013. It was only in 2016 that Eastern Visayas was able to not just return, but even surpass its pre-Haiyan GRDP per capita level at PHP37,144 (PSA 2019, n.p.).

However, it is not right to assume nor conclude that the 3.5% drop in the region's GRDP per capita is solely due to typhoon Haiyan. Knowing that the region also suffered from another major economic event in 2011, the fire incident in the PASAR smelting plant, the bases for such estimates are not very clear. Presumably, these values also include the lingering effects of the fire in 2011 misleading us from the true economic costs of the typhoon. As such, we should tease out the pure effect of typhoon Haiyan by constructing the region's counterfactual.

Estimating the counterfactual

In estimating the effects of any event or intervention, researchers usually use comparative analysis to compare the changes after the intervention happened (Abadie, Diamond, and Hainmueller 2010, 493). An easy solution would be to merely compare the "pre-intervention" and "post-intervention" scenarios (i.e., the *before* and *after*). However, analyzing an impact using this method makes it difficult to precisely determine whether the difference is due to the intervention of interest, or to a distinct event that happened during and/or before the intervention.

A more insightful method of comparison is done by analyzing the *control unit* in contrast with a *counterfactual*. The *control* is the actual unit/s that have been exposed to the intervention (in this case, typhoon Haiyan). A *counterfactual*, on the other hand, is one or more units that have not been subjected to the intervention of interest. The counterfactual must have the same characteristics as that of the control unit/s throughout the pre-intervention period—they must be able to mimic each another in the years before the intervention occurs (Abadie and Gardeazabal 2003, 116–117). Given this feature of a counterfactual, it becomes essential in estimating the impact of intervention because only then would we be able to identify the true effect of that intervention. It isolates the other possible impacts from variables that are unaccounted for in the analysis.

With these in mind, we employ the SCM in constructing a proper counterfactual for Region VIII. We do this by comparing the actual Region VIII with our counterfactual—the synthetic control (Abadie, Diamond and Hainmueller 2010).[28] The synthetic control is a systematically combined group of unexposed units (i.e., the non-Haiyan-affected regions) that approximates the characteristics of the exposed unit (i.e., Region VIII).[29] The resulting counterfactual (synthetic Region VIII) aids the researchers in estimating the impact of typhoon Haiyan on the region. Particularly, we estimate and compare the trend of the GRDP per capita of Region VIII with that of the synthetic Region VIII, our counterfactual, that did not experience typhoon Haiyan in November 2013.

The SCM, formalized by Abadie, Diamond, and Hainmueller (2010), is essentially an approach to estimate an impact of a particular intervention. Following this idea, the effect of Haiyan on Region VIII (α_t) at time t is evaluated as the difference between the GRper capita of the actual Region VIII (Y_t) and of the synthetic Region VIII (Y_t^{NY}), given by the equation:

$$\alpha_t = Y_t - Y_t^{NY}. \tag{1}$$

This impact is estimated through the synthetic control method using

$$\widehat{\alpha_t} = Y_t - \sum_{j=1}^{J} w_j^{*} Y_{jt} \tag{2}$$

with w_j^{*} referring to the weight assigned to region j included in the synthetic control group and Y_{jt} representing the GRDP per capita level of region j at time t.

In particular, SCM allows the researcher to construct a proper counterfactual for the unit(s) exposed to the intervention of interest. Instead of comparing a single unit alone, this approach systematically combines unexposed units to create the synthetic group which approximates the characteristics of the exposed unit(s) (Abadie, Diamond, and Hainmueller 2010; Mahoney and Barrenechea 2017). In developing a synthetic control, the values of w_j^{*} were obtained from the vector of optimal weights, $W^{*} = (w_1^{*}, \ldots, w_N^{*})'$, which minimizes the difference between the pre-Haiyan (2002–2013) output determinants of Region VIII and of the synthetic Region VIII as summarized in vectors X_1 and X_0, respectively.

Both said vectors contain q variables and the equation minimized is

$$\sum_{m=1}^{q} v_m (X_{1m} - X_{0m})^2 \tag{3}$$

with factor v_m as the weight assigned by the SCM to the variable m, indicating the relative significance of the particular output determinant in matching the pre-Yolanda GRDP per capita levels of Region VIII and of the synthetic Region VIII.

As stated in Abadie, Diamond, and Hainmueller (2010), the SCM has two key features which make it fare better relative to the traditional regression methods—transparency and safeguard against extrapolation. Since the SCM is a data-driven procedure, it specifies the similarities (or the lack thereof) between the units exposed and unexposed to the intervention through the weights given for each control unit included in the synthetic group (Abadie, Diamond, and Hainmueller 2010). Because of all these, various studies such as Coffman and Noy (2012), Barone and Mocetti (2014), and Cavallo et al (2013) have already adopted the said methodology in examining the effects of exogenous events on the respective variables of interest, such as the costs of natural disasters on economic growth.

But as expected, we found significant divergence in the resulting graphs of the actual and synthetic Region VIII even before 2013 (the year of the intervention) upon performing the SCM on our data. We attribute this to the aforementioned drop in Region VIII's GRDP per capita in 2012, which was due to the fire in PASAR's factory. Presumably, with the loss in production capacity, the effect on employment, income, and the multiplier effect on the rest of the region's economy, would have confounded the impact of typhoon Haiyan on the region. Hence, the observed discrepancy in GRDP between Region VIII and synthetic Region VIII may be due to both events.

Thus, we tease out the pure effect of typhoon Haiyan as follows: 1) we assume that the fire effects for all the post-Haiyan years are fixed and we subtract this fixed amount from the yearly losses due to typhoon Haiyan (2014 onwards);[30] 2) we take into account the discrepancy of the actual and synthetic Region VIII by getting the average of the differences from 2002 to 2011;[31] and 3) we omit the differences for 2012 and 2013 because we assume that this was already reflected in the value for the effect from the fire.

Overall, the validity of the SCM results rests on two assumptions: 1) the quality of the donor pool[32]—that is, the regions that will constitute the synthetic Region VIII; and 2) there are no major disturbances other than typhoon Haiyan that happened around or during the evaluation period.

Economic costs of typhoon Haiyan

Estimating the effects of typhoon Haiyan requires showing how the real GRDP per capita of Region VIII would have grown after 2013 if it was spared the wrath of the super typhoon. In doing so, we constructed Model 1 representing a synthetic control that is a linear combination of the non-Haiyan-stricken regions. Particularly, the synthetic Region VIII comes from the combination of National Capital Region (NCR), Ilocos, Central Luzon, South Cotabato, Cotabato, Sultan Kudarat, Sarangani, and General Santos (SOCCSKSARGEN), and Autonomous Region in Muslim Mindanao (ARMM).[33] Julian and Loja (2019) created two more models (Models 2 and 3), to further examine the true impact of the typhoon. Model 2 only includes those non-Haiyan-affected regions with economic characteristics (i.e., consumer price index (CPI), population density, income per capita, and poverty incidence among population) at par with Region VIII, while Model 3 consists of the regions spared from Haiyan that experienced a drop in GRDP from 2011–2012, specifically CAR and ARMM. For a thorough discussion on these models, please refer to Julian and Loja (2019). Among the three models, Model 1 has the simplest assumptions, with only non-Haiyan-hit regions as the criteria. In Table 9.1 we see the sum of the weights assigned to each region in the synthetic control for Model 1.

A good model, and hence a synthetic control, should be able to track Region VIII in the years before the onslaught of typhoon Haiyan. The Root Mean Square Prediction Error (RMSPE) is a measure of how well the synthetic control tracks the actual, overall, and before the intervention or treatment. The aim is to minimize RMSPE and this model obtained an RMSPE of 1352.852. Table 9.2 shows the summary of the economic growth predictor means for the actual and synthetic Region VIII of Model 1.

Additionally, the synthetic Region VIII should be able to mimic the actual Region VIII—both should have similar characteristics for the pre-intervention years. Figure 9.1 illustrates the trend line of the actual GRDP per capita of Region VIII in contrast to that of the synthetic Region VIII.

Table 9.1: Summary of weights

Region	Model 1
National Capital Region	0.013
Cordillera Administrative Region	0
Ilocos Region	0.201
Cagayan Valley	0
Central Luzon	0.045
Zamboanga Peninsula	0
SOCCSKSARGEN	0.557
Autonomous Region in Muslim Mindanao	0.185

Note: A cell containing "0" is different from an empty cell. Having a weight of "0" implies that the region was not assigned a weight even though it was part of the synthetic control group. On the other hand, an empty cell indicates that the region was excluded in the synthetic control group because it does not produce the best fit for the model.

Table 9.2: Summary of economic growth predictors of the actual and synthetic Region VIII, Model 1

Economic growth predictors	Actual Region VIII	Synthetic Region VIII
GRDP of the previous year	33,333.12	32,994.72
Population density	174.94	439.06
Average annual income	150.58	150.82
Inflation	5.46	5.47
Gross Capital Formation in construction	11,978.00	13,097.50
Gross Capital Formation in durable equipment	7,454.27	10,482.63
Gross Capital Formation in breeding stocks	6,838.46	4,918.82
Number of government personnel	63,401.00	56,484.92
Net commodity flow	35,900,000.00	12,100,000.00

It is noticeable that the synthetic Region VIII is somewhat able to closely replicate the actual Region VIII from 2007 up until the years before typhoon Haiyan, but a greater discrepancy is visible for the year 2012, which represented the economic impact of the fire in Leyte.

With this, we subtract the estimated fire effect, PHP5,359.54 in 2011 from the post-Haiyan values (2014 onwards). Consequently, the estimated poor tracking effect, PHP338.40 that was computed by getting the mean differences of the actual and synthetic Region VIII for the years 2002–2011, were also deducted. Figure 9.2 presents the trend lines of the actual Region VIII compared with the adjusted values of the synthetic Region VIII net of fire effects and poor tracking. Now the pre-Haiyan trend lines are closer together and both follow a similar trajectory. In the post-Haiyan scenario, we see that the difference is lesser in magnitude compared to the unadjusted values, but still has the same pattern of a lower difference for the year 2016.

Table 9.3 presents a summary of the yearly losses after considering the effect of the fire in 2011 as well as the effect of poor tracking of the synthetic Region VIII.

Immediately after the typhoon, there was a gross loss in GRDP per capita that amounted to PHP1,851.11 in 2014. On average, the GRDP lost for the first four years after typhoon Haiyan amounted to PHP2,263.17. The average annual difference is the mean of the yearly difference of the four years immediately after Typhoon Haiyan (2014–2017). The amount is equivalent to 5% of Eastern Visayas' average annual GRDP per capita. This indicates that the residents in Eastern Visayas earned 5% less than they should have without typhoon Haiyan.

Figure 9.1: Trends in the real GRDP per capita: actual vs. synthetic Region VIII, Model 1

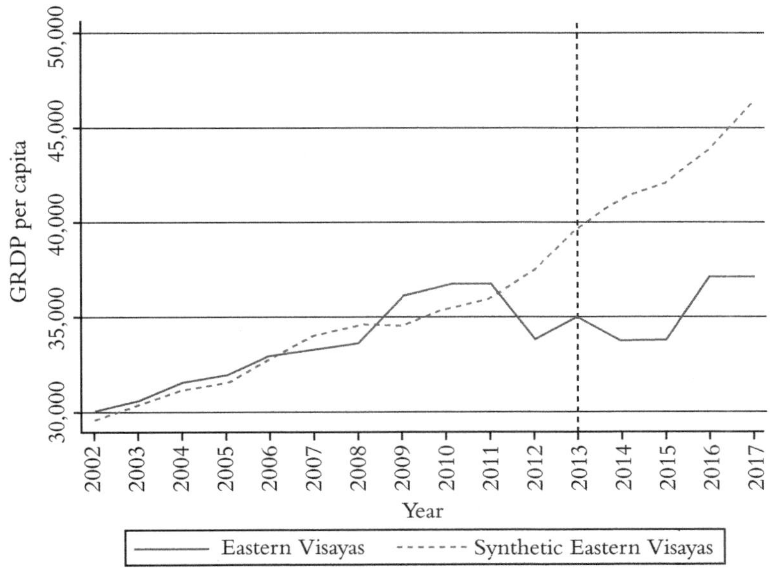

Figure 9.2: Actual Region VIII vs. Adjusted Synthetic Region VIII, Model 1

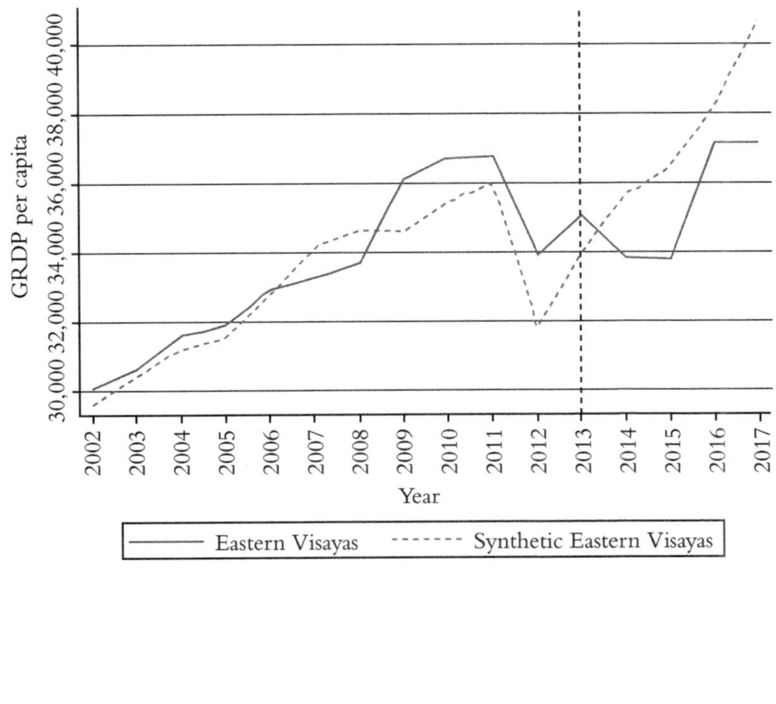

Table 9.3: Summary of gross losses in real GRDP per capita of Region VIII, Models 1–3

	2014	2015	2016	2017	Average annual impact
			Model 1		
Gross loss (in PHP)	(7,549.05)	(8,353.58)	(6,707.39)	(9,234.41)	(7,961.10)
Less fire effect (in PHP)	(2,189.50)	(2,994.03)	(1,347.84)	(3,874.87)	(2,601.56)
Less poor tracking (in PHP)	**(1,851.11)**	**(2,655.64)**	**(1,009.45)**	**(3,536.47)**	**(2,263.17)**

To make more sense of these values, we relate our estimates to losses in the food and poverty thresholds of Region VIII. According to PSA (2013), the food threshold is the minimum income required to meet the basic food needs and satisfy the nutritional requirements set by the Food and Nutrition Research Institute (FNRI), to ensure that one remains economically and socially productive. A similar concept is the poverty threshold which includes both the basic food and non-food needs such as clothing, housing, transportation, health, and education expenses. For 2015, the annual per capita food and poverty thresholds for Region VIII were PHP14,957 and PHP21,304, respectively. However, for 2014, only national (Philippines) values are available. Specifically, we only have the first semester 2014 Philippine food and poverty threshold values which are PHP6,125 and PHP8,778, respectively. To obtain the annual equivalents, we divided the monthly food and poverty thresholds by five, as these values are for a family of five. The resulting quotient becomes the monthly per capita values. Then, we multiply these by 12 to get the annual per capita values. The resulting values after this series of computations are PHP14,700 and PHP21,067.20, which become the annual per capita food and poverty thresholds, respectively. The value of the estimated loss translates into 27% of the food threshold and 19% of the poverty threshold. Put in another way, for those who are living in subsistence, losing PHP4,078.53 of their annual income means that they will have to make ends meet to survive and make it through each day. That amount, which was lost due to typhoon Haiyan, could have been used to buy the basic food and non-food necessities for one to live humanely.

Following a very large disaster such as super typhoon Haiyan, a great amount of damage is expected. Values higher than those we obtained are what most would hypothesize. However, attention must be drawn to

Table 9.4: Interpretation of the final value of estimated loss in PHP (converted into current prices)

	2014	2015	2016	2017	Average annual impact
Final value of estimated loss (in current prices)	(3,269.19)	(4,712.74)	(1,833.73)	(6,498.45)	(4,078.53)
As % of GRDP per capita	-5%	-8%	-3%	-10%	-6%
As % of the annual per capita poverty threshold**	16%*	22%	9%	31%	19%
As % of the annual per capita food threshold**	22%*	32%	12%	43%	27%

Note: * Computed using the threshold values for the first semester of 2014; the rest are computed using the annual threshold values for 2015

** Values for Region VIII only

the financial aid appropriated for post-Haiyan recovery efforts. To recall, the Philippine government, through its various agencies, immediately extended assistance following the onslaught of the typhoon. To name a few, over PHP200 million from the DSWD and PHP1 million-worth of medical assistance from the DOH. A Comprehensive Rehabilitation and Recovery Plan was also affected with a budget of PHP160 billion, released PHP52 billion-worth of funds within one year after Haiyan. The foreign community, likewise, has been more than generous with donations adding up to US$865.15 million (Financial Tracking Service 2016). As previous studies already showed (Coffman and Noy 2012; Barone and Mocetti 2014), the negative effects of a disaster are tempered by transfer payments and financial aid. These monetary assistances mitigate the impacts of a stimulus such as super typhoon Haiyan.

The results obtained from Model 1 were congruent with Coffman and Noy (2012) and Barone and Mocetti (2014). Monetary aid from both local and international sources cushioned the possible catastrophic economic impact of Haiyan. Although the negative impact was not totally eliminated, it is relatively lesser than what is to be naturally expected. Simply put, the results obtained already reflect the positive effects of financial aid relative to the GRDP per capita of the region.

Conclusion

Several studies have already been done to identify the effects of typhoon Haiyan on different sectors, such as rice production (Boschetti et al 2015), fisheries (Monteclaro et al 2018), and even mental health (Chan et al 2016), but never on the economy—much less on the economic growth of Region VIII, which suffered most from the wrath of the disaster. Studies conducted to measure such economic impacts only focused on the immediate direct effects of Haiyan (i.e., losses to human capital, livelihood, and productivity) but failed to estimate the medium- to long-term indirect losses of the super typhoon. They were not able to consider the counterfactual—the idea of how Region VIII's economy would have performed in the absence of typhoon Haiyan. Simply said, they did not look at it from the perspective of "What if typhoon Haiyan had not happened? Where would Region VIII's economy be today?" This is especially important in the case of Region VIII as it was already performing badly even before typhoon Haiyan struck. Hence, naive estimates of the economic losses after Haiyan may fail to consider the residual or lingering effects of another incident—for example, the fire in PASAR's plant—especially on the export production sector of the region.

Thus, this chapter provides the estimated economic effect of typhoon Haiyan in Region VIII. We used the real GRDP per capita (in constant 2000 prices) and employed the SCM developed by Abadie, Diamond, and Hainmueller (2010), to construct a synthetic Region VIII that represented Eastern Visayas without typhoon Haiyan. We estimated an average annual loss per capita amounting to PHP2,263.17, for the first four years (2014–2017) after the region was hit by the typhoon, which was equivalent to 6% of the average annual GRDP per capita of Region VIII. Likewise, this value is equivalent to 6% of the average income per capita of the residents of Region VIII. These estimates appear minuscule compared to the reported losses and damages of the typhoon, but we hypothesize that the post-Haiyan GRDP per capita used in the analysis already includes the effect of disaster relief and other assistance provided to the region. As in the study of Barone and Mocetti (2014), the presence of financial aid dampens the effect of natural disasters on the concerned economy.

It is important to note that the GRDP per capita of Region VIII is a measure of regional income and does not account for wealth or assets. Thus, the economic losses of typhoon Haiyan that we report here are in terms of *forgone regional income using GRDP*. The losses from Haiyan include both income and wealth, and, as such, there is no data for wealth that was used in the analysis. Additionally, our estimates do not account for the losses incurred in the lives lost, homes broken, and families separated by the tragic typhoon. Nevertheless, the method we used in the study provides another approach to identifying and estimating these indirect losses of natural disasters which moving forward, could serve as a guide on how policymakers and local leaders could form a better disaster risk management plan for their constituents. Building on the idea that financial aid weakens the economic effect of disasters, this should give light as to how much relief and assistance the government should be injecting into the affected region/s to mitigate economic loss.

Further studies regarding the true economic impact of disasters such as that of typhoon Haiyan are encouraged since Haiyan is not just a one-time phenomenon. With climate change, the chances that the Philippines will experience something as strong as this is getting more and more likely. The national government must strengthen current programs and services and its DRRM initiatives and the key to being better equipped to handle situations like this in the future is to understand the situation itself. These studies further the research on disaster impacts and could pave the way for the improvement of disaster preparedness of the Philippines and thus lesser vulnerability of the greater Filipino public.

Notes

1 Refer to Julian and Loja (2019, 5, Table 1) for the detailed list of the estimated damages and losses per sector of typhoon Haiyan in PHP.

2 Includes agriculture, fisheries, mining and quarrying, trade, industry and services, and tourism.

3 Comprises education, health, and housing.

4 Identifies governance and social impact assessment.

5 *Discount* in English; it means that the items are sold for a lower or reduced price.

6 PHP530 million—United Nations Flash Appeal; PHP 530 million—Australian NGOs and their partners on the ground (i.e. immediate life-saving assistance); and PHP82 million—Australian medical assistance team. Also refer to Official Gazette (n.d.).

7 Can$20 million – pledge through humanitarian partners; Philippine Crisis Fund.

8 US$52.1 million worth of aid with the following composition: 1) US$30 million—emergency grant aid; 2) US$0.6 million—emergency relief goods; 3) US$1.5 million—assistance through Japanese NGOs; and 4) US$20 million—Japan Fund for Poverty Reduction of Asian Development Bank (ADB).

9 £55 million for humanitarian response; including provisions of temporary shelters and access to safe water.

10 US$100,000—health, water, and sanitation support; US$20 million—additional humanitarian aid; US$10 million—additional humanitarian aid from USAID. Also refer to ABS-CBN News (2016).

11 For a more detailed list of post-Yolanda assistance, refer to Julian and Loja (2019).

12 Refer to Julian and Loja (2019, 10, Table 4) for the breakdown of the Quick Response Fund.

13 Damages to fixed assets and capital, losses of raw materials, crops, and extractable natural resources, and mortality and morbidity.

14 The economic activities lost due to the disaster—business interruptions, macroeconomic feedbacks, falls in demand, consequences for economic growth and health, social and community network disruptions, and impacts on poverty, security, and on stability.

15 National income accounts, fiscal accounts, balance of payments.

16 Households, firms, business, organizations.

17 Our sample period begins in 2002 because it is the earliest period for which Region IV was divided into Region IV-A and Region IV-B.

18 PSA-released constant GRDP per capita values for the years 2002–2008 were based on 1985 prices. Thus, we converted these values into 2000 prices based on our own calculations.

19 For a more detailed and complete list of the variables, refer to Julian and Loja (2019, 25–26, Table 5).

20 GRDP measures the economic performance of a region and covers the value of goods and services produced in the region for a given quarter/year (PSA 2019, n.p.).

21 Calculated by dividing the region's GRPD by its population.

22 Measures the minimum income required to meet the basic food and non-food commodities.

23 Measures the minimum income required to meet the basic food needs only.

24 Measures the average income required by the poor to get out of poverty, expressed relative to the poverty threshold (PSA 2013. n.p.).

25 Refers to the income shortfall (expressed in proportion to the poverty threshold) of families with income below the poverty threshold, divided by the total number of families (PSA 2013. n.p.).

26 Refer to Julian and Loja (2019, 5, Table 1) for the detailed list of the estimated damages and losses per sector of typhoon Haiyan in PHP.

27 All GRDP per capita values indicated in this paper are in real 2000 prices.

28 Refer to the Appendix in Abadie, Diamond and Hainmueller (2010), for details of the data and methods used in the analysis.

29 These are: 1) NCR; 2) CAR; 3) Ilocos Region; 4) Cagayan Valley; 5) Central Luzon; 6) Zamboanga Peninsula; 7) SOCCSKSARGEN; and 8) ARMM.

30 We obtain the fire effect using SCM but the treatment period is changed to 2011.

31 Example of computation: Actual Region VIII $_{2002}$ – Synthetic Region VIII $_{2002}$ = difference in GRDP for 2002.

32 The group of units considered to be included in the synthetic control.

33 Some of the non-Haiyan-affected regions were excluded in constructing the synthetic control since the SCM did not assign weights for these regions (i.e., CAR, Cagayan Valley, Zamboanga Peninsula).

References

Abadie, A., A. Diamond, and J. Hainmueller (2010) "Synthetic control methods for comparative case studies: Estimating the effect of California's tobacco control program," *Journal of the American Statistical Association* 105(490): 493–505.

Abadie, A. and J. Gardeazabal (2003) "The economic costs of conflict: A case study of the Basque Country," *The American Economic Review* 93(1): 113–132.

ABS-CBN News. (2013) "PASAR copper plant in Leyte damaged after typhoon," Reuters, November 12, https://news.abs-cbn.com/business/11/12/13/pasar-copper-plant-leyte-damaged-after-typhoon [Accessed July 20, 2020].

ABS-CBN News. (2016) "How much 'Yolanda' aid did PH get from US, EU?" October 8, https://news.abs-cbn.com/focus/10/08/16/how-much-yolanda-aid-did-ph- get-from-us-euengthen [Accessed July 20, 2020].

Anttila-Hughes, J. K. and Hsiang, S. M. (2013) "Destruction, disinvestment, and death: Economic and human losses following environmental disaster," *SSRN Electronic Journal*, February 18, doi:10.2139/ssrn.2220501

Barone, G.,and S. Mocetti (2014) "Natural disasters, growth and institutions: A tale of two earthquakes," *Journal of Urban Economics* 84: 52–66.

Berlemann, M. and D. Wenzel (2018) "Hurricanes, economic growth and transmission channels: Empirical evidence for countries on differing levels of development," *World Development* 105: 231–247. https://doi.org/10.1016/j.worlddev.2017.12.020

Boschetti, M., A. Nelson, F. Nutini, G. Manfron, L. Busetto, M. Barbieri, A. Laborte, J. Raviz, F. Holecz, M. R. O. Mabalaya. A. P. Bacong, and E. J. P. Quilang (2015) "Rapid assessment of crop status: An application of MODIS and SAR data to rice areas in Leyte, Philippines, affected by typhoon Haiyan," *Remote Sensing* 7(6): 6535–6557. https://doi.org/10.3390/rs70606535

Cavallo, E., S. Galiani, I. Noy, and J. Pantano (2013) "Catastrophic natural disasters and economic growth," *The Review of Economics and Statistics* 95(5): 1549–1561. https://doi.org/10.1162/rest_a_00413

Chan, C. S., K. N. S. Tang, B. J. Hall, S. Y. T. Yip, and M. Maggay (2016) "Psychological sequelae of the 2013 super typhoon Haiyan among survivor-responders," *Psychiatry* 79(3): 282–296. https://doi.org/10.1080/00332747.2015.1129874

Coffman, M. and I. Noy (2012) "Hurricane Iniki: Measuring the long-term economic impact of a natural disaster using synthetic control," *Environment and Development Economics* 17(2): 187–205.

Department of Agriculture (DA). (n.d.) "Agriculture and fishery modernization plan (2012–2017) Eastern Visayas," http://da08.da.gov.ph/index.php/da-nfrs/profile [Accessed July 20, 2020].

Department of Finance (DOF). (2015) "P52-B released to Yolanda victims since 2013: DBM," July 27, https://www.dof.gov.ph/p52-b-released-to-yolanda-victims-since-2013-dbm/ [Accessed July 28, 2020].

Department of Foreign Affairs (DFA). (2013) "Statement of the DFA spokesperson on the status of international assistance to the relief and rehabilitation efforts in areas affected by typhoon Yolanda," November 15, https://dfa.gov.ph/dfa-news/dfa-releasesupdate/1284-statement-of-the-dfa-spokesperson-on-the-status-of-international-assistance-to-the-relief-and-rehabilitation-efforts-an-areas-affected-by-typhoon-yolanda [Accessed July 20, 2020].

Department of Trade and Industry (DTI). (n.d.). "Profile of Region 8," https://www.dti.gov.ph/regions/region-8/profile/ [Accessed July 21, 2020].

Financial Tracking Service (FTS)- United Nations Office for the Coordination of Humanitarian Affairs (UN OCHA). 2016. "Philippines: Typhoon Haiyan - November 2013 (Rep.)," https://fts.unocha.org/appeals/441/summary [Accessed July 21, 2020].

Hanley, T., R. Binas, J. Murray, and B. Tribunalo (2014) "IASC inter-agency humanitarian evaluation of the typhoon Haiyan response," https://www.alnap.org/help-library/iasc-inter-agency-humanitarian-evaluation-of-the-typhoon-haiyan-response [Accessed July 20, 2020].

International Labour Organization (ILO). (2014) "Integrated livelihood recovery for super typhoon Haiyan-affected communities – Government of Japan," March 21, https://www.ilo.org/manila/projects/WCMS_378118/lang--en/index.htm [Accessed July 20, 2020].

Julian, S. A. O. and F. B. M. Loja (2019) "What could have been: Estimating the economic impacts of typhoon Haiyan in Region VIII," Unpublished bachelor's thesis in economics, University of the Philippines Diliman.

Mahoney, J. and R. Barrenechea (2017) "The logic of counterfactual analysis in case-study explanation," *The British Journal of Sociology* 70(1): 306–338. https://doi.org/10.1111/1468-4446.12340

Meniano, S. (2018) "NEDA confirms P729.6-M returned 'Yolanda' funds," https://www.pna.gov.ph/articles/1055300 [Accessed July 20, 2020].

Monteclaro, H., G. Quinitio, A. D. Moscoso, R. Napata, L. Espectato, K. Anraku, K. Watanabe, and S. Ishikawa (2018) "Impacts of typhoon Haiyan on Philippine capture fisheries and implications to fisheries management," *Ocean & Coastal Management* 158: 128–133. https://doi.org/10.1016/j.ocecoaman.2018.03.032

Mullen, J. (2013) "Super typhoon Haiyan, one of strongest storms ever, hits central Philippines," November 8, http://edition.cnn.com/2013/11/07/world/asia/philippines-typhoon-haiyan/index.html [Accessed July 20, 2020).

National Economic Development Authority (NEDA). (2015) "As year ends, several Yolanda rehabilitation projects near completion – NEDA," http://yolanda.neda.gov.ph/as-year-ends-several-yolanda-rehabilitation-projects-near-completion/ [Accessed April 14, 2019].

National Economic Development Authority (NEDA). VIII. (2012) "Eastern Visayas socio-economic performance report 2012". https://drive.google.com/file/d/0BxfUX4RrqnrhREFwa0kwRXl2UWM/edit. [Accessed July 20, 2020].

National Disaster Risk Reduction and Management Council (NDRRMC). (2013) "Final report re effects of typhoon 'Yolanda' (Haiyan)," http://ndrrmc.gov.ph/attachments/article/1329/FINAL_REPORT_re_Effects_of_Typhoon_YOLANDA_%28HAIYAN%29_06-09NOV2013.pdf [Accessed July 20, 2020].

National Disaster Risk Reduction and Management Council (NDRRMC). (2014) "Y (it happened): Learning from typhoon Yolanda," http://ndrrmc.gov.ph/attachments/article/2926/Y_It_Happened.pdf. [Accessed July 20, 2020].

Noy, I. and W. DuPont IV (2018) "The long-term consequences of disasters: What do we know, and what we still don't," *International Review of Environmental and Resource Economics* 12(4): 325–354. http://dx.doi.org/10.1561/101.00000104

Official Gazette. (2013) "Document: Reconstruction Assistance on Yolanda: GOVPH," December 18, https://www.officialgazette.gov.ph/2013/12/18/document-reconstruction-assistance-on-yolanda/ [Accessed July 28, 2020].

Official Gazette. 2014. "DBM: P52B released to Yolanda victims since 2013: GOVPH," November 7, https://www.officialgazette.gov.ph/2014/11/07/dbm-p52b-released-to-yolanda-victims-since-2013/ [Accessed July 30, 2020].

Official Gazette. (2015) "Yolanda donations used for aid and rehabilitation," Republic of the Philippines, https://www.officialgazette.gov.ph/2015/09/14/yolanda-donations-for-aid-and-rehabilitation/ [Accessed April 14, 2019].

Official Gazette. (n.d.). "RescuePH: A detailed list of government rescue and relief efforts before and immediately after Yolanda," Republic of the Philippines, https://www.officialgazette.gov.ph/rescueph-a-detailed-list-of-government-rescue-and-relief-efforts-before-and-immediately-after-yolanda/ [Accessed April 14, 2019].

Philippine Statistics Authority (PSA). (2013) "Despite rise in the number of families, extreme poverty among families remains steady at 1.6 million in 2012," https://psa.gov.ph/poverty-press-releases/nid/53187

Philippine Statistics Authority (PSA). (2002–2017) *Philippine Statistical Yearbook*, Quezon City: Philippine Statistics Authority, https://psa.gov.ph/products-and-services/publications/philippine-statistical-yearbook/2002.

Philippine Statistics Authority (PSA). (2017) "2015 FIES results for CAR," http://rssocar.psa.gov.ph/fies/2015%20FIES%20Results%20for%20CAR [Accessed July 20, 2020].

Philippine Statistics Authority (PSA). (2019) "Gross Regional Domestic Product," http://rsso01.psa.gov.ph/statisticalframeworksandindicatorsystem/GRDP [Accessed July 20, 2020].

Tuaño, P. P., M. S. Muyrong, and R. L. Clarete (2016) "Economic impact of typhoon Ondoy in Pasig and Marikina cities using a multiweek CGE model analysis," *Philippine Journal of Development* 43(2): 1–24.

10

Gender and Labor Market Regimes in Post-Disaster Haiyan Communities in Leyte

Glenda Tibe Bonifacio

Introduction

Labor is gendered; a fact that is known to societies (Leacock and Safa 1986; McCarl Nielsen 1990; Meade and Wiesner-Hanks 2006). Who does what work and how is socially marked in diverse economic activities ranging from households to industries (Baron 1991; Blumberg 1991). The types of work performed by women and men—including girls, boys, and gender-diverse individuals—could change through time and place. Conditions for doing work could also change with age, marital status, familial responsibility, access to technology, and the environment (Game and Pringle 2020). While gender and labor in paid and unpaid work have been extensively studied (Redclift and Sinclair 1991; Hatt 1997; Bettio, Plantenga, and Smith 2013; Milkman 2016; Yenilmez and Huyugüzel Kisla 2021), those that arise from disasters are still part of an emerging field of study (Enarson and Dhar Chakrabarti 2009; Takasaki 2012; Kinnvall and Rydstrom 2019; Akter 2021).

The aftermath of a disaster creates new realities in local economies and gender relations. Gender is a recognized aspect of the difference between men and women, for example, in survival rates and impact on their economic and social well-being during and after the occurrence of disasters (Enarson 2000; Neumayer and Plümper 2007). Members of the LGBTQ+ (lesbian, gay, bisexual, transgender, queer +) community have increased vulnerabilities during and after disasters such as isolation, discrimination, and violence (Reddy and Goodman 2019; Frank 2020). Displacement and migration often result in varying levels of social protection for members of affected households based on gender, class, and physical ability, among others (Oryema 2017). Post-disaster communities are unique in their ways of survival, resilience, and dynamics of intervenors in relief, reconstruction efforts, and development strategies. Governments, non-profit organizations, international aid agencies, and individual philanthropy work together or independently to address the humanitarian crisis and economic reconstruction, both in the short-term and long-term ventures. The interplay of these actors in creating labor market regimes in post-disaster communities affects labor participation of local populations; mainly, who benefits from it, who participates in it, and how they participate.

This chapter explores the relations between gender and labor market regimes that arise in post-Haiyan communities in Leyte. The worst-affected areas ravaged by super typhoon Haiyan were in Eastern Visayas, particularly in Tacloban City in Leyte. I conducted observational fieldwork in Leyte and Samar in 2015 followed by case studies in Tacloban City and Barugo in 2016–2017.[1] Interviews with local officials and focus group discussions among affected households in these two post-disaster communities were conducted in 2016–2017. The reminders of super typhoon Haiyan visibly remained two years after it made landfall on November 8, 2013. Washed-out cars lay piled up near the coastline in Magsaysay Boulevard in Tacloban City, and damaged roofs, collapsed houses, shattered windows, dilapidated buildings, and burnt debris were everywhere in the city and surrounding areas.

Going home to see the massive destruction in Tacloban City became a personal turning point as a scholar. Global warming creates extreme weather conditions, and the frequency of super typhoons looms large in the Philippines (Mongabay 2019; Climate Signals 2020) and elsewhere. Local realities provide important perspectives on how gender and labor markets become intrinsic to survival, rehabilitation, and reconstruction. The stories from 'ground zero' attest to our humanity and community resilience, of creative negotiations of survivors and hopes for a better tomorrow. In this chapter, the first section presents an overview of gender in the island economy; the second section examines gender and labor market regimes from ground zero, and the third section analyses gender and post-disaster economic aid. The concluding thoughts are inspired by intersectional feminist frameworks on gender and disaster.

Gender and island economy

Societies are fundamentally structured based on cultural contexts and particular histories. Gender is an organizing principle in society, as are race and class (Belkhir and Charlemaine 2007), which determines status, access to resources, and privileges. Roles and expectations are assigned based on sex and gender predominantly influenced by the traditional division between women/feminine and men/masculine (Lindsey 2015). Gender intersects with other systems of relations such as race and class and is multidimensional, or, according to Connell and Pearse (2015, 12), "[i]t is not just about identity, or just about work, or just about power, or just about sexuality, but all of these things at once." Communities and societies are also complex systems composed of interacting units, groups, and institutions, among others, (Sawyer 2005; Ball 2012) that give rise to their unique gender relations and responses to disaster.

Island communities such as Leyte and other islands in the Philippines develop socio-economic institutions, practices, and distinct ways of living with stark contrasts between rural and urban areas. For purposes of comparison, statistical indicators of Leyte represent the rural municipalities like Barugo, while those of Tacloban City depict the urban area. In this section, the geography, demography, and economy of Barugo and Tacloban City set a broader view of the representations of gender in the island economy in Leyte.

Geography

Leyte is the eighth largest island in the Philippines, situated on the eastern part of the archipelago. During the Second World War, the US forces led by General Douglas MacArthur landed in Leyte on October 20, 1944 (Encyclopedia Britannica 2013), using its strategic location directly facing the Pacific Ocean. The island of Leyte is linked to the island of Samar by the San Juanico Bridge, which was completed in 1973; both islands form part of Eastern Visayas or Region VIII together with Biliran Island (Department of Trade and Industry n.d.). Leyte has two cities, 40 municipalities, and 1,503 barangays as of June 2018 (Philippine Statistics Authority 2018). Barugo is a coastal municipality in Leyte with a total land area of 84.62 square kilometers and a population density of 387 persons per square kilometer across 37 barangays (Philippine Statistics Authority 2017a). The distance from Barugo to Tacloban City is about 51 kilometers with regular commuter transportation services.

Tacloban City is the largest city in Region 8 and the capital of the province of Leyte. It has its charter and is the first highly urbanized city (HUC) in the region. Located in the northeast part of Leyte, Tacloban City has a total land area of 201.72 square kilometers, 138 barangays, and a population density of 1,200 persons per square kilometers in 2015 (PhilAtlas 2021b; Philippine Statistics Authority 2017b). Since 1874, Tacloban City is a port with a "long deepwater wharf" and served as the "temporary capital of the Philippines until Manila was regained" by the US forces during the Second World War (Encyclopedia Britannica 2020).

Demography

The 2015 Census of Population recorded 1,724,679 people in Leyte with 889,054 males and 835,625 females (Philippine Statistics Authority 2018). Based on the household population of 10 years and over, Leyte recorded 646,284 women and 681,295 men in 2015, with 36.6% and 34.3% married, respectively (Philippine Statistics Authority 2020, 5). The youth population between the ages of 15 and 24 numbered 158,161 women (47.7%) and 173,519 men (52.3%) (Philippine Statistics Authority 2020, 9). Persons aged 60 years old and over were recorded as 83,447 women (54.7%) and 69,022 men (45.3%) in the province (Philippine Statistics Authority 2020, 10). In Barugo, the population reached 32,745 in the 2015 Census (PhilAtlas 2021a) with 16,980 males and 15,765 females (Philippine Statistics Authority 2017c, 13). Of the male population 10 years and older, 6,758 were single, 3,876 were married, 304 were widowed, 162 were separated, and 1,947 were in live-in relations. Females in the same population group recorded 5,201 single, 3,916 married, 944 widowed, 161 separated, and 1,962 in live-in relations. More women were married, widowed, and in live-in relationships compared to men in Barugo (Philippine Statistics Authority 2017c, 60).

As of 2015, the total population in Tacloban City was 241,089 with 50,547 households (Philippine Statistics Authority 2017b). Three barangays in the city registered the fastest population growth rate from 2005 to 2015—Santo Niño, New Kawayan, and Tagapuro—as a result of relocating displaced families in the aftermath of super typhoon Yolanda (Philippine Statistics Authority 2017b). Senior citizens or those over 60 years old numbered 9,297 women (57.1%) and 6,972 men (42.9%) in 2015 (Philippine Statistics Authority 2020, 11). Young people or those aged 15–24 years old were 27,001 women (49.8) and 27,183 men (50.2%) during the same census year (Philippine Statistics Authority 2020, 9). Of 94,482 women in Tacloban City, 43.9% were single, 31.7% were married, and 6.6% were widowed. Of 94,916 men, 50.8% were single, 31.2% were married, and 1.9% were widowed (Philippine Statistics Authority 2020, 6). Data showed that more women were married, widowed, and over 60 years old.

Economy

Barugo is an agricultural and fishing municipality rated as "third-to-sixth class in terms of annual income" but considered the "most improved" by the National Competitiveness Council (NCC) in 2016 (Gabieta 2016). Based on three indicators of "economic dynamism, governance efficiency and infrastructure", Barugo increased its ranking of similarly classified municipalities by the NCC from 436th in 2014 to 56th in 2015. Barugo has an annual income of PHP5 million (US$100,000) with 80 registered business establishments as of 2016 (Gabieta 2016). Males of all population age groups are mainly in skilled agricultural work, forestry, and fishery (Philippine Statistics Authority 2017c, 342). Barugo is famous for *tuba*, a coconut wine, and *roscas*, a local cookie. In partnership with Runggiyan Social Development Foundation since 2004, Barugo implemented the GREAT Women Project (GWP) to enhance women micro-entrepreneurs (Philippine Commission on Women n.d.). The Barugo Food Delicacies Producers Association (BFDPA) is a local cooperative (Gabieta 2016) composed mainly of women. Women entrepreneurs in non-farm activities contribute to local economic growth in Barugo and inspire women's economic independence from men (Cabanilla 2014).

Tacloban City is the major economic, educational, health, and administrative center in Eastern Visayas. In 2018, air traffic at the Daniel Z. Romualdez Airport reached 715,131 passengers with 21 daily flights from Manila, Cebu, Davao, and Clark (Meniano 2018). Major shopping malls such as SM, Robinsons, and Gaisano are located in the city which attracts shoppers from around the region. Tacloban City hosts six major hospitals, three state universities (City of Tacloban), and about eight private colleges. In 2015, there were 472,300 tourist arrivals in the city (Meniano 2018) with different types of hotel accommodation. *Pintados* and *Sangyaw* are the major festivals held in June during the Sto. Niño fiesta celebration. In terms of quality of life, the poverty rate among the population in the city reached 41.22% and poverty incidence among families was 32.95% in 2015. These poverty indicators, however, decreased to 30.86% and 23.99% in 2018, respectively (Department of Agriculture n.d.).

In general, the retail industry accounted for the largest proportion of employed women at 33.2% in Eastern Visayas in 2018 compared to men who are mainly employed in agriculture, hunting, and forestry at 35.2% (Philippine Statistics Authority 2020, 22). Major occupations are highly gendered with more women among the professionals, managers, clerical support workers, and service and sales workers at over 50% for each of these categories; men dominate in the armed forces, as machine operators and assemblers, craft and trade workers, skilled agricultural and forestry workers at over 80% for each of these categories (Philippine Statistics Authority 2020, 21). Labor force participation highly favors men at 75.3% compared to women at 46.5% in 2018 (Philippine Statistics Authority 2020, 18). Gender equality in income and economic opportunities remains a goal in the Philippines despite being the only Asian country with the closest gender gap index measured by the World Economic Forum (WEF). In 2018, the Philippines ranked 8th in the WEF rankings but slipped to 17th place in 2021 while remaining the only Asian country in the top 35 among 130 countries evaluated based on economic participation, educational attainment, health and survival, and political empowerment (WEF 2018, 2021).

From ground zero: gender and labor market regimes

Super typhoon Haiyan, locally known as Yolanda, struck Tacloban City in the early morning of November 8, 2013 with sustained wind speed at over 300kph, making it the "most ferocious storm in recorded history" (Liljas 2014). The Category 5 cyclone with a "tsunami-like-storm surge" left over 4 million displaced people (Liljas 2014) and over 8,000 dead (Mullés 2014), including members of my family. Tacloban City lay flat in ruins, about 90% destroyed (Reid 2018). It was ground zero.

In the aftermath of disaster and destruction, survival is key. Where to get food, clothing, and shelter? Who tends to the survivors? I asked these questions two days after realizing that my family in Tacloban City were among those missing and displaced after super typhoon Haiyan made landfall. The city government was incapacitated and unable to deliver the usual services during times of calamities. No means of communication from survivors in ground zero except posts coming out from Facebook from those with remaining charged cellphones. A local tragedy with national and global significance unfolded. In Canada, I watched the CNN coverage about Haiyan that was contradictory to the version presented by a Philippine TV anchor (Palatino 2013). I realized that the critical relief to Tacloban City became politicized at the costs of human lives (Salazar 2015). The political 'ping-pong' between the local government led by Alfred Romualdez and the national government under Benigno Aquino III, scions of political rivals, had impacted the management of disaster relief (Go 2016) and coordination of domestic and international donors.

This section examines the relations between gender and labor market regimes that arise in post-disaster communities in Tacloban City and Barugo. I use the term *regime* to emphasize the prevailing labor market conditions after a disaster. A regime is often associated with politics and the rule of government in power (Mainwaring and Pérez-Liñán 2013), and modes of control and regulation (Pulkowski 2014) that also reflect economic conditions (Herr and Kazandziska 2011; Magara 2014).. Markets form part of a complex system affected by politics and the environment, for example, the labor market. Broadly, labor regimes include "labour and skill availability and quality" as well as "labour market regulation, industrial relations and general labour market flexibility" (Myant et al 2012, 1). Ciccia (2015, 58) notes that labor regimes are conceptualized as the "intersection of two dimensions—selectivity and forms of regulation of employment relations—focusing on ways in which labour is mobilized in different political economies". Labor regimes also interact with the physical environment and its role in creating market conditions for women, men, and non-binary individuals. I offer an alternative definition of a post-disaster labor market regime as a form of the market created by conditions after a natural calamity in which labor participation in the community is contingent upon available resources, either domestic or foreign, and shifts accordingly from relief to reconstruction. These post-disaster conditions exacerbate labor demands and types of labor force participation whether paid or unpaid work.

While studies on labor market regimes are available (Bourdet and Persson 1992; Olofsson 2006), those that arise from meteorological and geophysical disasters are, arguably, few and emerging (Belasen and Polachek 2009; Mendoza and Jara 2020; Zhu et al 2021). According to Kirchberger (2017, 40), "less is known about how local labour markets in low-income countries adjust to such large shocks." Amidst the chaos and uncertainties, what was ground zero like with the coming of economic interveners from local, national, and international sectors? Who participated in the labor market at ground zero? Is the post-disaster labor market regime gendered? These questions are interrogated in this section using three economic spheres of activity that emerged from the qualitative case studies in 2016–2017: local, translocal, and international.

Reconstruction refers to the "full restoration of all services, and local infrastructure, replacement of damaged physical structures, the revitalization of the economy and the restoration of social and cultural life" (Aysan and Davis 1993, 10). This is a long-term phase that integrates capacity building, enhancement of resources and empowerment; in this case, the most vulnerable members of the community, such as women, are considered primary stakeholders in the process (Dominelli 2013). Reconstruction is viewed as part of the overall development goal of the community, in which economic and social well-being are connected to achieve a better quality of life. In the process of reconstruction in post-disaster communities, where various stakeholders from local, national to international agencies, and government and private interventions occur, certain types of labor market regimes are created.

Local

The *local* refers to the labor market regime characterized in Barugo and Tacloban City. At the end of the catastrophic storm surge, ground zero was an unimaginable wasteland with washed-out houses strewn with mud streaks and surrounded by mayhem. None were spared. All went down with super typhoon Haiyan, echoed by respondents in interviews and focus group discussions. No work to report to with destroyed buildings and no commercial establishments open for business for nearly a month in Tacloban City. In Albert Mullés' account of Haiyan, written in "candlelight for nearly six months" (2014, x), survivors "live off the land without food, water, and medical services in the face of surmountable odds" (back cover). Emergency relief came in too late with the politicization of disaster management response, and many survivors in the city, including my family members, scoured for canned goods under the rubble and relied on generosity. The physical conditions shaped labor flows, including evacuation and migration to nearby towns and islands.

At ground zero, barter of food-for-food or food-for-work market schemes seemed common among the survivors. According to respondents in Tacloban City, many households exchanged commodities for survival akin to a rudimentary form of 'sharing economy' (Sundararajan 2016). Rich and poor alike mingled in makeshift abodes or queued in the same line for relief, which somehow obscures differing social status based on previous material displays of wealth and comfort. Of course, those with higher income status had access to more goods and services available in other areas such as Ormoc, Leyte, and Catbalogan, Samar, with their private transportation and ready cash. Looting at the shopping malls and groceries was also reported (Marshall 2013; Ramzy 2013), and, according to field data, this was allegedly participated in surprisingly by people in the community known to be wealthy. Interestingly, some looted food items were shared by different families and households.

Perhaps the most observable mode of labor participation post-Haiyan disaster was volunteerism. Unlike in Western countries like Canada where survivors tend to be housed in facilities to undertake trauma counseling and support before returning to their devastated residences, survivors of super typhoon Haiyan had to do the clean-up themselves including the search for their missing and dead relatives. Neighbors and family members came together to assist those who needed them, for example, in fixing temporary shelters and cooking extra food. Men, women, and non-binary Filipino archetypes—*bakla* (born male with feminine gender identity) and *tomboy* (born female with masculine gender identity)—were fully engaged in these types of volunteer labor immediately after the disaster. Arguably, post-disaster communities rely on volunteer labor to support themselves and strengthen resilience (Gonsalves and Mohan 2011). A respondent in Barugo noted volunteerism at the nearby hospital to treat survivors with wounds, assist the laboratory, and secure supplies. Doctors arrived a week after the super typhoon. Post-disaster volunteerism in the form of unpaid labor reflects core resilient values of the collectivist orientation of Philippine culture that seems advantageous during times of crisis.

Because of the massive destruction of dwellings, offices, and commercial buildings, employed workers redirected or expanded labor from office work to repair and reconstruction in the community. Venn (2012, 4) points out that "[l]arge-scale natural disasters can have long-lasting effects on the labour market in affected areas in addition to their humanitarian and economic cost." Many workers lost their jobs as a result of collapsed business establishments but found themselves in demand from numerous reconstruction projects.

Translocal

Translocal refers to the labor market regime directed by post-disaster labor interactions in Leyte from those coming from elsewhere in the Philippines. Translocality is indeed common in labor markets; workers come from other islands or move farther between geographical borders (Freitag and von Oppen 2010; van Eerbeek and Hedberg 2021). Life in the Philippine archipelago deeply embeds translocality for work, education, and service. Centralized governance based in Metro Manila in Luzon participated by scattered island provinces makes translocality a fact of political administration and economic engagements.

In this section, my concern focuses on in-country labor market intervenors such as local government units (LGUs), government agencies, civic organizations, and non-resident individuals. LGUs have jurisdictional control and form part of various networks such as the League of Municipalities of the Philippines (LMP), League of Cities of the Philippines (LCP), and League of Provinces of the Philippines (LPP). Government agencies under the executive branch headed by the president of the Republic of the Philippines have regional or local offices. Civic organizations are diverse based on professions and interests, either local or national in scope. Essentially, these institutions and associations are translocal and often demonstrate solidarity and support to those impacted by typhoons and other catastrophes (Gomez 2020). Policy directives, whether by public or private groups, foster temporary labor infusion into disaster-affected communities.

In the case of post-Haiyan Leyte, the participants in the study well noted the role of selected government agencies such as the Department of Labor and Employment (DOLE), Department of Agriculture (DA), Bureau of Fisheries and Aquatic Resources (BFAR), and the Department of Social Welfare and Development (DSWD). The budget for disaster relief of the national government was transferred to local agencies in affected areas. According to focus group participants, *urhi na an gobyerno* (the government was late) and *waray umabat han gobyerno* (did not feel the presence of [national] government). Respondents also commented about the acts of some government employees based in other areas who allegedly took home foreign relief items intended for local distribution to survivors. These imported items were allegedly replaced with local sardines and corned beef that, according to them, "even dogs don't eat."

The immediate assistance provided by the LGU of Davao headed by Mayor Rodrigo Duterte was prominent in the narratives of the participants and consequently etched their votes in his historic bid to become the president of the Philippines in 2016. Mayor Rodrigo Duterte famously commented upon his return to Davao City from Tacloban City a few days after super typhoon Haiyan, "God must have been somewhere else or he forgot that there is a planet called Earth" (Espejo 2013). In the absence of coordinated national disaster response, Mayor Duterte provided much-needed medical and transport personnel, among others, in the ravaged communities as well as support to relocated survivors in Davao (Minda News 2013). For the national government, the ineffective typhoon response ultimately defined the Aquino presidency (Eadie 2013; Ramos 2013; Yee 2014; Agence France-Presse 2016).

Numerous civic organizations spearheaded relief operations and initiated the construction of temporary shelters and permanent resettlement of displaced families. Respondents remembered these non-government organizations such as Sagip Kapamilya/ABS-CBN Foundation, Tabang Bisaya, GMA Kapuso Foundation, and Caritas. Private initiatives were more immediate, visible, and coordinated with groups of volunteers and local labor from among the affected communities. In the northern barangays (village) in Tacloban City, according to respondents, the affected families helped construct the houses where they permanently relocated about a year, quite faster than those provided by the government (Amazona 2016). The private-household partnership utilized the availability of labor from members of the household while costs of materials were funded by donations.

International

International refers to the labor market regime created by interactions with international agencies, foreign or transnational civic groups and individuals in affected communities in Barugo and Tacloban City. Participants noted the presence of three types of international intervenors: bilateral, organizational, and civic groups. Bilateral refers to countries that provided emergency assistance, rehabilitation, and reconstruction support, namely Australia, Canada, China, Germany, Japan, Norway, South Korea, Switzerland, UK, US, and the Vatican. Organizational refers to international non-government organizations such as the Catholic Relief Services, Global Medic, Habitat for Humanity, International Red Cross, UNICEF, WHO, and Save the Children. Civic groups include the Circulo Barugeño (US), Tzu Chi Foundation (Taiwan), and Good Samaritan Society (Canada).[2] In 2014, the provincial government of Leyte held a donors' forum for Yolanda-affected areas with the following invited participants: United Nations Office for the Coordination of Humanitarian Affairs, Alliance for Safe and Sustainable Reconstruction (Assure Inc.), UNICEF, FAO, WFP, The World Bank, UNFPA, DFAT, The Asia Foundation, Oxfam, Sagip Kapamilya, Tabang

Bisaya, Assissi Foundation, GMA Kapuso Foundation, and others (*Manila Times* 2014).

Participants in the study vividly remembered the presence of international intervenors during the first phase of emergency relief and beyond. They hold positive reflections of international humanitarian assistance compared to those politically motivated deliveries of programs and services by the government. Respondents commented that local barangay chairmen prioritized those who were aligned to the same political party in the allocation of government financial aid. They referred to them as *malalakas an kapit* (strong political connections). There were claims that partial financial grants were distributed but with the remaining balance to be given after the national elections (cf. Maga 2015; *Manila Standard* 2015). However, some international assistance was also linked through the barangay such as from the Circulo Barugeño and the Good Samaritan Society. Support from Caritas was channeled through the local churches. International agencies were dispersed throughout the Visayas and not all were found in the communities under study.

Under extreme conditions of need and support post-disaster, local labor was critical and necessary. But the infusion of foreign capital resulted in doubled or tripled labor wages compared to pre-disaster rates. Increased rates persisted to the present day where labor costs from rural areas seem to be higher compared to those in the city. Labor availability dictates the apparent cost in affected communities that remained even after the completion of internationally funded reconstruction projects. Such change in labor wage dynamics was viewed differently by workers and employers, as an advantage to the former and a disadvantage to the latter. Migrant labor consequently became the source to repress the high local labor costs.

Cash-for-work programs were prevalent during the reconstruction of post-disaster communities in Leyte. International agencies provided cash payments in the clean-up drive in Tacloban City. In the San Jose area, according to respondents, the Tzu Chi Foundation paid PHP500 per person and was the first to arrive in the community. According to participants, international agencies arrived in the first two weeks in the devastated city. All interested residents regardless of gender worked under many cash-for-work schemes for livelihood provided by international organizations and selected agencies of the national government (Buhr, Almagro, and Tordesillas 2013; Asian Development Bank 2014).

Local, national, and international labor market regimes occurred simultaneously. Labor force participation was flexible based on availability and contract period. The international labor market regime offered the most attractive arrangement with higher wages. While respondents generally believe that "gender is not an issue" in any labor market regime, the type of work and paid beneficiaries were gendered. Clean-up drives and construction projects were open to all and gender-neutral, but men tended to take heavier physical work compared to women. These differing tasks were considered a complementary division of labor that enabled women and men to help each other akin to egalitarian relations quite common in agricultural societies like the Philippines (Quisumbing et al 2014). In terms of paid work, however, men were usually identified as beneficiaries for the household. No data exists for members of the LGBTQ+ community, but they are presumably included in the women and men categories.

Studies in labor economics suggest three explanations on the gendered differences in labor market performance and participation, or why men and women have different levels of employment and participation in labor activities: labor market discrimination, gender differences preferences, and productivity (Bertrand 2011). Gender discrimination in the labor market points to the gender-segregated economic activities and hiring practices resulting in wage differentials between men and women. Gender differences preferences in the labor market also demonstrate, for example, why men and women tend to choose particular economic activities, or how gender differences affect labor market outcomes. By and large, gendered labor market outcomes connect with power relations, structures, and norms that affect women and men in Filipino households and communities even post-disaster.

Gender and post-disaster economic aid

Economic aid is defined as the "unilateral transfer of resources from one economic entity (the donor) to another (the recipient)" (Abbott 1970, 1213). The flow of resources is usually from developed countries such as Canada to developing countries such as the Philippines. Economic aid encompasses support to local development programs, and often "donor ideology" impacts types of foreign aid (Brech and Potrafke 2014). The outpouring of humanitarian and economic assistance to the Philippines after super typhoon Haiyan was unprecedented, with pledges from 56 countries (Rappler 2013). In this section, the relations of gender and post-disaster economic aid in Leyte are explored based on insights from the narratives of respondents in 2016–2017. I limit my extrapolations to foreign economic aid by donor countries and the structures that impact gendered outcomes.

The Philippines has a unitary form of government with centralized bureaucracies and local government units. Foreign economic assistance by donor countries is channeled directly to the national government and, in a complex process, allocates funds to local agencies and communities. In 2015, the Leyte Representative at the House of Congress, Ferdinand Martin Romualdez, argued to declare the PHP17.2 billion in foreign donations as public funds and make the government accountable for its disbursements (Herrera 2015). Arguably, foreign economic aid supports existing programs of the recipient government, and disaster aid may not always benefit survivors. In the case of Haiyan, Romualdez noted that "no foreign-donated funds were downloaded to local government units" as of September 2015 (Herrera 2015).

However, the Philippines has commendable gender equality programs and services under the Magna Carta of Women (Republic Act No. 9710) which aims to remove discrimination and inequality in all spheres of life (Official Gazette of the Republic of the Philippines 2009). Laws, implementing policies, services, and programs in the Philippines are focused on binary gender identities and the promotion of women's empowerment. LGBTQ+ individuals, while culturally accepted and generally tolerated, remain vulnerable to discrimination and violence (Human Rights Watch 2017). Under a system based on gender binaries of the receiving government, foreign economic aid has the consequent effect to reproduce inequalities and further marginalizing vulnerable populations including persons with disabilities and seniors.

Respondents in the case studies overwhelmingly had positive recollections of international humanitarian assistance but not of foreign economic aid. Awareness of the billion-dollar pledges from different countries fostered claims of public accountability and led to mass protests (Salamat 2014). Rehabilitation of physical infrastructure, livelihood, and other programs to rebuild post-disaster communities appear gender neutral. But insights from local realities suggest that the existing structures could have reproduced the marginalization of Haiyan survivors, particularly women and LGBTQ+ individuals. Barugo, for example, has more men employed before and after the disaster compared to women. Micro-credit lending and livelihood programs facilitated more women's involvement, but these were noted to be privately initiated. According to Becerra, Cavallo, and Noy (2014, 561), the "efficacy of post-disaster foreign aid" by rich countries has not been systemically evaluated. It is surmised that post-disaster foreign economic aid is intrinsically connected to existing government programs without gender specificity to address economic gaps.

Concluding thoughts

Disasters could occur anytime, and the impact on human lives is gendered. Women, men, and members of gender diverse communities like the LGBTQ+ have different experiences during and after natural calamities. Data based on qualitative input from interviews and focus group discussions in 2016–2017 in Barugo and Tacloban City provided the basis to explore the relations of gender and labor market regimes created post-disaster. Gender was a key determinant in labor market participation under regimes facilitated by local, national, and international intervenors after super typhoon Haiyan destroyed Visayan communities. Men dominated as paid workers even if access to the labor market appeared open to all genders; however, the available tasks were subject to preferences based on gender differences.

Labor selectivity under different labor market regimes was affected by the intersections of gender, class, physical ability, and place. In the aftermath of a massive catastrophe in ground zero, the type of economic interventions and resources, either domestic or foreign, defined the labor market and who participated in it. Displaced workers and the unemployed tend to avail of cash-for-work programs funded by the national government and international agencies. Clean-up drives with cash incentives require physical abilities and complementary gender work relations to accomplish them. Rural and urban divide manifested in the access to opportunities in the labor market; more foreign economic assistance was located in Tacloban City compared to Barugo with donor and aid selectivity based on priorities.

Because men outnumber women in the paid labor market, women experience "slow recovery from economic losses" (Enarson 2000, n.p.) after a disaster. The formal avenues of employment afford more opportunities to men as male breadwinners in affected communities. Women consequently take the lead in the absence of men as household heads. In hindsight, about 64% of those killed by super typhoon Haiyan were women (Asia Pacific Forum on Women n.d.), which then would enable more men to participate in income-generating activities. Women participated more in non-agriculture cooperatives that were initiated before super typhoon Haiyan and were compelled to engage productively after the disaster.

The lack or absence of data about LGBTQ+ individuals in Barugo and Tacloban City post-Haiyan indicated the traditional gender binaries of reporting and recording. As well, respondents classify them as either belonging to male or female categories in the labor market. According to Arisa Junio at the 27th Foreign Correspondent Report, this essentialist "framing invisibilizes social groups who are outside the gender binary in disasters," and in Leyte, class impacted how LGBTQ+ persons were treated:

> Middle-income LGBTs socially benefited where their own sexuality was promoted and they were accepted by their own communities due to the foreign aid workers' presence on the ground. On the contrary, lower income sexual minorities do not have the same social acceptance treatment towards them as many of them have remained focused on earning a living. (Junio n.d., cited Ong et al 2015)

Ground zero of super typhoon Haiyan leaves many lessons for the accidental sociologist of disasters like myself—one of which is the realization that what comes out of destruction could transform and or reproduce gender relations. The lens provided by the rise of different labor market regimes post-disaster suggests that relief and reconstruction is a human endeavor in which systems of relations operate on usual terms pre-disaster. What makes them stand out was the immediacy of action to protect human welfare. Undeniably, gender intersects with other systems of domination like classism and ableism that limit inclusive participation. But consistent with collectivist cultural orientation among Filipinos, those abled tend to share the fruits of their labor to household members and beyond whether paid or unpaid.

Notes

1. The fieldwork for case studies in 2016–2017 was supported by the Prentice Institute Research Seed Grant and the Community of Research Excellence Development Opportunities (CREDO) grant at the University of Lethbridge.
2. Other international agencies and civic groups noted by respondents after super typhoon Haiyan were the American Refugee Committee, Care and Share Army (Australia), DAK Foundation (Australia), Family Service International, FAO (Food and Agriculture Organization), Heart to Heart International (US), Humane Society International (US), International Medical Corp, IOM (International Organization for Migration), JICA (Japan International Cooperation Agency), Medicins San Frontières, Nazarene Disaster Response (US and Canada), Project HOPE (US), Rotary Club Australia, Rotary Club Singapore, World Food Bank, World Society for the Protection of Animals, UNICEF (United Nation's Children Fund), UNHCR (United Nations High Commissioner for Refugees), UN Humanitarian Affairs, and USAID (United States Agency for International Development).

References

Abbott, G. C. (1970) "Economic aid as a unilateral transfer of resources," *Journal of Political Economy* 78(6): 1213–1227.

Agence France-Presse. (2016) "10 events that define president Aquino's legacy," *Inquirer.Net*, May 9, https://newsinfo.inquirer.net/784249/10-events-that-define-president-aquinos-legacy [Accessed August 4, 2021].

Akter, S. (2021) "Do catastrophic floods change the gender division of labour? Panel data evidence from Pakistan," *International Journal of Disaster Risk Reduction* 60: 102296. https://doi.org/10.1016/j.ijdrr.2021.102296

Amazona, W. (2016) "Tacloban starts mass relocation of Yolanda victims to resettlement areas," *CNN Philippines*, November 28, https://cnnphilippines.com/regional/2016/11/28/tacloban-starts-mass-relocation-of-yolanda-victims-to-resettlement-areas.html [Accessed August 3, 2021].

Asia Pacific Forum on Women. (n.d.) "Typhoon Haiyan claims lifetime loss and damages for women in the Philippines," https://apwld.org/wp-content/uploads/2014/12/flyer-amihan.pdf [Accessed August 8, 2021].

Asian Development Bank. (2014) "Cash-for-work program rekindles community life in typhoon-hit Philippines," *News*, March 25, https://www.adb.org/features/cash-work-program-rekindles-community-life-typhoon-hit-philippines [Accessed August 8, 2021].

Aysan, Y. and I. Davis (1993) *Rehabilitation and Reconstruction*, 1st ed, Disaster Management Training Programme, New York: United Nations Development Programme.

Ball, P. (2012) *Why Society Is a Complex Matter: Meeting Twenty-first Century Challenges with a New Kind of Science*, New York: Springer.

Baron, A. (ed) (1991) *Work Engendered: Toward a New History of American Labor*, Ithaca, NY and London: Cornell University Press.

Becerra, O., E. Cavallo, and I. Noy (2014) "Where is the money? Post-disaster foreign aid flows," *Environment and Development Economics* 20(5): 561–586.

Belasen, A. R. and S. W. Polachek (2009) "How disasters affect labor markets: The effects of hurricanes in Florida," *Journal of Human Resources* 44(1): 251–276.

Belkhir, J. A. and C. Charlemaine (2007) "Introduction: Race, gender and class as organizing principle," *Race, Gender & Class* 14(1/2): 153–156.

Bertrand, M. (2011) "New perspectives in gender," in O. Ashenfeller and D. Card (eds) *Handbook of Labor Economics*, Amsterdam: Elsevier, pp 1545–1592.

Bettio, F., J. Plantenga, and M. Smith (eds) (2013) *Gender and the European Labour Market*, Abingdon and New York: Routledge.

Blumberg, R. L. (ed) (1991) *Gender, Family, and Economy: The Triple Overlap*, Newbury Park, CA: Sage.

Bourdet, Y. and I. Persson (1992) "Youth Unemployment and labour market regime: The case of France and Sweden," Working Paper Series, Department of Economics, School of Economics and Management, University of Lund.

Bradshaw, S. (2013) *Gender, Development and Disasters*, Cheltenham, UK: Edward Elgar.

Brech, V. and N. Potrafke (2014) "Donor ideology and types of foreign aid," *Journal of Comparative Economics* 42(1): 61–75.

Burh, S., A. Almagro, and C. Tordesillas (2013) "DSWD, WFP to begin cash distributions for Yolanda survivors," United Nations World Food Programme, December 12, https://www.wfp.org/news/dswd-wfp-begin-cash-distributions-yolanda-survivors [Accessed August 8, 2021].

Cabanilla, G. A. C. (2014) "Targeting disadvantaged women – advancing women's empowerment through library entrepreneurship workshops," School of Library and Information Studies, University of the Philippines Diliman, http://library.ifla.org/id/eprint/980 [Accessed August 8, 2021].

Ciccia, R. (2015) "Classifying labour regimes beyond the welfare state: A Two dimensional approach," in S. O'Riain, F. Behling, R. Ciccia, and E. Flaherty (eds) *The Changing Worlds and Workplaces of Capitalism*, Basingstoke, UK: Palgrave Macmillan, pp 58–78.

City of Tacloban. (2019) "City profile". https://www.tacloban.gov.ph/city-profile/ [Accessed August 1, 2021].

Climate Signals. (2020) "Super typhoon Goni 2020: Philippines," https://www.climatesignals.org/events/super-typhoon-goni-2020 [Accessed August 1, 2021].

Connell, R. and R. Pearse (2015) *Gender in World Perspective*, 3rd ed, Cambridge: Polity Press.

Department of Agriculture (n.d.) "NEDA Says SAAD Contributes to Region 8's Decrease in Poverty," Special Area for Agricultural Development, Republic of the Philippines, http://saad.da.gov.ph/2020/02/neda-says-saad-contributes-to-region-8-decrease-in-poverty/ [Accessed August 1, 2021].

Department of Trade and Industry (n.d.) "Profile of Region 8," Republic of the Philippines. https://www.dti.gov.ph/regions/region-8/profile/ [Accessed August 1, 2021].

Dominelli, L. (2013) "Empowering disaster-affected communities for long-term reconstruction: Intervening in Sri Lanka after the tsunami," *Journal of Social Work in Disability and Rehabilitation* 12(1/2): 48–66.

Eadie, P. (2013) "Typhoon aftermath will be Aquino's legacy," *The Asia Dialogue*, November 28, https://theasiadialogue.com/2013/11/28/typhoon-aftermath-will-be-aquinos-legacy/ [Accessed August 4, 2021].

Enarson, E. (2000) "Gender and natural disasters," InFocus Programme on Crisis Response and Reconstruction, Working Paper 1, Recovery and Reconstruction Department, Geneva: International Labour Organization, https://www.ilo.org/employment/Whatwedo/Publications/WCMS_116391/lang--en/index.htm [Accessed March 15, 2020].

Enarson, E. and P. G. Dhar Chakrabarti (eds) (2009) *Women, Gender and Disaster: Global Issues and Initiatives*, New Delhi: Sage.

Encyclopedia Britannica. (2013) "Leyte," *Encyclopedia Britannica*, November 14, https://www.britannica.com/place/Leyte [Accessed August 1, 2021].

Encyclopedia Britannica. 2020. "Tacloban," *Encyclopedia Britannica*, July 24, https://www.britannica.com/place/Tacloban [Accessed August 1, 2021].

Espejo, E. (2013) "Duterte in tears after visiting Tacloban," *Rappler*, November 12, https://www.rappler.com/nation/duterte-in-tears-after-visiting-ravaged-tacloban [Accessed August 3, 2021].

Frank, T. (2020) "LGBTQ people are at higher risks in disasters," *Scientific American E&E News*, December 23, https://www.scientificamerican.com/article/lgbtq-people-are-at-higher-risk-in-disasters/ [Accessed August 5, 2021].

Freitag, U. and A. von Oppen (eds) (2010) *Translocality: The Study of Globalising Processes from a Southern Perspective*, Leiden and Boston, MA: Brill.

Gabieta, J. A. (2016) "Farming Town sprints on development track," *Inquirer.Net*, July 30, https://newsinfo.inquirer.net/801007/farming-town-sprints-on-development-track [Accessed August 1, 2021].

Game, A. and R. Pringle (2020) *Gender at Work*, 2nd ed, New York: Routledge.

Go, J. R. R. (2016) "On disasters and the Yolanda/Haiyan experience," Ballots & Bullets, School of Politics & International Relations, University of Nottingham, February 5, https://nottspolitics.org/2016/02/05/on-disasters-and-the-yolandahaiyan-experience/ [Accessed August 1, 2021].

Gomez, C. (2020) "League of cities conduct 'aggressive' fund drive for typhoon survivors," *Inquirer.Net*, November 25, https://newsinfo.inquirer.net/1364700/league-of-cities-conduct-aggressive-fund-drive-for-typhoon-survivors [Accessed August 2, 2021].

Gonsalves, J. and P. Mohan (eds) (2011) *Strengthening Resilience in Post-disaster Situations: Stories, Experience and Lessons from South Asia*, New Delhi and Ottawa: Academic Foundation and International Development Research Centre.

Hatt, S. (1997) *Gender, Work and Labour Markets*, London and New York: Macmillan Press and St. Martin's Press.

Herr, H. and M. Kazandziska (2011) *Macroeconomic Policy Regimes in Western Industrial Societies*, New York: Routledge.

Herrera, C. F. (2015) "Where did Yolanda aid go?" *ManilaStandard.Net*, September 24, https://www.manilastandard.net/news/-main-stories/187745/where-did-yolanda-aid-go-.html [Accessed August 8, 2021].

Human Rights Watch. (2017) " 'Just let us be': Discrimination against LGBT students in the Philippines," June 21, https://www.hrw.org/report/2017/06/21/just-let-us-be/discrimination-against-lgbt-students-philippines [Accessed August 8, 2021].

Junio, A. (n.d.) "Sexual minorities and disasters in the Philippines: Where are they?" The 27th Foreign Correspondent Report, http://www.kfaw.or.jp/correspondents/docs/27-2_Philippines_E.pdf [Accessed August 10, 2021).

Kahn, M. E. (2005) "The death toll from natural disasters: The role of income, geography, and institutions," *The Review of Economics and Statistics* 87(2): 271–284.

Kinnvall, C. and H. Rydstrom (eds) (2019) *Climate Hazards, Disasters, and Gender Ramifications*, Abingond and New York: Routledge.

Kirchberger, M. (2017) "Natural disasters and labor markets," *Journal of Development Economics* 125: 40–58.

Leacock, E. and H. I. Safa (1986) *Women's Work: Development and the Division of Labor by Gender*, New York: Bergin & Garvey.

Liljas, P. (2014) "Tacloban survived the world's strongest recorded storm, but only just," *Time*, November 7, https://time.com/3554112/tacloban-yolanda-supertyphoon-typhoon-haiyan-anniversary/ [Accessed August 1, 2021].

Lindsey, L. L. (2015) *Gender Roles: A Sociological Perspective*, 6th ed, Boston, MA: Pearson.

Maga, D. (2015) "Massive theft of 'Yolanda' donations," *Inquirer.Net*, August 1, https://opinion.inquirer.net/87201/massive-theft-of-yolanda-donations [Accessed August 8, 2021].

Magara, H. (ed) (2014) *Economic Crises and Policy Regimes: The Dynamics of Policy Innovation and Paradigmatic Change*, Cheltenham UK: Edward Elgar.

Mainwaring, S. and A. Pérez-Liñán (2013) *Democracies and Dictatorships in Latin America: Emergence, Survival, and Fall*, New York: Cambridge University Press.

Manila Standard. (2015) "Yolanda funds being used for polls," October 19, https://manilastandard.net/news/top-stories/189834/-yolanda-funds-being-used-for-polls-.html [Accessed August 8, 2021].

Manila Times. (2014) "Leyte hosts donors forum for Yolanda-affected areas," *Manila Times*, March 25, https://www.manilatimes.net/2014/03/25/regions/at-a-glance-regions/leyte-hosts-donors-forum-for-yolanda-affected-areas/85292 [Accessed August 3, 2021].

Marshall, A. R.C. (2013) "Desperate Philippine typhoon survivors loot, dig up water pipes," *Reuters*, November 12, https://www.reuters.com/article/us-philippines-typhoon-idUSBRE9A603Q20131113 [Accessed August 2, 2021].

McCarl Nielsen, J. (1990) *Sex and Gender in Society: Perspectives on Stratification*, 2nd ed, Long Grove, IL: Waveland Press.

Meade, T. A. and M. E. Wiesner-Hanks (eds) (2006) *A Companion to Gender History*, Malden, MA: Blackwell.

Mendoza, C. A. and B. Jara (2020) "Natural disasters and informality: Are local labor markets impacted after an earthquake?" *Regional Science Policy & Practice* 12(1): 125–157.

Meniano, S. (2018) "Tacloban airport passenger traffic up 21% H1 2018," *Philippine News Agency*, Republic of the Philippines, August 2, https://www.pna.gov.ph/articles/1043533 [Accessed August 1, 2021].

Milkman, R. (2016) *On Gender, Labor, and Inequality*, Champaign, IL: University of Illinois.

Minda News. (2013) "Duterte to Yolanda survivors in Davao City: 'stay here while Leyte rebuilds itself,'", December 24, https://www.mindanews.com/top-stories/2013/12/duterte-to-yolanda-survivors-in-davao-city-stay-here-while-leyte-rebuilds-itself/ [Accessed August 3, 2021].

Mongabay. (2019) "For the Philippines a warming world means stronger typhoons, fewer fish," *News & Inspiration from Nature's Frontline*, October 16, https://news.mongabay.com/2019/10/for-the-philippines-a-warming-world-means-stronger-typhoons-fewer-fish/ [Accessed August 1, 2021].

Mullés, A. (2014) *Typhoon Haiyan: The Untold Story*, Tacloban City: Tacloban Five Star-Grafica Printing Press.

Myant, M., J. Drahokoupil, M. Beblavy, S. Domonkos, and L. M. Kurekova (2012) "Linking labour regimes and technological innovation in Central and Eastern Europe: The case of automotive and software industries," NEUJOBS Working Paper No. 6.2 (July 1), http://dx.doi.org/10.2139/ssrn.2402728

Neumayer, E. and T. Plümper (2007) "The gendered nature of natural disaster: The impact of catastrophic events on the gender gap in life expectancy, 1981–2002," *Annals of the Association of American Geographers* 97(3): 551–566.

Official Gazette of the Republic of the Philippines. (2009) "Republic Act No. 9710," August 14, https://www.officialgazette.gov.ph/2009/08/14/republic-act-no-9710/ [Accessed August 8. 2021].

Olofsson, J. (2006) "Stability or change in the Swedish labour market regime?" Working Paper Series No. 2006:3, Lund University, https://portal.research.lu.se/portal/files/4628931/1453920.pdf [Accessed August 1, 2021].

Ong, J., J. Flores, and P. Combinido (2015) "Obliged to be grateful: How Local communities experienced humanitarian actors in the Haiyan response," *ALNAP*, May 30, https://www.alnap.org/help-library/obliged-to-be-grateful-how-local-communities-experienced-humanitarian-actors-in-the [Accessed August 10, 2021].

Oryema, J. B. (2017) "The Impact of social protection on resettlement of displaced persons: Evidence from post-conflict northern Uganda," *Journal of International Development* 29(6): 700–713.

Palatino, M. (2013) "Why CNN is getting praise and flak for its Philippine storm coverage," *Global Voices*, November 17, https://globalvoices.org/2013/11/17/why-cnn-is-getting-praise-and-flak-for-its-philippine-storm-coverage/ [Accessed August 2, 2021].

PhilAtlas. (2021a) "Barugo," https://www.philatlas.com/visayas/r08/leyte/barugo.html [Accessed July 29, 2021].

PhilAtlas. (2021b) "Tacloban City," https://www.philatlas.com/visayas/r08/tacloban.html [Accessed July 29, 2021).

Philippine Commission on Women. (n.d.) "Barugo and Runggiyan Social Development Foundation: Working together for WEE," Republic of the Philippines, March 9, 2011, https://pcw.gov.ph/barugo-and-runggiyan-social-development-foundation-working-together-for-wee/ [Accessed August 1, 2021].

Philippine Statistics Authority. (2017a) "Special release: Highlights of the 2015 census results for the municipality of Barugo," http://rsso08.psa.gov.ph/leyte/special-release/popcen/barugo [Accessed July 29, 2021].

Philippine Statistics Authority. (2017b) "Special release: Highlights of the 2015 census results for the City of Tacloban," http://rsso08.psa.gov.ph/leyte/special-release/popcen/tacloban [Accessed July 29, 2021].

Philippine Statistics Authority. (2017c) "2015 census of population – demographic and socio-economic characteristics in Leyte," Report No. 2, June, https://psa.gov.ph/sites/default/files/08_Leyte.pdf [Accessed August 1, 2021].

Philippine Statistics Authority. (2018) "Leyte quickstat," https://psa.gov.ph/statistics/quickstat [Accessed July, 31 2021].

Philippine Statistics Authority. (2020) *The Women & Men in Eastern Visayas (WAM).* 2020 Statistical Handbook, 9th ed, Tacloban City: Philippine Statistics Authority. http://rsso08.psa.gov.ph/sites/default/files/WAM%20 9th%20Edition_rev_06092020.pdf [Accessed July 31, 2021].

Pulkowski, D. (2014) *The Law and Politics of International Regime Conflict*, Oxford, UK: Oxford University Press.

Quisumbing, A., R. Meinzen-Dick, T. L. Raney, A. Croppenstedt, J. A. Behrman, and A. Peterman (eds), (2014) *Gender in Agriculture: Closing the Knowledge Gap*, New York: Springer.

Racioppi, L. and S. Rajagopalan (eds) (2016) *Women and Disasters in South Asia: Survival, Security and Development*, Abingdon and New York: Routledge.

Ramos, R. B. (2013) "Super typhoon 'Yolanda" defines Aquino presidency," *The Manila Times*, November 15, https://www.manilatimes.net/2013/11/15/featured-columns/columnists/super-typhoon-yolanda-defines-aquino-presidency/53296 [Accessed August 4, 2021].

Ramzy, A. (2013) "Philippine city's residents describe rifling of stores," *New York Times*, November 21, https://www.nytimes.com/2013/11/22/world/asia/philippine-looters-describe-frenzied-dash-for-goods.html [Accessed August 2, 2021].

Rappler. (2013) "Yolanda (typhoon Haiyan): Aid, donations from international community". *Rappler.com*, November 12, https://www.rappler.com/nation/yolanda-international-community [Accessed August 8, 2021].

Redclift, N. and M. Thea Sinclair (eds) (1991) *Working Women: International Perspectives on Labour and Gender Ideology*, London and New York: Routledge.

Reddy, H. and A. Goodman (2019) "Gender-based violence after a natural disaster," *Prehospital and Disaster Medicine* 34(1):s1333.

Reid, K. (2018) "2013 typhoon Haiyan: Facts, FAQs, and how to help," *World Vision*, November 9, https://www.worldvision.org/disaster-relief-news-stories/2013-typhoon-haiyan-facts [Accessed August 1, 2021].

Salamat, M. (2014) "Tens of thousands of survivors gather in Tacloban to demand justice," *Bulatlat*, November 11, https://www.bulatlat.com/2014/11/11/tens-of-thousands-of-survivors-gather-at-tacloban-to-demand-justice/ [Accessed August 8, 2021].

Salazar, L. C. (2015) "Typhoon Yolanda: The politics of disaster response and management," in D. Singh (ed) *Southeast Asian Affairs*, Singapore: Institute of Southeast Asian Studies (ISEAS)–Yusof Ishak Institute, pp 277–302.

Sawyer, R. Keith (2005) *Social Emergences: Societies as Complex Systems*, New York: Cambridge University Press.

Sundararajan, A. (2016) *The Sharing Economy: The End of Employment and the Rise of Crowd-Based Capitalism*, Cambridge, MA: The MIT Press.

Takasaki, Y. (2012) "Natural disasters, gender and handicrafts," *The Journal of Development Studies* 48(1): 115–132.

World Economic Forum. (2018) *The Global Gender Gap Report 2018*, Geneva: World Economic Forum.

World Economic Forum. (2021) *Global Gender Gap Report 2021*, Geneva: World Economic Forum.

van Eerbeek, P. and C. Hedberg (2021) "Chameleon brokers: A translocal take on migration industries in the Thai-Swedish wild berry business," *Migration Studies* 9(3): 830–851. doi:10.1093/migration/mnab030

Venn, D. (2012) "Helping displaced workers back into jobs after a natural disaster: Recent experiences in OECD countries," OECD Social, Employment and Migration Working Papers No. 142, November 29, https://www.oecd-ilibrary.org/social-issues-migration-health/helping-displaced-workers-back-into-jobs-after-a-natural-disaster_5k8zk8pn2542-en [Accessed August 3, 2021].

Yee, J. (2014) "1,000 'Yolanda' survivors set protest march during Aquino's sona," *Philippine Daily Inquirer*, July 8, https://newsinfo.inquirer.net/617895/1000-yolanda-survivors-set-protest-march-during-aquinos-sona [Accessed August 4, 2021].

Yenilmez, M. I. and G. S. Huyugüzel Kisla (eds) (2021) *The Economics of Gender Equality in the Labour Market: Policies in Turkey and Other Emerging Economies*, Abingdon and New York: Routledge.

Zhu, X., Z. Jin, S. Managi, and X. Xun (2021) "How meteorologial disasters affect the labor market? The moderating effect of government emergency response policy," *Natural Hazards* 107: 2625–2640.

11

No One Left Behind? A Case for Disability-Inclusive Disaster Risk Reduction in the Philippines

Kaira Zoe Alburo-Cañete, Corazon Clarin, Marivic Codiñera, and Katherine Velmonte

Introduction

The Sendai Framework for Disaster Risk Reduction 2015–2030 (United Nations 2015) that resulted from the Third UN World Conference in March 2015 articulated an important message for all participating countries, communities, and disaster risk reduction (DRR) stakeholders: *no one should be left behind*. Indeed, it has been a recurrent observation that disasters do not impact populations evenly (Bankoff and Hilhorst 2004; Blaikie et al 2004). Those who live in conditions of marginality are often those who suffer the most from disaster events. Yet, while this fact is established, conventional DRR practices have often excluded marginalized sectors from planning, strategizing, and implementing risk reduction measures (Peacock, Morrow, and Gladwin 1997; Peek 2008; Back, Cameron, and Tanner 2009; Bradshaw and Fordham 2013; Seager 2014).

In many cases, persons with disabilities are "the first to be forgotten, and the last to be remembered" (UNOCHR 2014), even from among the most marginalized groups. Recent examinations of disaster experiences all over the world—the Indian Ocean tsunami of 2004, hurricane Katrina of 2005, the Haiti earthquake of 2010, and the Great East Japan earthquake and tsunami of 2011, among others—have brought attention not only to how persons with disabilities are differentially affected by calamities but also how they are largely neglected and excluded in disaster risk management planning and programming (Suprobo 2011; Robinson and Kani 2014; Stough and Kang

2015). This invisibility has led to the further marginalization of persons with disabilities, which contributes to placing them at higher risk of injury, death, and loss of property during disasters (Twigg, Kett, and Lovell 2018).

A pioneering global study conducted by UNISDR (2014, 13) that looked into the intersection of disaster and disability revealed that 80% of persons with disabilities are unable to evacuate immediately without difficulty in the event of a sudden disaster, while 72% do not have personal preparedness plans, thus making it more likely for them to be left behind in the event of a calamity. Yet, the same study also shows that, if given sufficient time, the number of persons with disabilities who could evacuate without difficulty doubles. This points to the importance of putting in place risk reduction measures that ensure, among others, that timely life-saving information reaches all members of the community, despite any barriers to mobility or communication; and that at-risk groups, in particular, are provided with the means with which to act on such information (Robinson and Kani 2014). Against this backdrop, this chapter argues that DRR policies and frameworks cannot make good on their promise of *leaving no one behind* if the needs, conditions, and rights of persons with disabilities are not taken into account in DRR programming.

Given the imperative to 'leave no one behind' not just in DRR but also more broadly in development, as enshrined in the UN Sustainable Development Goals, the clamor for developing disability-inclusive DRR approaches has grown in recent years. But while the need for disability inclusion is acknowledged, there is a need to understand how it can be practiced on the ground.

In the Philippine context, data on the actual number of persons with disabilities is plagued with accuracy and reliability issues. It is often underestimated or not reported due to the stigma attached to disability, resulting in the exclusion of persons with disabilities in preparedness/contingency planning and relief efforts in times of disasters (Marella et al 2016). Even less research has been conducted on how disability shapes experiences and abilities to prepare for and respond to disasters. The experience of typhoon Haiyan in November 2013, in particular, highlighted the urgency of making visible the needs and conditions of persons with disabilities. In the 2014 UNISDR report on disability and disasters, the Secretariat for the UN Convention of the Rights of Persons with Disabilities (UN-ENABLE) noted that thousands of persons with disabilities were heavily affected by disasters, including during typhoon Haiyan. However, the rescue and relief operations by government task forces were not sensitive to the need to include persons with disabilities (UNISDR 2014, 4).

This statement of the UN Secretariat was supported by various studies conducted on disability and disasters in typhoon Haiyan-affected areas, which showed how disability remained largely invisible in disaster risk management efforts (Macalandag and Alburo-Cañete 2015; Leonardi et al

2016; Zayas et al 2017). Given that the Philippines has been considered as one of the countries with the highest risks of disaster (Mucke 2014; Welle and Birkmann 2016), and with a disability prevalence rate of 12% (Philippine Statistics Authority 2019),[1] disability inclusion is an urgent matter for DRR policy and practice in the country.

As the application of disability perspectives is a nascent but growing field in the Philippine disaster risk management landscape, this chapter aims to contribute to ongoing discussions and efforts to make disaster risk reduction inclusive—and sustainable development more broadly. Using the case of the Cebu Disability-inclusive Disaster Risk Reduction Network (CDiDRRN), a multi-stakeholder advocacy coalition formed after Haiyan by organizations of persons with disability (OPDs), the chapter highlights key lessons learned for disability inclusion. The CDiDRRN has been hailed as one of the pioneering initiatives for disability-inclusive DRR in the country and continues to gain wider attention both locally and internationally. From the experiences and challenges of establishing the CDiDRRN, the chapter presents a framework and practical strategies for undertaking disability-inclusive DRR in the Philippines.

To demonstrate the generative potentials for a more inclusive DRR through the experiences of building the CDiDRRN, we first provide an overview of disability and disasters in the Philippines. Here we present how we understand disability using a social and human rights lens and how this perspective is particularly useful for the practice of DRR. We also discuss the policy and operational environment for disability inclusion and risk reduction in the Philippine context. Next, we introduce the CDiDRRN by describing the conditions that led to its formation and the framework and approach that it adopts. This is followed by a discussion of the key activities and accomplishments of the network as well as the challenges encountered. Lastly, we conclude the chapter with the main lessons from the experience of building the CDiDRRN for promoting disability-inclusive DRR in the Philippines.

Disability and disasters in the Philippines: Understanding the links

We adopt the definition of the Convention on the Rights of Persons with Disabilities (United Nations 2006), which states that persons with disabilities "include those who have long-term physical, mental, intellectual or sensory impairments which *in interaction with various barriers* may hinder their full and effective participation in society on an equal basis with others" (emphasis added). While disability has been framed and conceptualized in many ways, we find this definition advanced by the UNCRPD as especially useful in understanding the direct relationship between functioning limitations (or impairments) and the environment—both physical and social. Here, our focus is not on the medical aspects of disability, but rather take on a social/ human rights approach that recognizes sociocultural and physical barriers preventing persons with disabilities from participating in society and claiming entitlements on an equal basis with other members of society (Priestley and Hemingway 2007). While we do not negate that persons with disabilities have real functioning limitations or 'impairments' that affect their everyday life experiences nor simply reduce their conditions as mainly 'social,' as some critiques of the social model of disability have raised (see Shakespeare 2013), we find that this approach is highly suitable to the context of disaster risk reduction.

The UNCRPD definition of disability becomes readily relevant when understanding the differential risks and impacts that persons with disabilities face in times of disasters. Persons with disabilities tend to embody higher disaster risks relative to the broader population specifically when their socio-physical environments provide *disabling conditions* for them to confront the impacts of such events (UNESCAP 2015, 30). Acknowledging that disasters have 'social roots' (Tierney 2014) that arise from similar processes that produce social inequalities, exclusions, and vulnerabilities, makes it clearer to see how disaster risk reduction initiatives must take on the task of addressing the various barriers experienced by persons with disabilities, which prevent them from accessing and acting on vital life-saving information and decisions if these are indeed to leave no one behind.

In the 2013 Census on Population and Housing, the disability prevalence in the Philippines was only put at 1.57% or about 1.44 million (Philippine Statistics Authority 2013). This falls way below the estimated worldwide disability prevalence. In its *World Report on Disability*, the World Health Organization (2011, 29) estimates that around 15% of any population is comprised of persons with disabilities. Using this estimate, the number of persons with disability in the country in 2010 should have been recorded at around 13.8 million.

The challenges in collecting disability data have been pointed out by the Philippine Coalition on UNCRPD in their 2013 Parallel Report on the Philippine implementation of the Convention on the Rights of Persons with Disabilities from 2008 to 2013 (Philippine Coalition on the UNCRPD 2013).[2] In the report, the Philippine Coalition on the UNCRPD (2013, 15) revealed that efforts for data gathering and documentation are lacking, and highlighted that different agencies have outdated and conflicting figures. The coalition stressed that the lack of systematic efforts for data gathering and documentation of vital information about Filipinos with disabilities (undocumented cases of violence and abuse) have caused continuing cycles of increasing invisibility, powerlessness, and marginalization (Philippine Coalition on the UNCRPD 2013, 14).

Problems with collecting disability data have often precluded a more accurate estimate of the prevalence of disability, especially in countries like the Philippines. The historical lack of data on persons with disabilities, according to UNCRPD may be attributed to several factors. These include: 1) the lack of a uniform definition and understanding of disability, which created confusion over what constitutes a 'disability,' particularly in cases of minor or moderate disabilities; 2) use of non-comparable methodologies in the measurement and collection of data; 3) reliance on self-reporting and identification; 4) cultural conceptions on disability; 5) use of the medical diagnostic testing to determine disability; and 6) distrust towards government offices tasked with registering persons with disabilities (UNCRPD 2015, 3–4).[3] At the local level, persons with disabilities may not have the means or access to representation, which might have facilitated their registration and allowed them to claim benefits and to articulate their voice (Suprobo 2011). In terms of 'standard' risk-assessment tools and disaster risk management processes in general, capturing disability data effectively is still an evident concern (Macalandag and Alburo-Cañete 2015; Alburo-Cañete 2016; Leonardi et al 2016; Zayas et al 2017).

Because of the efforts of OPDs and their advocates to improve the collection of disability data, the Philippines has gradually adopted more effective tools to measure disability prevalence. The recently released results of the National Disability Prevalence Survey (Philippine Statistics Authority 2019)[4] revealed that around 12% of Filipinos aged 15 and older experienced severe forms of disability, and one in every two Filipinos enumerated in the survey (47%) reported moderate forms of disability. In the said survey, females were more likely than males to experience severe to moderate disability, and about one in every three respondents who experienced severe disability (32%) were older people aged 60 and over. This demonstrates how disabilities intersect with other variables such as gender and age—intersectional factors that should be considered when assessing risks that specific social groups face (Twigg, Kett, and Lovell 2018).

Following typhoon Haiyan in 2013, a disability survey conducted in selected affected areas using the *WHO Disability Assessment Schedule 2.0* revealed that the disaster produced an increase in the prevalence of disabilities,[5] especially in urban contexts (Leonardi et al 2016, 268).[6] Indeed, according to the 2011 World Disability Report, people's environments have a huge impact on the prevalence and extent of disability. Major environmental changes such as those that accompany disasters "also affect the prevalence of disability not only by changing impairments but also by creating barriers in the physical environment" (World Health Organization 2011, 37). In other words, not only are persons with disabilities as a sector especially at risk of disasters given their functioning limitations and disadvantaged position in society, but *disasters in themselves also contribute to disability*. Despite this fact, considerations for providing reasonable accommodation[7] to persons with disabilities in risk reduction and disaster response remain the exception rather than the rule. The lack of sensitivity to disability was further exemplified in the experience of typhoon Haiyan, wherein serious concerns were raised by advocacy groups over the lack of transportation and inaccessibility of emergency services and healthcare (Robinson 2014).

While much is still yet to be done in terms of mainstreaming disability inclusion in the Philippines, institutional drivers nevertheless exist in the form of international commitments and national laws that would facilitate and serve as entry points to making disaster risk reduction and development disability inclusive. Among the most significant international commitments that the Philippines has declared adherence to are the UN Convention on the Rights of Persons with Disabilities, the ASEAN Bali Declaration on the Enhancement of the Role and Participation of Persons with Disabilities in the ASEAN Community, and the Biwako Millennium Framework for Action. Moreover, the Philippines has also adopted the Hyogo Framework for Action and the subsequent Sendai Framework, both recognizing the need to strengthen the resilience of at-risk sectors including persons with disabilities.

In addition to the adoption of international consensus documents and frameworks, the Philippines has also passed several relevant laws and policies which are useful for the promotion of disability inclusion in disaster risk management (see Box 1). Most notable is the passage of the Philippine Disaster Risk Reduction and Management (DRRM) Law, which identifies persons with disabilities as among the 'vulnerable'[8] groups that need to be accounted for and considered in DRRM programs. In establishing the CDiDRRN, knowledge of various laws, policies, and frameworks have been essential as these provided the entry points and enabling conditions for advancing advocacy for disability inclusion.

Learning from Haiyan

The establishment of the CDiDRRN was galvanized by the experience of typhoon Haiyan. The province of Cebu, located in the Central Visayas Region of the Philippines and approximately 587 kilometers (or 365 miles) south of Manila, was among the areas heavily affected by the typhoon. According to the Cebu Provincial Disaster Risk Reduction Management Office (PDRRMO),[9] the most seriously affected municipalities were located in the northern parts of the province, which impacted around 1,680,000 people in 36 municipalities. While Cebu is no stranger to typhoons, being centrally located in the country's typhoon belt, Haiyan had by far been the most powerful and destructive.[10]

The formation of the CDiDRRN was initially part of a project conceptualized and implemented by a local non-governmental organization

Box 1: Some Relevant National Laws and Policies in the Philippines

Republic Act 344 (1982) – Accessibility Law

"An act to enhance the mobility of persons with disabilities by requiring certain buildings, institutions, establishments and public utilities to install facilities and other device." It provides for minimum requirements and standards to make buildings, facilities, and utilities for public use accessible to disabled persons.

Republic Act 7277, 1992 – Magna Carta for Disabled Persons

This Law states that persons with disabilities must enjoy the same rights as any other citizens; and relevant departments and agencies are mandated to meet their rights-based needs such as rehabilitation, health, employment, and social participation through non-discriminating policies, programs, projects, provision of opportunities, and special benefits.

Republic Act 9442 – Amendment of the Magna Carta for Disabled Persons

Introduced the registration of persons with disabilities, which entitles them to discounted transport, medication, and other services. RA 9442 also legislated prohibition of verbal and non-verbal ridicule and vilification of persons with disabilities.

Civil Service Commission Memorandum Circular No.7, Series of 2014 –

Encouraging Government Agencies to hire PWDs pursuant to Republic Act No. 7277, as amended on Section 5 (Equal Opportunity for Employment) of the Magna Carta for Disabled Persons

Calls on all heads of government offices, agencies, and corporations to give PWDs an equal access to work opportunities, and support the implementation of at least 1% of positions being reserved for persons with disabilities.

Republic Act 9710 – The Magna Carta of Women (2009)

Women with disabilities are highlighted as a marginalized group, yet there is little mention regarding their specific needs or rights.

Republic Act 8425 – The Social Reform and Poverty Alleviation Act (1998)

Institutionalized the Social Reform and Poverty Alleviation Program. It was created to meet the objectives of the National Anti-Poverty Commission and includes the people with disability sectoral council; ensuring the full participation of people with disability in government, social, economic, civil, and political affairs.

DILG Memorandum Circular No. 2010–103 – Establishment of Persons with Disability Affairs Office (PDAO) in Every Province, City, and Municipality pursuant to RA 10070

Intended to formulate and implement policies, plans, and programs for the promotion of the welfare of persons with disabilities in coordination with concerned national or local government agencies, among others.

Philippine Plan of Action for the Asian and Pacific Decade of Disabled Persons 2003–2012

National priorities are directed towards the maximization of opportunities available to persons with disabilities in all aspects of life in the community. The goal of the Plan is to upgrade, strengthen, expand, and sustain activities aimed at realizing the full participation and equality of persons with disabilities.

Women's Empowerment, Development, and Gender Equality (EDGE) Plan 2013–2016

The Women's EDGE Plan serves as a guide for agencies in mainstreaming the gender dimension in their regular programs as well as in implementing gender-focused activities. The plan includes gender responsiveness in issues involving women with disabilities.

Republic Act 10121 – Philippine Disaster Risk Reduction and Management Act of 2010

In this Act, the aim of developing and strengthening the capacities of vulnerable and marginalized groups to mitigate, prepare for, respond to, and recover from the effects of disasters is clearly mentioned.

(NGO) in Cebu called A2D Project—Research Group for Alternatives to Development, Inc. (A2D Project) as a response to the call for more inclusive disaster risk reduction programming in the post-Haiyan context. A2D Project had been heavily involved in disaster relief operations in Northern Cebu when staff on the ground raised concerns regarding the difficulties persons with disabilities had been encountering in accessing life-saving assistance. To investigate these issues, A2D Project conducted a study in Tabogon, one of its partner communities, during the Haiyan response. Accounts of members of OPDs within Tabogon described how difficult it had been for them to evacuate from their homes when the typhoon struck, how evacuation centers were unable to cater to their particular needs, and how it was especially hard for them to reach relief distribution centers when transportation was

Figure 11.1: Map of typhoon Haiyan's actual track and summary of impact

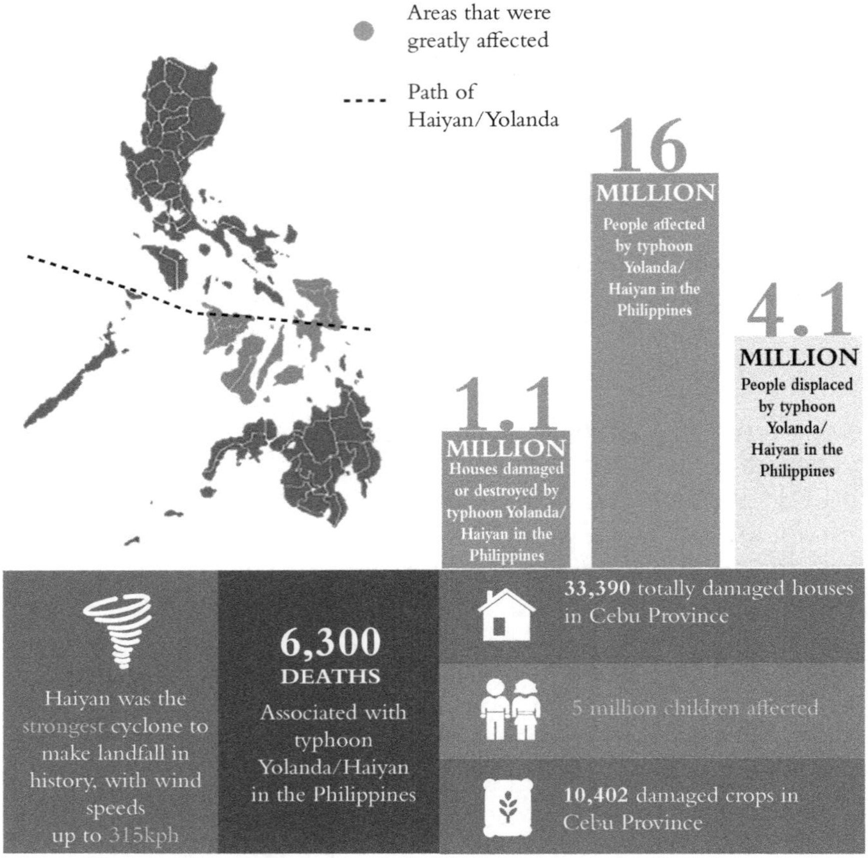

Source: Alburo-Cañete (2016).

not accessible to many. Box 2 highlights the key findings of the study (see Macalandag and Alburo-Cañete 2015).

This experience and insights gleaned from A2D Project's involvement in the Haiyan response opened an opportunity for the organization to develop a post-Haiyan program that addressed the lack of disability inclusion in DRR plans and systems. With the support of Arbeiter-Samariter-Bund, a German relief and social welfare organization engaged in a wide range of social service provisions in Germany and abroad, A2D Project launched an initiative, the ENABLED Project, to promote disability-inclusive DRR in Haiyan-affected communities in Northern Cebu. Among the project's notable accomplishments was the establishment of a network of OPDs, advocates from the local governments and agencies, and civil society organizations in the province. The main objective of this initiative was

Box 2: Key results of baseline study in Tabogon

General experience during Haiyan relief:

- Apart from lack of preparedness apparent in the community in general, persons with disabilities faced greater difficulties in terms of evacuation and access to relief goods and services. This is attributed to the general lack of consideration of specific needs of persons with disabilities, as well as other at-risk groups such as women and older persons, in carrying out disaster response and developing relief distribution systems.
- With the strict policy of "falling in line to claim relief goods and services," specific groups such as the older persons and persons with disabilities had no choice but to stand in line in the middle of crowded relief operations. Persons with difficulty in walking were unable to pick up goods from certain relief distribution centers.
- There were accounts that persons with disabilities received little to no assistance as most of them are not heads of the family. While some argue that they were able to avail of relief services and receive goods indirectly, persons with disabilities generally observed that there was not any organization which gave specific attention to their needs.

DRR participation and access to information

- While inclusion of persons with disabilities is acknowledged as important by municipal and barangay officials, there is still the need to deepen understanding on the roles and responsibilities of persons with disabilities from a rights-based perspective.
- Only 24% of the surveyed participants believed that persons with disabilities have the right to access to information.
- When asked what they think are the roles of persons with disabilities in helping the community become more prepared for disasters (i.e., help disseminate information, participate in making preparedness plans, be part of the DRRM committee), respondents expressed that they have largely no role in such kind of activities.
- Persons with disabilities have more difficulties achieving well-being and quality of life, and formal mechanisms that ensure inclusion and participation of persons with disabilities in existing LGU (Local Government Unit) structures, particularly DRRM, are not yet in place.

Barriers to participation:

- Person with disabilities expressed that they are not fully aware of their rights.
- There were prevailing biases that persons with disabilities are 'helpless' and need only be 'rescued' in times of disasters.
- Lack of representation in formal governance structures.
- Persons with disabilities were not confident to join community development activities because of prevailing biases against them.

to enable multi-stakeholder engagement (at community, municipal, and provincial levels) which would ensure that inclusive policies and programs are effectively lobbied for and advocated at different levels of disaster risk governance (Alburo-Cañete 2016).

We detail below how the network was organized as well as the key strategies and approaches employed. We demonstrate the important contributions that the network has made to the DRR landscape in Cebu in terms of institutionalizing disability inclusion in risk management processes. To do this, we highlight CDiDRRN's activities, accomplishments as well as challenges in pursuit of its goals.

The CDiDRRN

In recognition of the patterns of exclusion observed and in response to the need to mainstream disability perspectives in disaster risk reduction as a lesson learned from the typhoon Haiyan disaster, the CDiDRRN was formally organized in 2015 with a total membership of 15 OPDs and disability rights advocacy groups across the province of Cebu. It has been recognized as the first network of its kind in Cebu, and in the Philippines more generally. The network was represented by OPDs and community-based organizations in collaboration with government agencies and NGOs. A2D Project facilitated and served as the catalyst for the creation of the network.

The coalition has been envisioned as an advocacy platform and watchdog for the implementation of the Philippine Magna Carta for Disabled Persons, the Philippine DRRM Law, and the local implementation of the Sendai Framework. Its central mission is the inclusion of persons with disabilities in Disaster Risk Reduction (DRR) processes and structures to have OPD representation in local governments' Disaster Risk Reduction and Management Councils.

Organizing the network required undertaking exhaustive efforts to conduct a series of consultations, dialogues, and workshops among OPDs and stakeholders of advocacy groups dedicated to promoting the rights and welfare of disadvantaged, underprivileged, and marginalized groups, including persons with disabilities, and relevant local government structures, such as the Provincial Disaster Risk Reduction and Management Office of Cebu and the Local Disaster Risk Reduction and Management Offices of Mandaue City, Talisay City, Cebu City, Cordova, Catmon, and Lapu Lapu, among others. Through these various activities, the network began to promote its advocacy and laid down the foundations for establishing disability inclusion in formal DRR mechanisms and processes. As an entry point, the network leveraged existing policies and laws that supported disability inclusion in development programming and DRR, to argue for the need to pay attention to circumstances and needs of persons with disabilities. After establishing the network's common vision, which is "to create inclusive and resilient communities where persons with disabilities are secured and protected from disasters and calamities," the CDiDRRN crafted the following objectives to guide its actions:

1. Advocate for inclusive disaster risk reduction-climate change adaptation (DRR-CCA) policies and programs in all levels of governance achieve community resilience;
2. Promote the full implementation of relevant laws and international conventions thereby ensuring accessibility and inclusion;
3. Meaningfully participate and engage in mandated special bodies of government that work towards Disability-inclusive Disaster Risk Reduction—Climate Change Adaptation (DiDRR-CCA);
4. Actively engage with the private sector and other local and international humanitarian organizations to foster collaboration and partnership in DiDRR-related programs; and
5. Build the capacity of the network and its members in the area of DiDRR-CCA thereby providing for sustainability and development.

In pursuit of these objectives, the network applied the 'twin-track' approach with a rights-based perspective as its main organizing strategy. This approach was first conceptualized and developed by the United Kingdom's Department for International Development (DFID) for their gender equality program in 1996 as a response to the agreements and resolutions put forward in the 1995 World Conference on Women. The twin-track approach addressed gender mainstreaming and women empowerment objectives simultaneously (Watkins 2004). The framework has eventually been adopted to advance the rights of marginalized groups, most notably in the area of inclusion and disability rights advocacy. Applied to DiDRR initiatives, the twin-track approach combines disability-specific interventions aimed at empowering persons with disabilities on the one hand and societal interventions aimed at raising the resilience of all members of the community and ensuring that processes for risk reduction are barrier-free, on the other.

Following this twin-track approach illustrated in Figure 11.2, the CDiDRRN pursued a disability-specific agenda and interventions for empowering OPDs within Cebu's DRRM space (e.g., launching capacity development workshops and training for persons with disabilities and increasing their participation in formal mechanisms and institutional bodies). These include improvement of data on persons with disabilities, mainstreaming the application of universal design in government infrastructure, organizational strengthening of OPDs, promotion of inclusive DRR-CCA policies and programs, and establishment of persons with disabilities office in local governments in Cebu. At the same time, it engaged with broader risk reduction stakeholders in Cebu Province, to promote the

Figure 11.2: Twin-track approach to disability-inclusive development

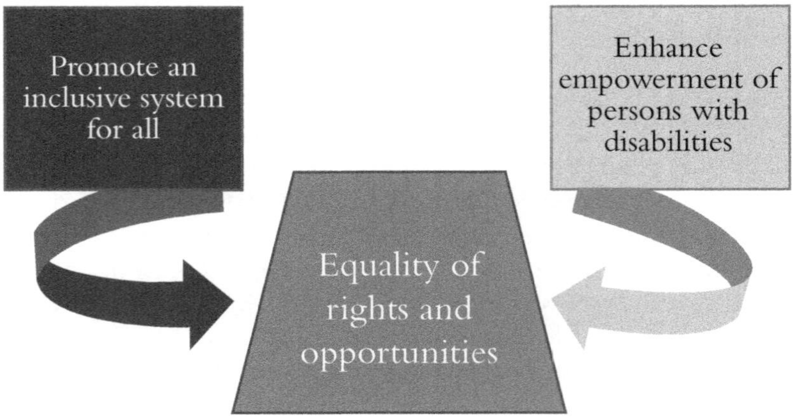

Note: Adapted from DFID (2000, 11).

adoption of a whole-of-society principle in DRR. This means that the network expanded its organizing and capacity building beyond OPDs to include government agencies and NGOs with a general end goal of building resilient and inclusive communities. With this strategy, the network has been able to forge a fruitful collaboration with government agencies and NGOs, which resulted in the enactment of policies and resolutions that specifically enabled OPDs to participate in DRR governance at the level of the province and selected municipalities of Cebu. Moreover, the discussions and dialogues organized through the network also triggered the review of practices and mechanisms using the lens of inclusion. In the next section, we discuss in more detail the results of the network's activities, and then lay out the continuing challenges being encountered.

Results of the CDiDRRN activities

Through the network, a platform for discussion and formulation of action plans was created which has, among others, facilitated capacity development activities for OPDs, and raised the awareness of the public regarding the roles of persons with disabilities in times of disasters to *de*-stigmatize persons with disabilities in society. We highlight below five of the major accomplishments of the CDiDRRN since its inception in 2015.

Established a pool of trainers on disability-inclusive DRR
As part of the twin-track strategy, training was conducted among members of OPDs to enable them to fully participate in DRR programming and even become valuable resources for making DRR inclusive. In partnership with Handicap International,[11] the CDiDRRN organized a training of trainers

on DiDRR and the use of the Inclusive Community-based Disaster Risk Reduction and Management Training Manual (or *Lahat Handa* Manual). The *Lahat Handa* Manual is a training guidebook, which supplements the Community-based DRRM Basic Instructor's Guide (BIG) Manual of the National Disaster Risk Reduction and Management Council – Office of Civil Defense (NDRRMC-OCD). It integrates inclusive approaches to community-based DRRM and promotes the rights and capacities of children, youth, older people, and persons with disabilities. This initiative resulted in creating a pool of trained resource persons from among the OPDs, who could *ably* assist local governments in conducting DiDRR programs, training, and planning. As an ongoing initiative of the CDiDRRN with the support of the Cebu Provincial Government, the *Lahat Handa* Manual is continuously being rolled out in different municipalities and cities in Cebu. Here, persons with disabilities take an active role in capacitating local governments to adopt an inclusive approach to DRR.

Training for 'self-rescue' during times of calamities is also a feature of the network's capacity development initiatives. The CDiDRRN seeks to change the perception that persons with disabilities are 'helpless,' thus their activities also involve conducting training and workshops for persons with disabilities in disaster preparedness and emergency self-help strategies. Such training includes conducting preparedness drills among persons with different forms of disability and coaching participants with techniques to address emergencies. The training, developed and conducted by network members, was recognized as a significant contribution to the typical ways preparedness drills have been performed. Because of this, the training design was included as an additional module in the *Lahat Handa* Manual's latest edition.

Influenced local government to integrate disability-inclusive risk management strategies in their DRRM and contingency plans

While local governments had extant disaster risk reduction and management plans, these generally did not consider the participation of, and issues confronted by at-risk groups, especially persons with disabilities. As a result, the network facilitated the review and enhancement of several DRRM and contingency plans of partner LGUs as well as facilitated the development of coordination mechanisms between Local Disability Affairs Offices and Local DRRM Offices. Notably, one of the key considerations in improving DRRM plans was establishing inclusive and community-based early warning systems and evacuation protocols.

In the Municipality of Tabogon in Northern Cebu, for example, local members of the CDiDRRN worked with the local government in setting up an inclusive early warning system. *Radyo baktas* (i.e., use of mobile loudspeakers), which has been the common method for information dissemination regarding anticipated hazards, was complemented by the use of colored flags (yellow, orange, and red) signifying different alert levels for incoming typhoons catering to those who might have hearing difficulties.[12] Moreover, communication protocols were developed to ensure that information is accessed by everyone. The protocols and plans were displayed in their village centers/meeting huts. Lastly, A2D Project and the CDiDRRN developed a module on Filipino Sign Language for Disaster and Emergencies. The barangay rescue teams were trained using the module to help them communicate advisories to deaf people. Lastly, family and caregivers of persons with disabilities were also involved in creating family preparedness plans as they play an important role in ensuring that information is communicated and discussed with family members who may need assistance.

The CDiDRRN continues to advocate among local governments, to take into account the differential access to information that persons with disabilities might have in specific localities and the kind of assistance that might be needed should an evacuation be required. This includes placing signs that someone, within the household, for example, has a hearing impairment. This indicates that risk communication to communities and households must accommodate the needs and rights of those who might not be able to hear auditory warnings during emergencies. Making evacuation routes and facilities safe and accessible are also important considerations that the network urges duty bearers to take seriously, including addressing specific accessibility and safety needs of women and children with disabilities. Indeed, in the study conducted by Zayas et al (2017), evacuation centers and bunkhouses housing internally displaced persons in Tacloban City after typhoon Haiyan lacked sensitivity to the needs of women with disabilities—shelters and facilities were not accessible to persons with mobility limitations and common latrines (apart from being inaccessible) were also perceived as unsafe by women with disabilities due to lack of lighting. This was also something observed in the Haiyan response in Cebu (Alburo-Cañete 2016). Thus, the CDiDRRN promotes addressing disability issues and their intersections with other conditions that might exacerbate disaster vulnerabilities such as gender and age.

Launched and organized the Provincial DiDRR Summit and Learning Events as a platform for the network's stakeholders to discuss the common vision, agenda, and commitments

The CDiDRRN launched the 1st Cebu Summit on Inclusive DRR and DRR Info Fair Exhibition on May 20–24, 2016. This was attended by over 300 participants. Among the highlights of the event had been the formulation of a Six-Point Inclusive DRR Agenda for the Province of Cebu, which was produced through a series of workshops attended by various stakeholders including persons with disabilities, children, and women. The Agenda advanced the following as action points for establishing a disability-inclusive DRR practice in Cebu:

1. Enforce the full implementation of the accessibility law and relevant laws on disability and DRR through consultation, collaboration, and establishment of linkages for greater participation of persons with disability in development and nation-building.
2. Empower the sector of persons with disabilities including their resilience through the provision of skills training for self-rescue, values formation, sustainable livelihood programs, and assistive devices as criteria for an inclusive community.
3. Make accessible disaggregated data on persons with disabilities, provide access to such information, organize and conduct a holistic and sustainable information drive including disability sensitivity on inclusive DRR, as well as rights of, and impacts of disasters on, persons with disabilities to bring about behavior change in the community and different levels of governance.
4. Ensure the active participation of persons with disabilities, their families, and groups that support them in governance and decision-making bodies, so they become productive members of society and leaders and implementers of good governance.
5. Guarantee continuity of participation of persons with disabilities organizations through accreditation, implementation of laws, provision of sign language interpreters for the deaf, and education of communities on basic sign language to ensure good communication and easy access to information.
6. Warrant sustainability of the persons with disabilities organizations and their representation in DRR committees through formulation and implementation of the city/municipal/barangay ordinances and resolutions, utilization of allocated funds, provision of capacity building activities, and institutionalization of the Person with Disability Affairs Office in every municipal/city local government unit in Cebu.

This six-point agenda was approved by acclamation during the summit and was monitored for its implementation. The success of the summit in galvanizing a movement for inclusion in DRR at different levels of governance and among a wide range of stakeholders led to the institutionalization of the summit as an annual event. It became a platform for reviewing commitments and crafting plans of action for DRR—with the active involvement of persons with disabilities. Furthermore, the network organized roadshows to localize these agenda, and to encourage more local governments to adopt inclusive approaches to DRR.

Contributed to the enhancement of Cebu local governments' capacity to collect disability data using the Washington Group questions for disability data collection

Low prioritization of data collection on disability not only contributes to the low awareness regarding persons with disabilities, but it also contributes to their further marginalization and invisibility in development programming. From a DRR perspective, disability data is extremely important in terms of assessing community risks and vulnerability, access to information, infrastructure, and services, and capacities to address identified risks. Although effective collection and utilization of disability data is considered a work-in-progress, the network has supported the use of the Washington Group (WG) short set of questions to address functioning limitations that need to be considered in addressing the issue of disaster risk.

The WG set of questions is named after the Washington Group on Disability Statistics, which is a voluntary technical working group comprised of "over 100 National Statistical Offices as well as international, non-governmental and disability organizations" coordinated by the UN Statistical Division (Madans, Loeb, and Altman 2011, 1). To deal with the challenge of multiple definitions of disability and ways of measuring prevalence, the WG developed a short set of questions that can be used on censuses, surveys, and other data collection formats across a wide range of cultural contexts. The questions comprise six core functional domains: seeing, hearing, walking, cognition, self-care, and communication (see Table 11.1).

Applications of the WG questions and their adaptation to the context of disaster risk reduction have been considered a way forward for enhancing disability data collection (Sloman and Margaretha 2018). In 2016, members of the CDiDRRN participated in the training of trainers (TOT) on using the WG questions for DRR. After this training, the network incorporated a discussion on applying the WG questions to the *Lahat Handa* training is carried out to LGUs in Cebu Province.

Following the *Lahat Handa* training, particularly, for the municipalities of Tabogon and Cordova, their DRRM and Persons with Disabilities Offices carried out a barangay- and municipality-wide survey using the WG questions. The result of the survey was used to improve their current DRRM and contingency plans. For Cordova, the survey results influenced the LGU appropriation for DRRM funds for persons with disabilities, including stockpiling of assistive devices.

Institutionalized the CDiDRRN programs at the province and one municipality

One of the key achievements of the network, apart from securing representation at the Provincial DRRM Council, was its successful negotiation with the Provincial government to allocate DRRM funds to support the network's activities and initiatives. In particular, the Province of Cebu allotted more than PHP3 million (approximately US$60,000)[13] to sustain the network's operations throughout 2015–2019. This meant that

Table 11.1: The Washington Group short set of questions

	No difficulty	Some difficulty	A lot of difficulty	Cannot do it at all
a. Do you have difficulty seeing, even if wearing glasses?	1	2	3	4
b. Do you have difficulty hearing, even if using a hearing aid?	1	2	3	4
c. Do you have difficulty walking or climbing steps?	1	2	3	4
d. Do you have difficulty remembering or concentrating?	1	2	3	4
e. Do you have difficulty (with self-care such as) washing all over or dressing?	1	2	3	4
f. Using your usual (customary) language, do you have difficulty communicating (for example, understanding or being understood by others)?	1	2	3	4

Source: http://www.washingtongroup-disability.com/washington-group-question-sets/short-set-of-disability-questions/

the CDiDRRN initiatives were recognized as part of the Cebu Provincial Government's regular DRR program. Moreover, the network has also succeeded in institutionalizing its programs in some partner municipalities. The municipality of Cordova, for example, has started to allocate funds (approximately PHP90,000 or US$2,000) from its local budget to support the network's activities for DiDRR. One of the network members was also appointed as the Municipal Disability Affairs Officer and, at the time of writing this chapter, sits at the local development council as a representative of persons with disabilities.

In sum, the activities and initiatives of the CDiDRRN have produced a range of tangible results in Cebu's DRR landscape at various levels of engagement. Figure 11.3 presents the results of the Cebu DiDRR Network's rights-based initiatives through the twin-track approach to advance disability-inclusive DRR in the context of Cebu.

Although this list of achievements is not exhaustive, these are arguably among the most significant contributions to promoting DiDRR not only within the province of Cebu but also at a national scale. Members of the CDiDRRN have been invited as participants and resource speakers at national learning events to propagate the lessons and achievements from

Figure 11.3: Results of the CDiDRRN's rights-based initiatives reflect the application of the twin-track approach

building the network. Beyond the Philippines, network members have also been invited to speak at international learning platforms to share their experiences of advocating for disability inclusion in disaster risk reduction.

We need to emphasize, however, that these accomplishments have been the product of years of intensive advocacy work with local government officials and NGOs. While the accomplishments of the network may appear to have been seamless undertakings, on the contrary, network members have expressed the struggles they have had to take on to advance their vision of an inclusive society. In the next section, we discuss some of the challenges confronted by the CDiDRRN as they navigated to overcome barriers embedded in their social, cultural, and political environments.

Challenges

Marginalization of persons with disabilities in DRR is largely a result of pre-existing prejudices and discrimination within the sociocultural environment. In the Philippines, disability is often regarded as an "incapacity to do things" that so-called 'normal' people can do, or sometimes as a "trial from God," where disability is viewed as a test of perseverance and faith (Zayas et al 2017, 25). These views about disability have often resulted in the social stigma that deters persons with disabilities from participating in community activities fully, on the one hand, and paternalistic programs that frame persons with disabilities as passive recipients or beneficiaries of various projects, including DRR interventions, on the other.

Prejudicial and discriminatory attitudes and practices continue to be a challenge that members of the CDiDRRN confront in their work. In some cases, local government officials were hesitant to include persons with disabilities in DRRM initiatives. This is due to prevailing ideas that persons with disabilities do not need to participate but rather need only to be 'rescued' (see Macalandag and Alburo-Cañete, 2015). However, as disability perspectives do not usually inform DRR planning as a result of these attitudes toward participation, the capacities of DRR personnel are often severely limited in addressing concerns and conditions of persons with disabilities during emergencies. For example, one network member who was also an officer of the Blind Association of Lapulapu City in Cebu commented how rescue staff are not trained to handle the needs of persons with disabilities during disasters. During several trainings using the *Lahat Handa* Manual which he helped facilitate, he recounted his observation (personal communication, December 4, 2019):

> There were many cases wherein rescue workers could not appropriately work with persons with disabilities… sometimes causing more harm or injuries. We had a case wherein a rescuer during a drill was not able to

> communicate properly with a visually impaired person and therefore was unable to successfully lead the person safely to evacuation. Also, the rescue teams always used pick-up trucks for evacuations which are too high for many people with disabilities to use. Just a simple ramp or lift from the pick-up would make things so much better in times of rescue.

Apart from the *disabling* sociocultural environment posing a challenge to members of the CDiDRRN as they advance their agenda and advocacy, problems also arose from the political environment. One particular issue that the network has had to deal with is the constant change in local government leadership every three years. This means that after elections, especially when new officials are elected in to office, advocacy work has to be repeated to convince leaders regarding the need to prioritize disability inclusion—and officials and key decision-makers are not always committed to championing disability issues, DRR, or both. Navigating tensions between opposing parties and political rivals is always a complicated endeavor, which could drastically affect the initiatives of the network if not dealt with carefully.[14] However, the local governments have gradually become receptive of the DiDRR advocacy—although results and uptake remain uneven, because of the ongoing awareness-raising campaigns conducted by stakeholders from OPDs, NGOs, and government offices who are active in the network; and the continued leveraging of existing laws and policies that support the network's efforts.

Although the CDiDRRN has had apparent success in terms of opening spaces for participation in DRR, from a more critical perspective, it is still important to situate disability inclusion within broader issues of development. In many cases, network members from community-based OPDs have expressed that while they are generally appreciative of the opportunities to participate in DRR planning and implementation, much still needs to be done in addressing exclusions that they experience in everyday life. Because of social stigma, for example, persons with disabilities are often deprived of opportunities for education and employment. "Without adequate livelihood," one member of an OPD said, "participation in DRR is good, but not enough to uplift our conditions." Thus, one of the key challenges that the network is facing, is the need to address structural issues more holistically, and beyond—but still largely relevant to—DRR.

(Without) conclusion

Rather than provide a 'conclusion,' we have chosen to highlight lessons learned as guideposts for future work on DiDRR in the Philippines. This is because we consider the CDiDRRN as a *work in progress*, necessitating constant and reflexive learning-by-doing. It must be understood that addressing the marginalization of persons with disabilities in DRR as well as in broader development processes is a gradual undertaking. This involves implementing strategies to better identify persons with disabilities and collecting vital information to understand the enabling and hindering factors affecting their participation in society and enjoyment of human rights on an equal basis with others. More importantly, this entails changing widely held perspectives, biases, and practices that require painstaking and consistent efforts and (un)learning.

By taking on a rights-based perspective on disability and by striving to make inclusive disaster risk reduction a concern of everyone in society, the network has produced several good practices and practical strategies that practitioners and advocates can take as entry points as well as build on—taking into account varying contexts in which DiDRR initiatives are to be undertaken. These good practices include but are not limited to:

1. Mainstreaming disability inclusion in different levels of governance to shift approaches to disability from charity and medical models that view persons with disabilities as passive recipients of aid to a rights-based model that recognizes them as persons who are entitled to equal access to society's valued resources and who are capable of contributing to disaster risk reduction and the attainment of their respective communities' development goals;
2. Training OPD leaders on data collection to enhance disability data using and/or adopting tested tools in disability research such as the WG short set of questions, Rapid Assessment of Disability toolkit, or other instruments designed to increase awareness about persons with disabilities;
3. Going 'twin track' in the approach to disability inclusion rather than just creating separate or special programs for persons with disabilities;
4. Developing leaders and trainers among persons with disabilities for them to represent themselves as they are the best resource for promoting disability inclusion across various processes of disaster risk reduction, and development in general;
5. Connecting with like-minded institutions and organizations to gain support for disability-inclusive DRR as well as to expand the resource base for carrying out advocacy for DiDRR; and,
6. Providing an enabling environment or continuous dialogue and engagement between communities of practice working in disability rights promotion and disaster risk management.

One of the important messages that the CDiDRRN imparts to all its members and partners is that persons with disabilities are not passive, helpless, or incapable. It challenges charity-based frameworks that render persons with disabilities as simply 'recipients' or 'beneficiaries' of assistance and welfare. Instead, the network advances an alternative view: persons with disabilities are entitled to equal access to society's valued resources and are capable of contributing to disaster risk reduction and the attainment of their respective communities' development goals.

Moreover, rather than create solely disability-specific interventions, the network has been effectively able to promote its advocacy by applying the twin-track approach and working with a range of stakeholders at different levels to open spaces for participation and promote the capabilities of persons with disabilities. However, as the experience of the network shows, it is important to situate the application of the twin-track approach in a broader set of power relations that are rooted in structural inequalities which produce disaster vulnerabilities (Bankoff and Hilhorst 2004; Oliver-Smith 2004). By keeping in mind the role of power, we avoid the pitfalls of many DRR interventions that tend to reduce vulnerability as a technical rather than an inherently political condition (Hewitt 1983; Alburo-Cañete 2014; see also Ferguson 1994 on depoliticization).[15]

Finally, persons with disabilities are not a homogeneous group. Thus, their experiences and perspectives are shaped by a range of social, cultural, and economic factors that need to be taken into account, such as gender, class, age, and ethnicity. As disability intersects with these other axes of identity and experience, differential vulnerabilities and capacities are also produced. As Twigg, Kett, and Lovell (2018) argue, disability inclusion cannot be achieved if broader marginalization, exploitation and institutional discrimination based on multiple intersecting identities are not challenged. Thus, leaving no one behind, as the Sendai Framework and Sustainable Development Goals have underscored, means acknowledging that inclusion is a complex undertaking that requires addressing the root causes of exclusion in society.

Notes

1. This prevalence rate is for severe forms of disability, according to the National Disability Prevalence Survey.
2. The Philippine Coalition on UN CRPD was formally established in 2011. The core group is comprised of over 20 OPDs and NGOs covering various disability constituencies and several national federations.
3. Item 5(c) of the provisional agenda during the Eighth Session of the Conference of States Parties to the Convention on the Rights of Persons with Disabilities.
4. The National Disability Prevalence Survey or Model Functioning Survey (NDPS/MFS) was conducted in 2016. It adopted the Model Disability Survey (MDS) developed by the World Health Organization (WHO) and the World Bank (WB).
5. The Disability Assessment Schedule is a generic assessment instrument for health and disability utilized by the World Health Organization. The WHODAS instrument may be accessed through https://www.who.int/classifications/icf/whodasii/en/ [Accessed December 2, 2022].

6 After interviewing 1,982 respondents in Haiyan-affected areas, the survey concluded the following result: "that age above 65, being employed or a student, living out of one's own house, having one or more health condition, and rating one's own health as moderate or very poor are the most relevant drivers of disability 6 months after the Haiyan/Yolanda typhoon" (Leonardi et al 2016, 268).

7 Reasonable accommodation refers to adjustments made within organizations, institutions, or systems to ensure that an individual, based on one's need, is able to participate in key activities as well as access opportunities fairly in relation to others (e.g., providing sign language interpreters for persons with hearing impairments).

8 When using the term 'vulnerable' groups, we highlight the need to consider how vulnerability is also socially produced. Rather than attaching the notion of vulnerability as something inherent in persons with disabilities, we adopt the conceptualization of Bankoff and Hilhorst (2004) that vulnerabilities must be situated within differentiating processes found in specific social contexts. Thus, we consider vulnerabilities as fluid, shifting, dynamic, and political.

9 This data was made public during the disaster response coordination meetings which some of the authors have attended.

10 Among other powerful typhoons that also caused much damage to property, infrastructure, and livelihoods in Cebu was typhoon Mike (Ruping) in 1991, which caused over PHP10 billion in damages (roughly US$200 million). Prior to Haiyan, typhoon Mike was considered the most destructive to have affected Cebu.

11 Handicap International now goes by its new name Humanity & Inclusion.

12 Determining the kind of early warning methods in Tabogon was informed by a prior baseline study conducted by A2D Project which assessed the kinds of functioning limitations that community members had, which precluded them from receiving and acting on information being disseminated using the usual protocols. Recommendations were then made regarding appropriate means of spreading information.

13 Exchange rate of PHP51 = US$1.

14 Entrenchment of state power in local politics is an attribute of Philippine political organization and relations (Abinales and Amoroso 2017). Political culture is often described as operating through kin-based ties, (inter)personal relationships, patronage, and clientelism (Hutchcroft 1991, 1994; Bulloch 2018), thus loyalties are often demanded by local politicians in exchange for their support and assistance.

15 In his *Interpretations of Calamity*, for example, Kenneth Hewitt (1983) highlights how prevailing paradigms of disaster management tend to privilege technocentric approaches and elide the ways that disasters are inherently political. Alburo-Cañete (2014), in a study on sexual and reproductive rights in post-disaster response in Iligan City after tropical storm Washi, made the observation that assistance and programs employed a highly techno-managerial approach which regarded sexuality and reproduction not as political issues but as problems to be managed. Neglecting the role of power—in its polyvalent forms and locations (Foucault 1980)—runs the risk of perpetuating the inequalities that play a central role in the production of disaster vulnerabilities.

References

Abinales, P. N. and D. J. Amoroso (2017) *State and Society in the Philippines*, 2nd ed, Lanham, MD: Rowman & Littlefield Publishers.

Alburo-Cañete, K. Z. K. (2014) "Bodies at risk: 'Managing' sexuality and reproduction in the aftermath of disaster in the Philippines," *Gender, Technology and Development* 18 (1): 33–51.

Alburo-Cañete, K. Z. K. (2016) "Breaking the barriers to inclusion: Promoting disability-inclusive disaster risk reduction through the Purok system in Tabogon, Northern Cebu," Cebu City: A2D Project.

Back, E., C. Cameron, and T. Tanner (2009) "Children and disaster risk reduction: taking stock and moving forward," https://www.preventionweb.net/publication/children-and-disaster-risk-reduction-taking-stock-and-moving-forward [Accessed December 2, 2022].

Bankoff, G. and D. Hilhorst (2004) "Introduction: mapping vulnerability," in G. Bankoff, G. Frerks, and D. Hilhorst (eds) *Mapping Vulnerability: Disasters, Development and People*, London and Sterling, VA: Earthscan, pp 1–9.

Blaikie, B., T. Cannon, I. Davis, and P. Wisner (2004) *At Risk: Natural Hazards, People's Vulnerability and Disasters*, 2nd ed, London: Routledge.

Bradshaw, S. and M. Fordham (2013) "Women, girls and disasters – a review for DFID", https://www.eldis.org/document/A72649 [Accessed December 2, 2022].

Bulloch, H. C. M. (2018) *In Pursuit of Progress: Narratives of Development on a Philippine Island*, Honolulu: University of Hawaii Press.

Department for International Development (DFID). (2000) "Disability, poverty and development", https://hpod.law.harvard.edu/pdf/Disability-poverty-and-development.pdf [Accessed December 2, 2022].

Ferguson, J. (1994) *The Anti-Politics Machine: Development, Depoliticization, and Bureaucratic Power in Lesotho*, Minneapolis: University of Minneapolis Press.

Foucault, M. (1980) *Power/Knowledge: Selected Interviews and Other Writings, 1972–1977*, New York: Pantheon Books.

Hewitt, K. (1983) *Interpretations of Calamity from the Viewpoint of Human Ecology, The Risks and Hazards Series 1*. London and Sydney: Allen and Unwin Inc.

Hutchcroft, P. D. (1991) "Oligarchs and cronies in the Philippine state the politics of patrimonial plunder," *World Politics* 43(3): 414–450.

Hutchcroft, P. D. (1994) "Predatory oligarchy, patrimonial state: The politics of private domestic commercial banking in the Philippines," in J. F. Eder and R. L. Youngblood (eds) *Patterns and Politics in the Philippines: Implications for Development*, Tempe: Arizona State University Press, pp 77–102.

Leonardi, M., R. Talampas, S. Chatterji, N. F. Kostanjsek, C. Regadio, M. C. Tarroja, S. Schiavolin, N. Naidoo, A. Raggi, M. Pindog, J. Zayas, and N. Zagaria (2016) "Measuring functioning and disability after a disaster," *International Journal of Rehabilitation Research* 39(3): 267–271.

Macalandag, R. and K. Z. Alburo-Cañete (2015) "Post-Haiyan: Enhancing resilience through disability inclusive community-based disaster risk management in Tabogon, Cebu, Philippines – baseline study", Cebu City: A2D Project.

Madans, J. H., M. E. Loeb, and B. M. Altman (2011) "Measuring disability and monitoring the UN Convention on the Rights of Persons with Disabilities: The work of the Washington Group on disability statistics," *BMC Public Health* 11 (Suppl 4): S4, https://bmcpublichealth.biomedcentral.com/articles/10.1186/1471-2458-11-S4-S4 [Accessed December 2, 2022].

Marella, M., A. Devine, G. F. Armecin, J. Zayas, M. J. Marco, and C. Vaughan (2016) "Rapid assessment of disability in the Philippines: Understanding prevalence, well-being, and access to the community for people with disabilities to inform the W-DARE project," *Population Health Metrics* 14(1): 26.

Mucke, P. (2014) "Urbanization trends and risk assessment," World Risk Report 2014. United Nations University Institute for Environment and Human Security.

Oliver-Smith, A. (2004) "Theorizing vulnerability in a globalizing world: A political ecological perspective," in G. Bankoff, G. Frerks, and D. Hilhorst (eds) *Mapping Vulnerability: Disasters, Development and People*, London and Sterling, VA: Earthscan, pp 10–24.

Peacock, W. G., B. H. Morrow, and H. Gladwin (1997) *Hurricane Andrew: Ethnicity, Gender, and the Sociology of Disasters*, London and New York: Routledge.

Peek, L. (2008) "Children and disasters: Understanding vulnerability, developing capacities and promoting resilience – an introduction," *Children, Youth and Environments* 18(1): 1–29.

Philippine Coalition on the UNCRPD. (2013) "2013 UNCRPD Parallel Report," https://www.slideshare.net/akkapppasig/2013-uncrpd-parallel-report-of-the-philippine-coalition [Accessed December 2, 2022].

Philippine Statistics Authority. (2013) "Persons with disability in the Philippines (desults from the 2010 Census)," https://psa/gov.ph/content/persons-disability-philippines-results-2010-census [Accessed January 16, 2023].

Philippine Statistics Authority. (2019) "Disability spares no one: A new perspective," https://psa.gov.ph/ndps/disability-survey-id/138567 [Accessed December 2, 2022].

Priestley, M. and L. Hemingway (2007) "Disability and disaster recovery," *Journal of Social Work in Disability & Rehabilitation* 5(3–4): 23–42.

Robinson, A. (2014) "Inclusive disaster risk reduction: Reflecting on Haiyan," AADMER Partnership Group, http://www.aadmerpartnership.org/wp- content/uploads/2014/02/Inclusive-Disaster-Risk-Reduction.pdf [Accessed July 29, 2015].

Robinson, A. and S. Kani (2014) "Disability-inclusive DRR: Information, Risk and practical action," in R. Shaw and T. Izumi (eds) *Civil Society Organization and Disaster Risk Reduction*, Tokyo: Springer, pp 219–236.

Seager, J. (2014) "Disasters are gendered: What's new?," in A. Singh and Z. Zommers (eds) *Reducing Disaster: Early Warning Systems For Climate Change*, Dordrecht: Springer, pp 265–281.

Shakespeare, T. (2013) "The social model of disability," in L. Davis (ed) *The Disability Studies Reader*, New York: Taylor and Francis, pp 214–221.

Sloman, A. and M. Margaretha (2018) "The Washington Group short set of questions on disability in disaster risk reduction and humanitarian action: Lessons from practice", *International Journal of Disaster Risk Reduction* 31: 995–1003.

Stough, L. M. and D. Kang (2015) "The Sendai Framework for disaster risk reduction and persons with disabilities," *International Journal of Disaster Risk Science* 6(2): 140–149.

Suprobo, N. (2011) "Lessons learned from the project mainstreaming disability into disaster risk management initiatives in Indonesia and Philippines," Yogyakarta and Makati: Handicap International Philippines.

Tierney, K. (2014) *The Social Roots of Risk: Producing Disasters, Promoting Resilience*, Redwood City, CA: Stanford Business Books.

Twigg, J., M. Kett, and E. Lovell (2018) "Disability inclusion and disaster risk reduction: Overcoming barriers to progress," https://interagencystandingcommittee.org/iasc-task-team- inclusion-persons-disabilities-humanitarian-action [Accessed January 1, 2020].

United Nations. (2006) "Convention on the rights of persons with disabilities," https://www.un.org/development/desa/disabilities/convention-on-the-rights-of-persons-with- disabilities/convention-on-the-rights-of-persons-with-disabilities-2.html [Accessed August 1, 2020].

United Nations. (2015) "The Sendai Framework for disaster risk reduction 2015–2030," http://www.preventionweb.net/files/43291_sendaiframeworkfordrren.pdf [Accessed April 16, 2019].

United Nations Convention on the Rights of Persons with Disabilities (UNCRPD). (2015) "Improvement of disability data and statistics: Objectives and challenges," 8th Session of the Conference of States Parties to the Convention on the Rights of Persons with Disabilities, New York, https://www.un.org/development/desa/disabilities/conference-of- states-parties-to-the-convention-on-the-rights-of-persons-with-disabilities-2/8th-session-of-the-conference-of-states-parties-to-the-convention-on-the-rights-of-persons-with-disabilities-9-11-june-2015.html#docs [Accessed December 2, 2022].

United Nations Economic and Social Commission for Asia and the Pacific (UNESCAP). (2015) "Disability-inclusive disaster risk reduction," www.un.org/disabilities/documents/desa/3WCDRR/Sendai_2015- [Accessed January 1, 2020].

United Nations International Strategy for Disaster Reduction (UNISDR). (2014) "Living with disability and disasters: UNISDR 2013 survey on living with disabilities and disasters – key findings". http://www.unisdr.org/2014/iddr/documents/2013DisabilitySurveryReport_030714.pdf [Accessed December 2, 2022].

United Nations Office of the High Commissioner for Human Rights (UNOCHR). (2014) "Statement on disability inclusion in the third world conference on disaster risk reduction and beyond," https://www.ohchr.org/EN/NewsEvents/Pages/DisplayNews.aspx?NewsID=15184&LangID= E [Accessed January 2, 2020].

Watkins, F. (2004) "Evaluation of DFID development assistance: Gender equality and women's empowerment DFID's experience of gender mainstreaming: 1995 to 2004 evaluation department", https://www.oecd.org/countries/bangladesh/35074862.pdf [Accessed December 2, 2022].

Welle, T. and J. Birkmann (2016) "The World Risk Index 2016," *The World Risk Report 2016*. United Nations University Institute for Environment and Human Security, pp 42–51.

World Health Organization. (2011) "World report on disability," Malta, https://www.who.int/publications/i/item/9789241564182 [Accessed December 2, 2022].

Zayas, J., L. Lacsamana, F. Garcia, J. Garcia, and K. Z. Alburo-Cañete (2017) "Build back better: Making inclusion work in disaster recovery in the aftermath of typhoon Haiyan," https://www.preventionweb.net/publication/build-back-better-making-inclusion-work-disaster-recovery-aftermath-typhoon-haiyan [Accessed December 2, 2022].

12

Climate Change and Applied Theater as Post-Disaster Response

Dennis D. Gupa

Being born and raised in the Philippines, I became aware of environmental disasters and destructions early on. I have witnessed images of mining activities that have devasted Indigenous lands. I have seen wretched pictures of the ocean with plastic and oil pollutants, landslides after typhoons, and flooded cities which constantly occupy the intermedial world of broadcast and print communications. Most of the time, these images of disasters pervade our lived lives. The words of Sheila Coronel (2014, 88), "our country's twin" aptly capture the intensity of Filipinos experience with disasters. I cite her words here as living echoes of my narratives of disasters that I experienced in a highly urbanized community constantly engulfed by perennial flooding. In an urban community, environmental destructions and weather events are entangled with socio-economic challenges. Growing up in an urban poor community in Quezon City, I witnessed how the lack of economic mobility forced certain communities to become impuissant and marginalized from civic political dialogues and social agency. Due to my lived experiences of poverty and lack of equal opportunities for people situated in the periphery of society, I began asking this question: "*Bakit mahirap ang buhay sa lugar namin?*" (Why are people poor in my community?). Like a ghost, this question in my young mind appeared intermittently in between direness and moments of possibilities. Amidst despairing situations, I was already curious about the possibilities of human empowerment through collective work. As someone who was introduced to theater at the age of nine, I have witnessed the potency of performance in gathering entities that contain their imaginations in a particular temporality and site. Provoked by my personal stories of disasters as critical positionality in examining climate change, applied theater, and local ecological knowledge, in this chapter,

I illustrate the processes of applied theater as a post-disaster response in an island community that has a recent history of devastation by a strong weather event. Taken from my dissertation, "Applied Theatre as Post-Disaster Response: Re-Futuring Climate Change, Performing Disasters, and Indigenous Ecological Knowledge," I aim to render stories of disasters and narratives on super typhoon Yolanda (known internationally as Haiyan)[1]—the strongest typhoon ever recorded to hit the Philippines—that my collaborators from my field site entrusted me as a performative articulation of possible futures. In my doctoral project, I engaged autoethnography, participatory action research, and applied theater as methods of research in rethinking disaster experiences and stories to mobilize an agentic process of collective creation of community-based-theater performance on acute violence of climate crises in an island community in the Philippines. By using the process of applied theater, I and my collaborators performatively examined climate change in Tubabao Island, Eastern Samar, Samar Province, by theatrically performing our disaster stories.

Tubabao Island is located in the Municipality of Guiuan where Haiyan first entered from the Pacific Ocean and left the entire municipality in a parlous and unimaginable state of destruction. To reflect on the cultivation of social cohesion in the wake of climate crises in an island community, I ask this question: how can applied theater mobilize local disaster stories to animate futurity?[2] This chapter provides a narrative inquiry on the practice of applied theater that aims to recenter grassroots participation as a post-disaster recovery program. A central theme of this chapter is the renewal of creativity within the discussion on social relationality through collective creation of community-based theater performances that aims to underscore an attentiveness and assiduousness in shared empowerment through applied theater's multivalent possibilities in mobilizing creativity, relationality, and solidarity.

Storm surge

The early warning forecast of the Philippine Atmospheric, Geophysical and Astronomical Services Administration (PAGASA) released information that Haiyan could cause the uprooting of trees. Robert Sawi, officer-in-charge of the PAGASA said: "*Sobrang lakas na ito baka halos wala nang matirang puno sa lugar if ever ganun kataas ang signal*" (This is ferociously powerful and no trees may be left in place, if ever the signal is raised higher" (Bacani 2013). Five years after Haiyan, I did my field research in Guiuan, Eastern Samar, the first of the ground zeros where the typhoon made its landfall. In my daily interactions with the community members, I was swamped with stories of the enormity of its impact. Stories, for example, of the huge satellite 'ball' of PAGASA's meteorological weather radar in Guiuan which was blown away by the wind and how it rolled down from the hill toward the municipality proper. Oral stories like these record the consciousness and agency of nature.

Community members in Guiuan's Barangay San Pedro heard from the weather forecast that Haiyan could propel storm surges. The storm surge was used as a pre-warning term that describes the projected magnitude of Haiyan. Found in the same article published by the *Philippine Star*, quoted above, we read this: "Storm surges are also expected in the coastal areas of Catanduanes, Albay, Sorsogon, Eastern Visayas, Dinagat Island and Siargao Island as early as Thursday evening and over the seaboards of Visayas and southern Luzon area by Friday and Saturday morning, respectively" (Bacani 2013). When I asked one of the local elders of the village, Francisco 'Tatay Frank' Abuyen, what he thought of the term, "storm surge," he made several guesses with silences in between his sentences:

> "*Mga balod? … It nga pagdako hit dagat? … Aw hit balod?*"
>
> ["The waves? … The time that the sea becomes enormous? … Oh, the waves rather."] (F. Abuyen, personal communication, June 2, 2019)

His response expressed a quandary and honesty:

> "*Hindi. Eh kasi English yun, pero kun ha waray-waray balod nga tikang han bagyo nga balod nga dako tungod han bagyo.*"
>
> ["No. Because it is English, but if it is Waray-Waray, it would be a large wave from the typhoon."] (F. Abuyen, personal communication, June 2, 2019)

The term 'storm surge' raised confusion for many. Assistant weather services chief of PAGASA, Ma. Cecilia Monteverde, confessed, "We weren't able to tackle that. It had more to do with the signals and in delivering the forecasts and warning disseminated to the public. But the storm surge wasn't explained there […]" (Bernal 2013). On the other hand, PAGASA Chief Engineer based in the Visayas, Alfredo Quiblat Jr., has a different narrative, "Five years before Yolanda, they [Western Visayas] already experienced a storm surge and PAGASA was already educating the public about storm surge. But people tend to forget as it did not happen often. But we keep on reminding the people about storm surge" (Abatayo, 2018, para. 13). For, Tatay Frank 'storm surge' is an image of destruction:

> "*Aw oo sugad hiton pagkuku-an hiton, sugad hiton mga dagko nga bagyo gin kuku-an man iton hiton mga tv, sugad hiton mga radyo nga asya na ngay-an ine ku-an nga maabot sugad han kan Yolanda nga maabot. Nga makaharadlok ine*

> *hin umabot ine, madako it ku-an madako iton tubig maku-an iton balod paro hiya hiton mga ano kapa-ine mga paro hiton mga nagtitidal wave sugad hiton tidal wave sugad iton klase hiton. Waray pak hiton pakakita kay largo man la ano kapa hine. Pero dida hiton mga palabas hiton telebisyon bisan han pagbaha han Japan nga han pagtidal wave han Japan nga pagkardamo han mga nagkakakuku-an nga mga sasakyan, mga kotse nga pinanganganod nga nawawasak nga mga sarakyan.*" (F. Abuyen, personal communication, June 2, 2019)
>
> ["Yes, just like when big storms are being reported on TV, or through the radio. Like when reports say that this is what's about to come, much like Yolanda. That it will be so alarming if it landfalls, how tidal waves, those kinds of waves. I haven't seen those kinds [of waves] first-hand because I immediately leave to evacuate. I have only seen them during newscasts on TV, like flooding in Japan due to the tidal wave and how so many vehicles were toppled over, how so many cars were washed ashore and destroyed."]

A plethora of stories of Haiyan and the confusion around disaster warning terms have elicited both confusion and discourse on climate crises in the Philippines. And this storied afterlife of Haiyan gave birth to performance processes and performances that mobilized dialogues on climate crises in an auspicious theatrical site of creativity and collaboration.

Haiyan

The first time I arrived in Leyte and Samar in February 2018, and upon hearing the distressing stories of the locals of Guiuan, I kept questioning why I was studying the stories of disaster that Haiyan had brought to the lives of my potential collaborators. En route to Guiuan, I asked Amado 'Arjhay' Babon— who became my artistic collaborator— to accompany me in the mass graves in Tacloban City and Palo municipality in Leyte Province where thousands of people who died from Haiyan were laid. We walked on the grassy pathways lined up with white *lapida* (gravestones) while the clouds were pouring heavy rains. As an outsider, I read the names on the *lapida* of those who perished from the typhoon. It was a critical moment for self-reflexivity as I began my doctoral research. This visitational act might be limited to grasp the weight of the magnitude of climate crises to those who have lost their lives in the wake of Haiyan and the trauma inherited by those who survived the catastrophe. Those thousands of names etched on the *lapida* manifest a lamentation of survival from untimely demise and unjust destruction caused by the severity of climate crises in Samar and Leyte Provinces. Taking some fragments of 'The Haiyan Dead,' a poem by University of the Philippines Visayas professor emeritus Merlie Alunan

(2016), I continue to subject my intention of doing field research and creating a performance on Haiyan. Some lines of the poem merit a place here:

> THE HAIYAN DEAD
> do not sleep.
> They walk our streets with us
> climb stairs of roofless houses
> latchless windows blown-off doors
> they are looking for the bed by the window
> ...
> the Haiyan dead will walk among us
> endlessly sleepless— (Alunan 2016, 170)

Recalling from memory, I once heard Alunan say: "When we walk in the streets of Tacloban, we are walking with the ghosts [...]" (Merlie Alunan, personal communication, April 1, 2018). The words she uttered and the lines of her poem, forced me, once again to ask myself: What conversations would I offer and what kind of co-presence would I share with the spirits of those who perished from storm surges that engulfed their villages while I was occupying their local spaces? I left Tacloban with no answers. But, perhaps, it might be enough, for the time being, that this symbolic act of intersubjective co-presencing was carried through as an ethical approximation in navigating the streets of Tacloban into the island of Tubabao.

In my peripatetic inquiry on disasters along the streets where the Haiyan dead were also walking, I ask one last question: to what extent can my disaster stories help in the affective embodiment of critical empathy as conceptualization and exploration of social justice from climate crises in local communities where the dead and the living continue to converge in zones of precarity? Ivan Brady (2009, xiii) in his foreword to the book, *Poetic Inquiry: Vibrant Voices for the Social Sciences*, asserts that "There is more than one way to see things, and therefore to say things, each inviting different points of entry into the research equation." I write this chapter to rethink disaster memories and reconceptualized them as agentic sources in rendering a possible future that rejects the dichotomies of "me and them," "dead and alive," and "human and the non-human" in performing climate disasters.

Outsider

I first lived in Tubabao Island's Barangay San Pedro in the summer of 2018 to conduct my field research. Early that year, I visited my father's birthplace in Balangkayan for a week where I was introduced to my relatives and learned about ancestral stories of my great-great-grandfather, Cesnero Contado. I came to these communities five years after Haiyan. Although the villages where I visited and lived have rebuilt their houses, the memories of destruction were still fresh among its community members. During this initial period, I listened to the community members' stories on the impact of Haiyan; observed the everyday activities of its elders; participated in the rituals; and fishing practices in the community. When my relatives learned that I was studying Haiyan they encouraged me to go to Guiuan, a municipality two hours away from Balangkayan. After a few months of visiting Balangkayan, I returned to Samar and did my field research in Guiuan. From Guiuan, I was assisted by the local government office staff to visit Barangay San Pedro where I met my applied theater collaborators: elders, public school teachers, and other community members. The first time I arrived in Tubabao Island in March 2018, I tried to explain to the community the goals of my doctoral project. It was challenging at first to introduce to them how theater can make an investigation of the processes of climate crises. Subsequently, through a workshop that led to performances that I organized among the public schoolteachers, I was able to explain to the community the rationale and goals of my doctoral project. Together with my Manila-based collaborators,[3] the theater artists from Sirang Theatre Ensemble, local elders from Barangay San Pedro, and the public schoolteachers of Guiuan, we created community-based theater performances on climate crises in island communities in the Philippines. Although I have received a warm welcome from the community in conducting my field research, it was a circuitous and complex journey of building trust and relationships.

Heartbeat

"*Ay, ano yung applied theatre, Sir Dennis.*[4] *Parang yung ginagawa ni Lea Salonga sa Manila?*" ["Huh? What is applied theater, Sir Dennis? Is it like what Lea Salonga is doing in Manila?"], one of the women in the community asked me. As a response to questions about my doctoral project by the community members, I co-organized an applied theater workshop with Sir Robert Yodico[5] and Angge Dayao. My objectives in this workshop were two-fold: 1) to present my dissertation to the community, and 2) demonstrate to them what applied theater is by performing stories of climate change.

In the early phase of the research, I distributed written consent forms to elders and their families before I engaged them in an interview and applied theater workshops. Even though I had translated my consent forms and dissertation script in Tagalog, I was surprised that some members of the community would feel alienated by written material that asked for their consent to participate in my study. I received joking remarks, such as: "*Hindi naman kami makukulong dito kapag kami ay pumirma, tama ba?*" ["You wouldn't want us to go to prison for signing this form, right?"]. Certainly, the complexity of cross-cultural encounters demands an ethical examination in conducting research and embedding oneself in the community. I knew that this jocular tone communicated a surreal relationship with legal contracts. A formal contractual relationship involving the signing of written documents seemed to be looked on with skepticism, if not anxiety, among the community members. I found myself sympathetic to this concern, and hence, decided to conduct an applied theater workshop that would explain my research goals to my potential collaborators. It aimed also to introduce myself and my Manila-based collaborators to the community members. Thus, the decision to 'perform' my research with the help of public schoolteachers came into being. When I consulted with Sir Robert about this idea, he observed that it would also help their schoolteachers learn about art theory and acquire skills in teaching theater.

Teachers in public elementary schools are required to deliver lectures on multiple subjects. They are expected to have a general knowledge of different courses including the fundamentals of music theory. But many of them have little knowledge of Western art theory. Sir Robert told me:

> "*Na-skip namin ang music theory sa lecture at theatre. Wala naman kasing nagtuturo sa aming nyan.*"
>
> ["We usually skip teaching music theory and theater in our lectures. No one is teaching us that."] (Yodico, personal communication, March 20, 2018)

It turned out that this workshop was also a way to reciprocate the generosity of Sir Robert and his family in accommodating me on the island. This six-day applied theater workshop resulted in the first community-based theater performance, attended by 12 Guiuan public schoolteachers. The workshop took place from April 14–19, 2018, at Barangay San Pedro Elementary School. We would begin at 10.00am and end at 4.00pm. The participants were engaged in various applied theater processes including theater games, breathing and voice projection, body movement, and, most significantly, improvisational music/sound creation. The workshop and performance were held in one of the classrooms that was destroyed by Haiyan.

When Haiyan made landfall on the island, the buildings of this school were destroyed. In providing the participants with a space for creativity and a place that would hold them safely, Angge used the framework of *Pintigan* (heartbeat) in her music improvisational workshop buttressed from the principles of shared agency, equity, and inclusivity where "people… see themselves as actors… with agency and control" (O'Connor and Anderson 2013, 12). "Pintigan" is a biological condition of blood circulation that helps generate a heartbeat: "*ritmikong paggalaw ng mga ugat dahil sa pagdaloy ng dugo mula sa puso, karaniwang nadaramá sa pulsúhan at leeg*" ("rhythmic movement of veins because of the blood that flows from the heart, usually felt through the pulsation of the wrist and the neck") (Diksionaryo 2020). Another definition relates to non-biological terminology: "*ang katulad na galaw ng tunog, liwanag, koryente, o musika*" ("a similar movement in the sound, light, electric current, or music") (Diksionaryo 2020). As a facilitator, she brought forth the creativity of the participants through the *Pintigan* Music Improvisational Workshop, a process in which the participants were encouraged to bring out their sense of rhythm. When Angge instructed the participants during the workshops: "*Pintigan nyo*" (allow the heartbeat to flow) she was instructing the teachers to bring to life the objects they found in the schoolyard by improvising sounds using the materials they gathered. The goal of this *Pintigan* Workshop was to develop an appreciation of objects that the participants have recovered as improvised musical instruments for collective sound creation. Angge wanted to stir up the curiosity of the teachers by facilitating them into realms of creativity where ordinary objects retrieved in the surrounding environments were re-utilized as instruments for music creation with "emotions," agency, and performative use. Music creation, "… may not be expensive", she tells the teachers after they created a collective soundscape from the objects they picked up outside the room. When Sir Robert shared his thoughts with the audience about his experience with *Pintigan* workshop he declared that as performers they became the *Pintig*. In his words:

> "*Mga waray na mga gamit karon gintagan ng buhay, nilagyan ng bawat isa ng tunog diba, nagkaroon ng buhay an mga waray na mga gamit pareho an akon chairs nakikitan tala it dinhi, lokal, oh kun ano paman it dida nga waray nira gineekspektaran ano it tunog ini? Ano ba it tunog hini? Pero binigyan ng buhay nira Dennis ngan Ange kun nakakabati la kamo han kada presentation, nira an lecture ano, may ada ngayan nira kinabuhi o sariling tunog. Ano an iyo napapansin han aton instruments? Mga waray na gamit liwat, pero kun iyo ig-uuntog-untog kada usa hiton iba iba it tunog hiton. So, iba an iya, iba liwat an akon [Kun diin gintutudlok niya an iya kausa], iba liwat an ganito, hi kami waray namon hini nga parti musika, pero nakuha kami tikang han napakaupay, napakagandang pagtuturo ni Angel, so, pintigan tayo pintigan. Hi ako diri gud ako nakakatiming hito. So thank you so much Dennis, damo na Salamat.*" (Yodico, speech, 43.21, 2018)
>
> ["We used bits of garbage that litters our community and found life in them. [We found that] it is imbued with sounds. Life prevails in that leavings and scraps we thought have no value just like those we have here. Or those things we just see around us: we can ask ourselves, what is the sound of this thing? Dennis and Angel helped us give life to these and if you heard the lecture of Dennis and Angel, you would have realized that they do have their own life or distinct

sound. What did you observe from these instruments? They are things that you just want to throw away. But when you put them together, they sound different from each other. His instrument is different from mine [referring to his co-participant], we are not musicians so basically we are not experts in music, but we learned from an excellent lecture by Angel. So, we are the beat-makers, sound-makers."]

Sir Robert's heartfelt declaration of becoming a *Pintig* or life-giver to the discarded materials did not come from a vacuum. It was a consequence of collective labor involved in co-organizing with Angge and Sir Robert. One becomes the *Pintig* by being an ensemble member, a creative participant who can improvise using the debris left by catastrophic events or the litter disposed around the community. The teachers have discovered the beat in those objects through *Pintigan* and animated the agency of those scraps by giving them sounds. In this workshop, Angge's *Pintigan* was used as a technique for original ensemble music creation using various organic and non-organic materials (e.g., shells, bamboo, tin cans, glass), that focused on rhythm as an essential element of the improvisational composition. According to Angge, rhythm is a vital element in *Pintigan* improvisational music composing. In her *Pintigan* workshop, she highlights the value of rhythm by illustrating how a series of sounds can create patterns through a recurring musical tone. She instructs the participants to think of one beat by using the object and each one of them offers this beat to the group.

Once a beat is created by a workshop participant it will be followed by another beat from another participant until the whole ensemble develops a full orchestral sound. She asked the participants, "What word can you think of to describe our technique in music making?", Angge prompted the teachers to reflect on their orchestral *oeuvre* after the group presented their output. By drawing out some words from the teachers, she was allowing free thoughts to come to the fore instinctively without too many cerebral interventions. Her process was spontaneous and devoid of censure, like a stream of consciousness that she allowed everyone to share continuously. She responds:

"*Ano pa? Sige po baka may maidadagdag pa yung iba… Iba pang mga salita. 'Found objects.' Pag sinabi pong found objects, ano pong pagkakaintindi ninyo doon? 'Mga local materials' … 'Indigenous materials' sabi ni Willmer.*" (Dayao 2018)

["Natural … 'natural' … What else? Just go on, perhaps there are more words that some of you can add to what has been said. Other words.

Figure 12.1: Pintigan sound workshop

Note: Participant Michelle Esquierdo enjoys the playfulness of this music improvisation. In front of her is her *tulipsay* (scallop shells) that she used for improvising her sound creation. In this improvisational sound workshop, the facilitator explains to the participants the difference of rhythm from other elements of music (pitch, timbre, and duration) through Pintigan sound exercises. Photo by Dennis Gupa, 2018.

> 'Found objects.' If we say found objects, how do you understand that? 'Local materials' … 'Indigenous materials', says Wilmer (*Natural*'… '*natural*' …"]

After Angge's *Pintigan* music session, I conducted a workshop on poetry writing that used their retrieved objects as themes to generate their imagination. This was followed by a storytelling workshop, tableaux, and creative bodily movement activities. These activities were designed to elicit the unhindered expression of the teachers by demonstrating their non-verbal and gestural performative abilities. Similar to the Angge's workshop the activities aimed at producing bodily movements most spontaneously without the intervention of heavily cerebral effort. Both Angge and I were immersing the participants in creative processes that made them focused, active, and animated.

Storytelling

Amidst the intrusions of modernity, Barangay San Pedro has maintained practicing some of its cultural traditions including its customary ways of storytelling. In one of our workshop activities, we once again went out of the workshop venue for a storytelling activity where I and Angge borrowed the practice of traditional storytelling in Tubabao Island, the *iru-istoryahan*. *Iru-istoryahan* is a Waray term that refers to a gathering where a small group of people recounts and listens to stories. From the root word 'istorya' or 'story,' *iru-istoryahan* is an oral narration practice of local community members who engage in an informal conversation on different topics. While local folks in Barangay San Pedro find in the collective sharing of stories their form of entertainment and passing the time, marine anthropologist Cynthia Zayas thinks that storytelling in the aftermath of disaster induces some cathartic effects (Zayas, personal communication, March 9, 2018). Community members in Barangay San Pedro gather under the shade of a tree where chairs or benches are set up, or in front of a *sari-sari* store to share tales on different topics in a relaxed manner.[6] It is carried out with a sense of friendliness and ease. While one person shares a story, another one listens or witnesses the narrator of the story.

Iru-istoryahan entertains both the narrator and the listener through a shared encounter of storytelling and listening. It enables a sense of expression and empathy between entities involved. In one of the late morning *Pintigan* sessions, we all headed directly to Dionicio 'Tatay Doning' Abuyen's house. Since I had been interviewing with him as my collaborator-informant, I thought of having him as our workshop guest speaker to an *iru-istoryahan*. And without a second thought, he accepted my invitation. Usually, at this time, Tatay Doning would be outside his house, unwinding. Tatay Frank and Tatay Doning are brothers. They are the oldest *Panggal* fishers in the community.[7] Being among the last few *Panggal* fishers on the island, his knowledge of the tradition would make him an excellent *iru-istoryahan* narrator. Under the *Talisay* tree,[8] we gathered and listened to Tatay Doning's stories. He shared with us how he learned fishing with *Panggal* from his father. During this visit, his grandchildren were beside him who also shared some personal observations on Tatay Doning's ways of *Panggal* fishing. Having lived in Barangay San Pedro for weeks and being exposed to Tatay Doning almost every day, I had gained a certain idea of his good-natured *persona*.

It was a special time to be with Tatay Doning and to listen to his stories. I sensed Tatay Doning's joy in sharing his stories with the teachers. To be the center of our attention with most public schoolteachers listening to a fisherman was a unique picture of knowledge production generated by a local elder for the schoolteachers. The teachers' curiosity was piqued when he told them about his exploit of accidentally recovering a metal receptacle from the seabed of Leyte Gulf. While we were listening to him, the teachers' attention was caught momentarily by the high-pitched sound of a bird from a nearby house. Beside this house was another *Talisay* tree where nested a brown billed-heron. He shifted his story from one topic to another immediately with ease. Certainly, he knew how to improvise, and his ability to extemporaneously tell stories was evident.

He enacts his memories of these topics through oral narration with vivid imagery of the past and sudden laughter that he imbues in between pauses.

> "*Ha Lapay kasama na namin yan sa aming pamilya.*"
>
> ["That is Lapay and he is part of our family."]

Tatay Doning told us amusingly that never fails to capture the attention of his listeners. According to him, one of his nieces found Lapay when it was a fledgling in the *bukid* (farm). The bird was brought into their house since it seemed abandoned by its mother. The family agreed to build a treehouse for Lapay. Tatay Doning continued,

> "*Nahulog yan si Lapay sa kanyang kanyang treehouse noong bata pa sya. Kaya napilay.*"

Figure 12.2: Theater games

Note: Our early afternoon workshop is spent with theater games. Certainly, a favorite part of the workshop, as one teacher would describe it, as our 'energizer.' Theater games spark liveliness of participation especially when the summer can be unbearably hot. In this photo, we are playing 'Chill Chill Bang Wha.' Right to left: San Pedro Elementary School teachers, Sonny Awa-oa with Rachel Cano, Perlita Machica, Michelle Esquierdo, Robert Yodico, Dennis D. Gupa, and Emelita Jadulan on who is looking at Robert from the far right. Photo by Angge Dayao, 2018.

> ["Lapay fell from the treehouse when he was still young. That is why he is lame now."]

We learned later that Lapay became an endeared bird in the community. The bird was especially remembered by the community as a survivor of Haiyan who kept himself safe under the piles of debris during the height of the typhoon. We ended our *iru-istoryahan* with Lapay's story at around noon. During the workshop, we played typhoon-themed theatER games and created tableaux that rendered their Haiyan experiences. After these games, we sat on the floor and continued our *iru-istoryahan*. This time it was the teachers who played as narrators. After our *iru-istoryahan*, I divided them into groups and asked them to select any stories they remembered from Tatay Doning and collectively we translated them into their groups' tableaux. One group was assigned to create the improvised sound using their found objects as musical instruments. The goal was to re-create stories that they had learned from Tatay Doning into sets of performances by using the technique of *Pintigan*. These performances were created for an intergenerational audience in the community. The teachers decided to translate the stories of Lapay and the discovery of the metal receptacle into a series of tableaux. Some of the more skilled dancers of the group choreographed a movement piece that thematized the danger of a fishing expedition during the typhoon season. They also crafted poems about the objects they had found, which became the prelude of the final performance on experiences of the participants on Haiyan. After the performance, Ariely Go Pinton, a participant from Camparang Elementary School, shared to the group that: "Acting is creation, creation is from within," a statement that demonstrates the authenticity of storytelling drawn from one's environment, self, and community.

Dependency

During my first few weeks in the village before conducting the two performances, the community members kept asking about the purpose of my research. The biggest challenge was to answer their questions, "*Ngano nga kita?*" ["Why did you choose us?"] and "*Ano nimo pagkasabot han among barangay?*" ["How did you know about our *barangay*?"], which came my way constantly. I was mistaken to interpret these questions as implicit articulations of material reciprocity expected from me by the community members. It must be recalled that non-profit humanitarian organizations surged in Guiuan after Haiyan hit the village. Local and international non-profit organizations (INGOs) dispatched relief goods and set up recovery programs including monetary support for cleaning up the devastated areas in Tubabao Island. While the community members were appreciative of the help, some were curious as to why these humanitarian efforts stopped after two years. "*Marami nang pumunta dito para tumulong pagkatapos ng Yolanda. Pero, kayo lang Sir ang tumagal...*" ["There were many people and organizations that came here after Yolanda. But you are the only one who stayed longer..."] said a young mother one late morning while I was observing the fishers dock their boats in the *baybay* (seashore). The young mother was waiting for her young husband, a fisher, who also carries a baby in her womb.

Reflecting on this encounter, I think of development aid that potentially can spur dependency. This prompted me to examine my presence as an outsider in the community. That morning of encounter with the young mother marked my second week in the village. And that time, I was not yet fully sure of what I was doing on the field site. The young mother continued telling her stories to me. The influx of emergency aid and the overflow of cash benefits for cleaning up their community allowed some families on the island to buy television sets and get access to cable networks. Before Haiyan struck their village, there were only a few families who owned TV sets in the community. But the young mother was thankful for the INGO. Her family was a direct recipient of the monetary aid of the INGO after the typhoon. She told me that she was able to purchase her TV set and she is now a regular watcher of daily soap operas. The young mother recalled that she was a few weeks pregnant when Haiyan happened. During the peak of Haiyan's wrath, her family evacuated to the 'Building,' the only concrete two-story house in the village. The day Haiyan hit the village she was one of 200 villagers who camped inside it. It was the house of a prominent woman in Barangay San Pedro who migrated to New York. The morning the typhoon entered the village, the young woman saw how the roof of the 'Building' was blown off by the strong wind. "*Nakakatakot talaga*" ["It was extremely scary"]. Fear was the only word that could encapsulate the experiences of the villagers inside the 'Building,' she told me. She was ready for her death. Mothers who were devotees of Sto. Niño were leading the prayers for the safety of 70 people inside the house composed of various age groups including elders and children. This was happening while the wind was blowing. After six years, the memories of Haiyan were still fresh to her. Her story sheds light on the complexity of social dependency after the typhoon and the religious value of reliance to Sto. Niño when the forces of nature strike.

Performing disaster

We culminated our workshop with a performance that transpired in the classroom.[9] Inside an elementary classroom, there were around 25 community members who attended the performance. Among them were schoolchildren who were seated on the floor and some of them watched from behind the windows. I saw how they were taking pleasure in witnessing their teachers perform stories about Tatay Doning and Lapay. After the performance, Sir Robert facilitated a discussion where the community shared their thoughts about it. Tatay Doning was present in this performance. "*Maupay*" or "beautiful," he told me after the performance. During the performance, some audience members would steal a glance at him and observe his reactions. At one point, I saw him smiling when his story of the missile shell that he recovered from the depths of the Leyte Gulf was being performed in an animated romantic comedy style. The performers used this genre to render a love story of a desperate "missile shell" who talks like a human. In their tableaux, the "missile shell" was looking for the right fisher to find him. Even before the performance, many members of the community were aware of Tatay Doning's story of how he recovered the "missile shell." They called Tatay Doning's missile shell, "torpedo." Sir Robert assumed that this was debris from the Battle of Leyte Gulf in World War II that took place October 23–26, 1944. Historical accounts tell of the battle between American and Japanese forces that included aerial bombardment and heavy destruction of the military fleet. Torpedoes were used, causing the fleets to sink. Today, Tatay Doning's torpedo is an important object of his daily life that he uses as a container for catching rainwater. For the participants of the workshop, Tatay Doning's ability to improvise war debris into everyday implements of survival is worthy of a theatrical performance. After the performance, I see the torpedo quietly placed on the corner of his house symbolizing Barangay San Pedro's capacity to cultivate creativity after a disaster.

When the teachers learned that the local government was willing to host the performance in the *bayan* (town), they began scheduling a pick-up rehearsal a week after the show in Barangay San Pedro. The Mayor's Office offered us an auditorium in one of the local government offices. But I negotiated for the performance to be held in the old Gabaldon Building inside the Guiuan Elementary Central School. The Gabaldon Building was destroyed by Haiyan and it was in a state of total wreckage when I found it as a possible site for our performance. The week that I organized the performance, I met with the principal of Guiuan East Elementary School to ask his permission for using it as our venue for the show, but he hesitated at the idea of using it as our performance site. The building was dirty, and the pillars were in total disarray. But the Mayor's Office saw the merit of the performance and helped me persuade the principal in allowing us to use the Gabaldon Building for the performance. Through the help of the Tourism Office, the local community members cleaned up the space and eventually we used it for the performance. On the night of the performance, there was an intergenerational audience of around 20 who came to see the creative output of the schoolteachers.

Figure 12.3: Gabaldon Building after Haiyan

Note: Gabaldon Building after cleaning for the applied theater performance. The performance started at around 5.30pm and ended at around 6.45pm. Through the benevolence of the Mayor's Office, the show received a sponsorship for lighting equipment which helped illuminate the performance inside the rubble edifice. The event was attended by local officials. Photo: Dennis Gupa, 2018.

Future

In this chapter, I illustrated the processes of applied theater as disaster response that I employed in my field research. This theatrical practice mobilized creativity among the community members in an island community recently impacted by a strong weather event. Guided by the principle of Indigenous self-determination, in conducting my research in Tubabao Island, I was inspired by the research principles of Jelena Porsanger (2004, 107–108) who contends that "[t]he main aim of indigenous methodology is to ensure that the research on indigenous issues can be carried out in a more respectful, ethical, correct, sympathetic, useful and beneficial fashion, seen from the point of view of indigenous peoples." Applied theater as disaster is a task of re-centering the voices of the community who have been peripheralized in socio-political dialogues around climate justice. Theatrical and performance processes must adapt to the situations of local communities if it intends an agentic exercise of reimagining a possible future. Within the local sites that constantly receive the impact of climate change, applied theater's capacity for community enablement is relevant in this creative gesture of enacting this future. Nora Quizon, another schoolteacher, shares:

> "*You are thinking beyond the surface of each theme. Sa loob ng classroom, may mga pupils ako, I think they will learn from me, if I do the same just like what I am doing ngayon. Halimbawa, ang dagat, na-imagine ko, 'pwede ba itong magsalita?' (referring to an object.) Sa ilalim nito, maraming nag uusap usap.*"
>
> ["Inside the classroom, I have these students who can learn from me, only if those students do the same thing that I am doing now. For example, I am imagining, 'can this speaks'? [referring to an object]. Under the depth of this ocean, there are many things who are speaking to each other."] (Quizon, April 27, 2019)

For Quizon, the processes of creativity in theatrical exploration can unleash the creativity of her students. On the other hand, we heard Cheryl Lanzaderas' proposition of drawing stories from real stories that they hear and the experiences of people in the local community. She bravely suggested:

> "*Yung mga tao na gumagamit ng dinamita sa dagat… mas maganda yun Sir kung ipapakita sa performance natin. Kasi makikita ng mga tao kung ano ang nangyayari sa mga isda… Fe-ni-feel mo ha hiya. Pinakikiramdaman mo.*"
>
> ["Those fishers that use fishing dynamite… we can represent as a story on that. This will show what is happening to the fish when they employ fishing through dynamite. You make them feel what the fish feels. You will need to feel them."] (Quizon, April 27, 2019)

During our post-discussion session, Lanzaderas expressed her final words on localizing theater and the power of specificity in storytelling: "By using the experience, you are empowering self and empower others. Because it is relevant to the experience. Through creativity and imagination, we can go deeper. There is no limit on that."

For these teachers, climate change is a lived experience. It is not an abstract idea. Climate emergency is an everyday negotiation entangled within the economic, political, and social spheres of their lives. Performing disasters in island communities with recent traumatic memories of Haiyan demands collective process and care. We did this by innovating our theatrical practice and vernacularizing the process of performance making. In rendering Haiyan as a subject of performance, we think of theater as a process that gives agency to local community members, mends the wounded memories, creates new solidarity, and looks into the future.

Notes

1 See World Vision (2013).

2 Cervera et al (2019) introduced me to the concepts of multiple futures, futuring, and futurity. The introduction to the online journal written by performance studies scholars, provides an insightful discussion of the thematics, logics, and subjectivities of the "future" that sets my initial meditation and dialectic on the agentic theater in the Dramaturgy of the Islands (DIs).

3 Francis Matheu, Marie Angelica Dayao and Jon Lazam were my volunteer-artists who joined me in my field research in Tubabao Island. Their exposure and participation in my doctoral project in Samar and Leyte prepared them in conceiving their artistic outputs to my doctoral performance project.

4 While the term, "Sir" denotes an honorific address in the village, it is also an endearing word attached to a name.

5 Robert Yodico is a public teacher at San Pedro Elementary School. His family hosted me and my artistic collaborators while we were conducting my field research in Tubabao Island. Marie Angelica 'Angge' Dayao was the third Manila-based artist who joined me in my field research in Tubabao Island. When I had my ocular visitation in Eastern Samar, I was accompanied by Francis who eventually became part of my doctoral performance project in Canada. During my first field research, Jon Lazam, an experimental filmmaker, assisted me in documenting the ritual performances in Tubabao.

6 *Sari-sari* store is a community/convenience store. A neighborhood convenience store that sells various basic items for everyday use. Usually, a small *barangay* of a thousand people would have two to three *sari-sari* stores. *Sari-sari* is a Tagalog word meaning mix-mix. This small family-run convenience store sells a variety of goods including but not limited to rice and other ingredients (salt, sugar, milk, onion, ginger, etc.), school supplies, snacks, and sometimes toys. A bigger *sari-sari* store would offer fresh vegetables, meat, and cold cuts. The *sari-sari* store is usually named after the owner (example: Adelina's Sari-Sari Store).

7 *Panggal* is a traditional fishing method that uses a *Panggal* fishing trap made of bamboo strands.

8 *Talisay* is commonly known as the beach almond tree (*Terminalia catappa*). It is a tropical tree found usually on the shoreline of Philippine beaches.

[9] For a complete description of what transpired in the process and how the performance was exhibited please refer to Gupa (2021).

References

Abatayo, R. O. (2018) "Yolanda: A wake-up call for Pagasa," *Cebu Daily News*, 5 November, https://cebudailynews.inquirer.net/202220/Yolanda-a-wake-up-call-for-pagasa [Accessed November 21, 2019].

Abuyen, Francisco. Interview by Dennis D. Tubabao Island, 2 June 2019.

Alunan, M. M. (2016) *Our Memory of Water: Words after Haiyan*, Naga City: Ateneo De Naga University Press.

Bacani, L. (2013) "PAGASA: 'Yolanda' could be strongest typhoon to hit Phl this year," *The Philippine Star*, 6 November, https://www.philstar.com/nation/2013/11/06/1253694/pagasa-yolanda-could-be-strongest-typhoon-hit-phl-year [Accessed August 15, 2017].

Bankoff, G. (2015) *Cultures of Disaster: Society and Natural Hazards in the Philippines*, London and New York: Routledge.

Bernal, B. (2013) " 'Storm Surge' not explained enough-PAGASA official," *Rappler*, November 14, https://www.rappler.com/move-ph/issues/disasters/typhoon-yolanda/43735-yolandaph-haiyan-preparedness-philippines [Accessed August 15, 2017].

Brady, I. (2009) "Foreword,". in M. Prendergast, C. Leggo, and P. Sameshima (eds) *Poetic Inquiry: Vibrant Voices in the Social Sciences*, Rotterdam: Sense Publishers, pp i–xiii.

Cervera, F., S. Chua, Y. Demetriou, A. Jeong, E. Laine, A. Shafiri, E. Wan, and A. Warren (2019) "Orientations: Where is the future now?" *GPS: Global Performance Studies* 2(2). DOI: https://doi.org./10.33303/gpsv2n2a1.

Coronel, S. (2014) "Unnatural disasters," in R. Abuyuan (ed) *Agam: Filipino Narratives on Uncertainty and Climate Change*, Quezon City: Institute for Climate and Sustainable Cities.

Gupa, D. (2021) "Applied theatre as post-disaster response: re-futuring climate change, performing disasters, and Indigenous ecological knowledge," Unpublished doctoral dissertation, University of Victoria.

Macasa, Marianito A. Interview by Dennis D. Gupa. Guiuan Eastern Samar, March 2018.

O'Connor, P. and M. Anderson (2013) *Applied Theatre: Research Radical Departures*, London: Bloomsbury.

Quizo, Nora. Discussion Workshop by Dennis D. Gupa. 27 April 2019.

Pinton, Ariely G. Discussion Workshop by Dennis D. Gupa. 27 April 2019.

Porsanger, J. (2004) "An essay about Indigenous methodology," *Nordlit* 15 (July): 105–120. DOI: https://doi.org/10.7557/13.1910

World Vision. (2013) "2013 typhoon Haiyan: Facts, FAQs, and how to help," https://www.worldvision.org/disaster-relief-news-stories/2013-typhoon-haiyan-facts [Accessed December 2, 2022].

13

Beyond *Bayanihan*: Overcoming Myths of Community Resilience in Typhoon Haiyan Post-Disaster Recovery

Yvonne Su, Ladylyn Mangada, and Ara Joy Pacoma

Introduction

On November 8, 2013, Typhoon Haiyan made landfall in the central islands of the Philippines. The typhoon affected over 14 million Filipinos and displaced thousands of families from their homes. In total, approximately 6,300 individuals were reported dead and 28,688 were injured, but some suspect these numbers are much higher (OCHA 2013; NDRRMC 2014; IBON 2015). Tacloban City, the biggest city on Leyte Island, experienced the greatest amount of physical damage and causalities (NDRRMC 2014). Tacloban is a highly urbanized city and is the administrative center of the Eastern Visayas region (NDRRMC 2014). While Tacloban is considered an economic hub of the Philippines, there are pockets of deep poverty in the city, especially in vulnerable coastal areas. As an effect of the typhoon, Tacloban experienced a deadly storm surge that damaged the vast majority of houses and infrastructures. Taclobanons, the residents of Tacloban City, recount seven harrowing days of search and rescue, burying the dead, looting for food and searching for fresh water as they waited for international humanitarian assistance to arrive (BBC 2013). Several sources have criticized the government for its inadequate risk reduction and disaster preparedness, evidenced by its insufficient disaster budget (Heydarian 2013; IBON 2015). Yet, despite the dangerous and traumatic environments described by many survivors, the national and international media coverage in post-disaster recovery tended to focus on extraordinary narratives of resilience and survival, particularly stories of communities coming together under the banner of *bayanihan*.

Bayanihan (pronounced as buy-uh-*nee*-hun) is the Indigenous Filipino spirit of communal unity and cooperation (Sarmenta and Hirano 1999, 675). Gertrudes R. Ang (1979, 91) defines *bayanihan* as "the ancient Filipino custom of group work," describing it as "a system of mutual help and concern which has become the backbone of family and village life through the Philippines." Ang has explained that *bayanihan* is "derived from the Tagalog root word bayan, a town or nation" so *bayanihan* means "being in a *bayan*" (Ang 1979, 91). This is about the Filipino tradition of helping families move their houses, where "[t]he able-bodied men would insert bamboo poles under the locally-made house to facilitate the task of lifting the entire house for relocation" (Adviento and de Guzman 2010, 106). Thus, one interpretation of the term is that it refers to the spirit of being in a community, together with cooperating to achieve a common goal. Similar definitions can be found in the literature. For instance, this sentiment is captured by Bankoff's (2007, 26) definition of *bayanihan* as "toiling on another's behalf and assuming another's burden." But as more Filipinos move away from rural areas into urban cities, the value of *bayanihan* has evolved from Indigenous ideas around collective labor and house relocation to more contemporary definitions around cooperation, volunteerism, and just "helping each other out" (Su and Mangada 2016). Eadie and Su (2018) liken *bayanihan* to bonding social capital, such as support from close family and friends. The IBON Foundation's publication (2015, 19) defines *bayanihan* as cooperative labor, labor pooling, or collective work in the context of describing survivors helping each other with reconstruction. Thus, cooperation is a more versatile and reflective definition of the practice of *bayanihan* in urban settings.

It is now a common practice to see NGOs, governments, and the media in the Philippines evoke the value of *bayanihan* after major disasters, to demonstrate the resilience of the Filipino people. Indeed, *bayanihan* is often seen as the source of Filipino resilience. But what about the accounts from local people who survived typhoon Haiyan? Did they also witness and practice *bayanihan* repeatedly? And what does *bayanihan* look like in theory and practice? Does this Indigenous value serves to increase community resilience in urban communities suffering from widespread poverty and frequent disasters? Moreover, what does resilience look like for the urban poor who are often forced to be self-reliant?

Our analysis of *bayanihan* is informed by two separate but overlapping case study research projects. The first research project examined three barangays in coastal communities in Tacloban City that were heavily affected by typhoon Haiyan. Short surveys were conducted with 465 households across the three barangays from November to December 2017. Utilizing systematic random sampling, the research team knocked on every two or four doors depending on the size of the barangay, to achieve a representative sample of each barangay.

The second research project, 'Project Yolanda,' was an Economic and Social Research Council and Department of International Development funded three-year collaborative project between the University of Nottingham and the University of the Philippines. The project examined 20 barangays of comparable size across the jurisdiction of the three local government units (LGUs): Tacloban City, Palo, and Tanauan. These areas under the three LGUs are all coastal communities that were also heavily affected by Haiyan. It was a multi-year study with 800 households surveyed once a year throughout 2015–2017. In each of the barangays, 40 households were surveyed. The 40 households were composed of the 20 most affected and the 20 least affected by Haiyan, to establish the factors that helped or hindered rehabilitation. For this chapter, we focus on community resilience in the form of *bayanihan* in post-Haiyan Tacloban City. Thus, for this chapter, the survey data of the 320 households from the eight barangays collected in 2017 from Tacloban City were used. It is important to note that the two studies looked at different barangays so no barangays were surveyed by both projects. The results and analysis of the study are based on most of the data drawn from the first project with supporting evidence provided by the second project. By combining the data from the two projects in Tacloban City, we were able to triangulate multiple sources of survey data and enrich our understanding of community resilience after typhoon Haiyan. We argue that the combination of the two research projects generated an even more robust account of the narratives of community resilience as well as the frequency of *bayanihan* occurring across the household and community levels.

We also draw on the 'insider-outsider' dimensions of doing disaster research. The methodological theory of insider and outsider is a useful tool for fieldwork analysis. This approach adds depth and nuance to research involving vulnerable groups. Insider research refers to when a researcher conducts research with a population in which they are also members, sharing an identity, language, and experiential base with participants. This insider status lends more legitimacy to the research, leading to rapid and more complete acceptance of the research project by the participants. Outsider research refers to when an individual conducts research with a population in which they are not a member. They share no identity, language, and have limited connections with the participants. The positionality of the members of the two teams—both being a mix of foreign researchers and local academics—allowed the research teams to benefit from being both insiders as well as outsiders. Local academics ensured that our research was grounded in local knowledge and allowed the surveys to be conducted in the local language of Waray-Waray. Having local academics conduct the surveys allowed the Haiyan survivors to be more open when sharing their experiences during the typhoon because they know that the academics were also survivors of Haiyan; the local academics understand deeply what they are feeling. At the same time, the foreign researchers on the teams help to bring international legitimacy to the projects.

Drawing on these two research projects in the province of Leyte after typhoon Haiyan, and drawing on the 'insider-outsider' dimensions of doing disaster research, we put forward three main arguments in this chapter. The first is that media articles, NGO reports, and government statements in the Philippines have overstated the practice of *bayanihan* after Haiyan. Secondly, what *bayanihan* looks like in practice is drastically different than what has been described in theory. Lastly, we argue that evoking *bayanihan* following disasters minimizes the gravity of the issue and resources needed to effectively lead communities toward post-disaster recovery. As such, we call for a more critical examination of the potential of using *bayanihan* as a post-disaster coping mechanism in the Philippines. The emphasis on *bayanihan* obfuscates the retrenchment of the state's role in post-disaster recovery, shifting accountability and responsibility away from governmental institutions and humanitarian organizations and outing them onto the disaster-affected people themselves. Moreover, such a singular narrative robs the ability of affected communities to tell their own stories and represent the diversity of experiences. As geographer Chaya Go explains, "people may be strong and happy, but they may also be tired, angry, and fearful. People are rarely defined by a sole characteristic" (Su, Mangada, and Turalba 2018). In this chapter, we try to go beyond the singular narrative and illustrate the variety of experiences people had, in an attempt to present a more comprehensive view of community post-disaster recovery after Haiyan.

Background on *bayanihan*

Before typhoon Haiyan was about to make landfall in central Philippines on November 8, 2013, President Benigno Aquino III called on all Filipinos to practice *bayanihan*, stating that "*Alam nating walang bagyong maaaring magpaluhod sa Pilipino kung tayo'y magbabayanihan*" ["No storm will make the Filipino get down on his knees as long as we help each other"] (Cupin 2013). The popular use of this Filipino value has resonated across the world, even making its way into the White House. In President Barack Obama's reaction statement to Haiyan, he mentioned that "I know the incredible resiliency of the Philippine people, and I am confident that the spirit of *bayanihan* will see you through this tragedy" (White House 2013). Such common use of the term has inspired others such as the local

NGO, Gawad Kalinga (2013), to make the statement "Let us declare a state of *bayanihan* not just a state of calamity." As Su and Mangada (2017, 104) have observed, "for the population that 'goes the extra smile,'[1] the idea of survivors working collectively fits the narrative," of a resilient and cooperative nation.

In addition to the mentions of *bayanihan* by President Benigno Aquino III and President Barack Obama, numerous news agencies referred to the *bayanihan* spirit about the typhoon Haiyan local recovery efforts and the resilience of Filipinos (Christian Science Monitor 2013; GMA News Online 2013; Sevastopulo and Kerr 2013; Abenoja and Lacamiento 2014; Viray 2014). Some articles focus on the spirit of *bayanihan* on the ground and among the diaspora (GMA News Online 2013; Sevastopulo and Kerr 2013; Abenoja and Lacamiento 2014; Viray 2014). Others described resilience and *bayanihan* using very strong and sensational language and the two are often treated as synonymous in meaning. Filipino-American Anthony Advincula (2013) claims that "Resilience seems innate in Filipino people. … Accustomed to hardship, we are naturally self-reliant." Similarly, an article by the Christian Science Monitor (2013) describes *bayanihan* as Filipinos' weapon against disasters, stating that "Every nation has its own shock absorbers for such calamities, such as proper zoning and close social ties. Filipinos have their own spirit of *bayanihan*, or neighborly deeds of empathy, as well as a tropical abundance of materials to rebuild houses." The Salvation Army Vision Network (2014) has published a film entitled *Bayanihan: The Spirit of Community*, which followed two Salvation Army cadets on their journey back home to find their families affected by typhoon Haiyan. In the film, they defined *bayanihan* as "a word that expresses, we are a community caring for each other, and that's what I saw, it's not just a word, it's a way of life" (Salvation Army Vision Network 2014). In all of these publications, *bayanihan* is described as something innate, instinctive, and intrinsic to Filipinos. This sentiment is fully captured by the then Chief of the United Nations High Commissioner for Refugees (UNHCR), Bernard Kerblat's statement about the spirit of *bayanihan* and resilience after Haiyan (Quismundo 2013):

> What I've seen in Ormoc, in Tacloban, in all these barangays, in this landscape of devastation, with survivors completely hazed as if they've been shell-shocked after bombardment … I must admit I was pretty moved. The other thing that equally moved me—and we need to reinforce and sustain this—is this incredible national mobilization, *bayanihan* on the move in support of their "kababayan." That is also very impressive. And I would venture in saying that Filipino resilience is unique. It is a typical Pinoy trait. It's part of your genes, part of your culture. You're born with it.

Such a strong narrative of inherent resilience and organic community support by people in positions of power combined with similar stories from NGOs, humanitarian organizations, government officials, and even survivors themselves, has perpetuated the myth of *bayanihan* and resilience that consistently placed the responsibility to survive on the affected communities themselves. In fact, at a press briefing less than two weeks after typhoon Haiyan, Bernard Kerblat, stated "We have to focus, all of us. Bickering and finger-pointing are not needed at this time, we have to focus, the *bayanihan* spirit must be preserved for our kababayan" (Lozada 2013). This statement captures perfectly how *bayanihan* can be used as a spectacle that diverts responsibilities away from government officials whose actions or inactions contributed to increasing the risk of communities to disasters in the first place. Instead, the narrative of the innate and instinctive *bayanihan* spirit places the solutions to post-disaster recovery wholly on the inherently resilient Filipinos, who have been practicing cooperation to rapid recovery in their communities.

Practice of *bayanihan*

Communities that are frequently described as tight-knit (Ong, Flores, and Combinido 2015), could logically have strong foundations of community support where *bayanihan* is a value commonly practiced after the crisis. The Filipino value of *bayanihan* after Haiyan is explored as we examine data about community engagement before Haiyan, whether respondents witnessed the *bayanihan* or community help, and how long they thought the *bayanihan* spirit lasted after Haiyan.

Community engagement: pre-Haiyan

Before Haiyan, 49% of the total respondents of the first project said that their households have frequently joined together with others in the barangay to address a common issue, while 28% have responded "never." In the second project, 48% of the total respondents have cited that before Haiyan, the people in their community supported each other through credit, loans, babysitting, labor, if the household income was not enough. However, the majority (52%) of the respondents have revealed that before Haiyan, the people in the community had no support for each other if the household income was not enough. It appears that less than 50% of the households observed *bayanihan* when it was most needed to address common issues and support each other when household income was insufficient before typhoon Haiyan.

Timeframe of bayanihan *spirit: post-Haiyan*

The experiences of disaster-affected households become clearer when the timeframe is examined. The first project asked the households "How long did the *bayanihan* spirit last after Haiyan?" and the majority of the total respondents (48%) cited one week to a month, while 12% said less than one week. Our findings suggest that while the spirit of *bayanihan* existed, mutual assistance and other forms of *bayanihan* were experiences that were short-lived in the urban context of Tacloban City. Previous research demonstrates that while disasters can present unique opportunities for change that encourages transformations in the social fabric of society, the change is often temporary and social divisions within the community return after the immediate danger passes (Oliver-Smith 1979). Similarly, Davis and Alexander (2016) found that while the short-term reactions to sudden-onset disasters can increase social cohesion, the longer-term consequences in many cases lead to worsened social tensions and further-divided communities. These observations appear consistent with our findings among Tacloban respondents who often offered more examples of problems with their neighbors during the response and relief distribution period than stories of *bayanihan*. Their stories temper the myth that there was an overwhelming flow of mutual assistance or that *bayanihan* was a consistent thread throughout the post-disaster experience of Tacloban, especially when the experiences of mutual help were described as short-lived.

In terms of *bayanihan*, community well-being was secondary or serves as a 'positive' side effect to securing self-interest or family's welfare in the immediate aftermath of the disaster. Study results show that 40% of the total respondents from the first project agreed with the statement that "as a whole, people here look out mainly for the welfare of their own families and they are not much concerned with barangay welfare." Most respondents believed that *bayanihan* is driven more by a sense of familial or clan interests rather than the altruistic promotion of community welfare. This fine distinction matters again in challenging the overly positive portrayal of *bayanihan* in Haiyan-affected communities. Rushing to mutually help one another is not at the expense of protecting their own families first and may specifically arise in response to the competitive distribution of resources in post-disaster recovery. In the case of extreme disasters, while outsiders may see the devastation as opportunities for the significant expansion of mutual help, solidarity, and social exchange, such as the outpour of international donations, Haiyan-affected households often see the scarcity of resources as a reason to be wary of others and to rightfully prioritize their own family's security. That is, mutual assistance did not always equate to mutual trust. This is evidenced by the 73% of the total respondents in the first project who revealed that people "need to be very careful" regarding trusting others in the barangay and 21% mentioned that most people in the barangay can be trusted. Issues of trust may have also been exacerbated by reports of widespread looting from stores and stealing from neighbors, as well as stories of people hoarding relief goods in the immediate aftermath of Haiyan (Evangelista 2013; IBON Foundation 2015; Su and Mangada 2017).

If community support and joining together did not seem to be something a majority of what the communities did before the disaster, was it something they did afterward? The first project asked the respondents if "After Haiyan, did you witness forms of *bayanihan* spirit in your barangay?" The results show that 73% of the total respondents answered yes, while 27% of them said no. The second project asked a slightly different question with more options for answers. They asked the respondents "To what extent did your community help itself immediately after the disaster?" The majority of the respondents (43%) replied "some but could have done more." Interestingly, 22% of the total household respondents said that the community "did not help itself" and 24% mentioned "not much," which means almost half of the households did not observe the community helping itself in a significant manner immediately after the disaster. This is contrary to the narrative among the media, NGOs, and governments that community members overwhelmingly banded together immediately after Haiyan to help each other (Christian Science Monitor 2013; GMA News Online 2013; Sevastopulo and Kerr 2013; Abenoja and Lacamiento 2014; Viray 2014).

Bayanihan *as collective mobilization, not a social safety net*

Two of the questions this chapter aims to understand are, "what is *bayanihan* in theory and practice?" and "does this indigenous value really serve to increase community resilience in urban communities suffering from widespread poverty and frequent disasters?" The literature largely defines *bayanihan* based on an Indigenous and traditional understanding of it as "group work," "mutual help," and "shared labour". As Su and Mangada (2016) have found in the modern and urban context, *bayanihan* is defined more along the lines of cooperation, volunteerism, and just helping each other out. To understand how *bayanihan* is perceived and defined in urban settings after a disaster, we explore the forms of *bayanihan* that were observed by respondents in the first project.

The most common forms of *bayanihan* observed after Haiyan was the community clean-up (44%); sharing food, water, or shelter (24%); and cash for work (13%). Community clean-ups were frequent events that took place after Haiyan, organized by barangay officials or local and international NGOs as part of their post-disaster programming. For example, the Taiwanese organization Tzu Chi was often cited as a source of inspiration as they frequently encouraged people to come together to clean up their communities. But the involvement and encouragement of NGOs and local government officials complicate the situation by confusing the incentives for community clean-ups. Did community members engage in community clean-ups because of the spirit of *bayanihan* and mutual help or did they get involved because they believe an NGO would pay them or give them other benefits as a result? The latter explanation is supported by the number of respondents who believed that "cash for work" was also a form of *bayanihan*. Cash for work is a program whereby locals are paid a certain amount of money, often a fixed daily rate, for specific post-disaster work. Cash-for-work programs have become very popular as ways for NGOs and governments to provide income in the form of cash to affected families or individuals in exchange for their labor, often toward a goal that would contribute to the public good. The common cash-for-work program after Haiyan was removing debris and clearing the roads. Yet, the paid nature of this collective action does not resemble the traditional conceptualization of the *bayanihan* as shared labor or mutual help—all of which were conceptualized as unpaid. Thus, the popularity of cash for work as a form of *bayanihan* among respondents demonstrates that certain aspects of the traditional conceptualization of *bayanihan* have been lost.

While understanding the common forms of *bayanihan* listed by respondents helps unpack their idea of *bayanihan*, an examination of the less common forms observed is even more interesting. For example, while stories of local men forming ad hoc security groups to protect the community was a common example of *bayanihan* shared by barangay captains and officials, saying that the volunteer barangay police, or *tanods*, were mobilized to keep communities safe at the night (Eadie and Su 2018), two of the total respondents in the first study have identified security as a form of *bayanihan* they witnessed after Haiyan. Similarly, while heroic stories of people helping others out with evacuation, search and rescue, and burying the dead were mentioned by barangay officials, they were barely cited as examples by respondents. One respondent has shared that his brother, a tall man who knew how to swim, saved hundreds of small children in the barangay as he swam through the storm surge to evacuate children to his concrete, two-story house. He was awarded a medal of honor by the barangay for his heroic act. Another respondent has shared that about 100 people had evacuated to her concrete home, but she also complained that they had eaten all of her food and stolen her furniture. But given the narratives of *bayanihan* and collective action in the face of challenges, one would assume that there would be many more stories of this nature. However, these stories of mutual help that fit the altruistic and positive description of *bayanihan* were the exception, not the norm.

Many respondents shared that they stayed in their homes and did not evacuate due to a fear of their house being looted or damaged by the typhoon. A study by Dalisay and De Guzaman (2016) on evacuation decisions after

Haiyan, found that men refused to leave their houses because the house and their possessions represented their life's work and sacrifices. The authors also found that "being family breadwinners, the men worked for these material possessions, and they were vulnerable because they would readily risk their lives to protect their possessions from being ravaged by typhoons" (Dalisay and De Guzman 2016, 709). When responding to the survey, a middle-class respondent shared a vivid story of when the storm surge came, he heard loud knocks on his door. There seemed to be many people outside, so he feared that his nice home with flat-screen TVs and other luxuries would be targeted for looting. He said,

> "*Nakabati ako hin aringasa ha gawas nga nahadlok ako bangn kami pangawatan. Nak ginbuhat kinuha ko an sundang ha kusina ngan kumadto ako ha purtahan panginano hino an nakadto ha luyo.*"
>
> ["I heard many voices outside and I was scared they were going to rob us. So, I grabbed the biggest knife in the house and I went to the door to see who was on the other side."] (Benben)

In his re-enactment of the events, he showed us how he had the knife raised and he was ready to strike whoever tried to enter. When he opened the door, he saw some small children and women who were staying at a small hotel nearby which was flooded. They came to evacuate to his concrete home, not to loot or hurt him. This resonates with the earlier survey finding that the majority of respondents agreed that you "need to be very careful" regarding trusting others in the barangay.

Low trust in cities and inequality in relief distribution undermine bayanihan

The issue of trust or mistrust among community members is a theme considering that 5% of the second project's respondents have observed the *bayanihan* spirit in relief distribution-related activities—such as sharing information on when relief would be distributed, sharing relief goods, and physically helping to distribute the goods. One respondent has shared that they would "relay information or inform our neighbors that there is relief distribution." One would assume that given how simple it is to share this information and the spirit of *bayanihan* alive and well in the communities, many more respondents would have observed relief distribution activities as a form of *bayanihan*. Instead, the respondents have shared stories of neighbors withholding information about relief distribution from them and hoarding relief goods for themselves as common survival tactics. Several respondents also shared that their neighbors had gone as far as reporting them as middle-income households to NGOs in an attempt to exclude or disqualify them from receiving aid. As one female, middle-class respondent stated,

> "*Waray ako makakarawat bulig ha Red Cross kay nam sapit nagyakan han empleyado ha Red Cross na medyo may kaya ako, labi na an ak balay sementado, so diri ak dapat tagan hin bis ano na bulig. Napa-usa ako kay nagpinanbulig ak ha ira na akon mga sapit/amyaw, gin pakadi ko hira ha balay, ginpakaon ko pa.*"
>
> ["I did not receive any assistance from Red Cross because one of my neighbors had shared with their staff that because I am middle class and have a concrete house, I should not be eligible for assistance. I was very surprised because I had provided so much support to my neighbors during Haiyan, letting them evacuate to my home and eat my food."] (Evelyn)

The respondent shared her shock with us and added that not only did evacuees eat all of her food, they also stole her furniture as well as other goods in her home. She suspected that her neighbors envied her and did not have much sympathy for her as she did not suffer a significant loss due to typhoon Haiyan. Her story is a reminder that in times of crisis, one's kindness does not necessarily cover up or completely dissolve social tensions that may have existed in the community before the disaster.

In addition, several studies have noted that relief distribution after Haiyan was a source of conflict and community tension (IBON Foundation 2015; Ong, Flores, and Combinido 2015; Su and Mangada 2016; Eadie and Su 2018). Ong, Flores, and Combinido (2015) documented many incidences of 'relief envy,' whereby different households in the same heavily affected communities got different amounts and types of aid. Those who got less, such as middle-class households, were openly bitter about being excluded during the emergency phase because they felt equally vulnerable and impacted in those early days. Su and Mangada (2016) found that community members that had become friends, helping each other during the emergency phase, quickly resorted back to selfish attitudes and behaviors when relief operations started. Competition over relief goods led people to self-preservation and conflicts between community members had emerged. Eadie and Su (2018) have presented evidence that the haphazard and inequitable distribution of relief goods after Haiyan generated discontent within communities and that this inequality undermined community-level cooperation. Lastly, IBON Foundation (2015) observed that uneven distribution of relief within communities disrupted the community culture of *bayanihan*. They have cited that humanitarian relief targeting "created conflict and division among community members and has raised the issue that enduring solidarity within affected communities is just as important as need" (IBON Foundation 2015, 45). These same patterns of behavior were observed in the two projects and go to undermine the narrative that *bayanihan*, an expression of cooperation, was common after Haiyan.

Filipino brand of resilience silences criticisms of government and humanitarian failures

One of the main struggles faced by typhoon Haiyan survivors was the long time it took for any form of external aid to arrive. It took one week for this to happen, leaving survivors to fend for themselves (BBC 2013). The lack of aid led to one of the unique urban forms of *bayanihan*, which may be looting, or the act of stealing goods during a crisis. Two respondents have said looting is a form of *bayanihan*. Many others who have mentioned sharing food as a form of *bayanihan* stated that it was the sharing of looted food. A respondent has also shared that in addition to looting, he informed others about where they could go to loot as well. He said:

> "*Nagsusumat ha kada miembro han pamilya/kaurupdan kun diin may malolooting para makakuha makakaon/gamit.*"
>
> ["We inform family members/relatives where looting was possible or happening to get more food/things."] (Raul)

It is very interesting to consider why looting was considered a form of *bayanihan* as it is often seen as a criminal or negative act that goes against the positivity that surrounds the *bayanihan* spirit. Some scholars argue that looting is simply a necessity in the context of a disaster. Lukes (2005) highlights a television clip during the hurricane Katrina coverage whereby a looter was asked by the reporter "Why are you looting?" and the looter replied, "Can you see anyone to pay?" Curato and Ong (2015, 9) highlight that during a disaster, the familiar, everyday structures of social life break down, so deviant behaviors like looting cannot be understood through normative expectations constructed during normal times. While the respondents have shared positive stories of looting and the sharing of the looted goods, reporter Patricia Evangelista, who was on the ground immediately after Haiyan struck Tacloban, described very different conditions. This was how she described the looting situation immediately after Haiyan:

> They climbed the broken wall [of the warehouse]… and jumped down to fight it out with a hundred other thieves. Sardines, water, whatever they could carry. It was not the stealing that was dangerous. It was protecting what they had stolen. (Evangelista 2013, 123)

While there would have been more food than they could have carried in these warehouses, the protective nature of looters immediately after Haiyan shows the fear that inhabited the survivors. Evangelista's portrayal of the conditions on the ground is similar to that of journalist Anderson Cooper who described the situation immediately after Haiyan as a "miserable, miserable" one and that these are "very, very difficult conditions for the people here on the ground" (CNN 2013a). Their descriptions of the misery and difficulties on the ground highlight that, in times of disaster and crisis, some choose to conserve resources for themselves instead of sharing with others in need. This is contrary to the spirit of *bayanihan* described by Bernard Kerblat, who spoke of the survivors who shared the very little that they had, stating, "They open their doors, they don't have much, they tighten up their belts and they share their rice, they share their hospitality, they support each other. I mean, hats off. This is remarkable" (Quismundo 2013). While some survivors did show remarkable kindness to others, these stories should not cover up or displaced the other stories of selfish, protective, and sometimes violent behavior that takes place after disasters.

Similarly, crude accounts of human survival can be found in the many descriptions of the hostile conditions at evacuation centers. Dalisay and De Guzman (2016) noted that some respondents shared that the lack of privacy in evacuation centers led to concerns related to sexual harassment. Nguyen (2014) has described the conditions in evacuation centers as "often poor, with overcrowding, limited sanitation, and reports of crime and abuse." The United Nations Population Fund (UNPFA) estimates that 5,000 women were subjected to sexual violence in the month following Haiyan and the head of the Philippines anti-trafficking unit described the wake of Haiyan as a "feast for human traffickers" (Morales 2017). Dr. Babatunde Osotimehin, the UNPFA Executive Director, mentioned that "In the aftermath of natural disasters, chaos and a breakdown of law and order lead to increased risk of abuse" (Morales 2017).

Looting, fighting, and sexual violence are seen as common post-disaster experiences, thus hearing such stories in Tacloban City was not surprising. What is surprising is that positive headlines and praises of *bayanihan*, resilience, and heroism often trump stories of misery, exploitation, and failure of governments and humanitarian organizations to provide adequate aid. Some of these are the results of direct manipulation by government officials. The Centre for Media Freedom and Responsibility (CMFR), an NGO dedicated to press freedom, completed an analysis of the media coverage (print and broadcast) of Haiyan. In their analysis, CMFR (2013) found that after CNN's Anderson Cooper's reporting from Tacloban City said there was a lack of search and rescue missions, feeding centers, and organized relief, President Aquino requested that the Association of Philippine Broadcasters feature stories that "move(d) others to action" and "uplift(ed) spirits of the Filipino people." The President said news organizations should find stories of "resilience, hope, and faith, and show the world just how strong the Filipino people are" (CMFR 2013). The broadcasters complied by reporting uplifting stories focused on survivors and people's actions to help each other (IBON Foundation 2015). The pressure on media to show positive stories also made its way back to Anderson Cooper. Shortly after his coverage was aired, Korina Sanchez, a radio broadcaster and the wife of then-Interior Minister Secretary Manuel Roxas II who oversaw the relief efforts on the ground, accused Anderson Cooper of wrongly claiming that there was no government presence in Tacloban. In his response to Sanchez, the journalist responded with a clip entitled "In every report we have shown how strong the Filipino people are" in which he pays tribute to the strength of the people he met in the Philippines (CNN 2013b).

President Aquino and the government's intervention and involvement in the media coverage of Haiyan, his successful request to show uplifting stories, and Anderson Cooper's acknowledgment of Sanchez's criticism (CNN 2013b), demonstrate the strength of the Filipino brand of resilience. It shows how there are many pressures from politicians to uphold this brand and a psychological desire to reinforce the image of the strong Filipino people. Yet, the strong influence politicians have over the media coverage of typhoon Haiyan shows just how dangerous using success stories such as *bayanihan*, can be for silencing criticism of inadequate risk reduction and preparedness by national and local governments as well as slow response and recovery by humanitarian organizations.

Taclobanons feel less resilient after typhoon Haiyan

It is important to discuss how *bayanihan* and resilience have become synonymous with each other. As noted earlier, *bayanihan* is often seen as the source of Filipino resilience, referring back to the news articles and publications cited earlier (Christian Science Monitor 2013; GMA News Online 2013; Sevastopulo and Kerr 2013; Abenoja and Lacamiento 2014; Viray 2014). But interrogating the meaning and practice of resilience shows that it too was a concept that was often overstated and misused by media, NGOs, and governments. The findings from the surveys of both the studies found that the much-celebrated virtue of resilience comes from painful negative coping strategies involving high-interest debt, skipping meals, and living in substandard rebuilt shelters with no basic services. While NGOs were quick to publish glossy reports on resilience and Haiyan such as the Food and Agriculture Organization of the United Nations' *Portraits of Resilience* (FAO 2015) and the International Organization for Migration's *Portraits of Recovery* (IOM 2014), our findings reveal that behind the smiles on the covers are less camera-worthy stories of recovery. FAO (2015) states, their book is a "tribute to their [farmers and fishers'] resilience and our work together to build back better their agricultural livelihoods after suffering such devastating losses," while IOM's (2014) book is described as "a powerful advocacy tool celebrating the resilience of the affected communities." Both NGOs portrayed resilience as something to be celebrated, yet for our respondents, survival after the disaster does not necessarily elicit positive reactions or experiences. In the Philippines, a society known for low trust among strangers and limited state support, the urban poor of Tacloban have adapted by being extremely self-reliant. An intense self-reliance is a form of individual resilience borne out of necessity. It is not the same as the positive community resilience that is implied in the humanitarian and development mantras of 'building back better' or 'bouncing forward' (Su and Le De 2020). Irrespective of whether survivors are individually resilient, this reality cannot be extrapolated as a positive outcome of community support or government policies nor is 'being resilient' the sole responsibility of survivors. Focusing on the resilience of the people masks the need for the whole system to be resilient and particularly the need for the government to implement policies and invest in infrastructure that can prepare the city for future mega-disasters.

Our research shows that the affected populations developed coping mechanisms, as opposed to relying on their pre-disaster resiliency. This is showcased by comparing the responses of the survey questions examining respondents' views of their resilience and that of their community. When respondents were asked, "Would you describe yourself as disaster-resilient pre-Haiyan?" Of the 318 respondents, 81% said yes and 19% said no. Given the abundance of stories on how resilient Filipinos were in the face of Haiyan, one may assume that they would only feel more resilient after surviving Haiyan. Yet, that was not the case. When asked "Would you describe yourself as disaster-resilient post-Haiyan?", the percentage of people who said yes dropped to 69%, or a 12% decrease from pre-Haiyan. These figures appear to indicate that while the majority of respondents considered themselves to be resilient, this feeling was weaker after Haiyan, alluding to the reality of the many negative coping strategies that were employed for households to be able to undergo rapid self-recovery. Lastly, when asked "Do you think you or your community can deal with the problems caused by Haiyan?", 58% of the respondents said yes, and 42% said no. With just over half of the respondents confident in their own and their community's ability to deal with the problems caused by typhoon Haiyan, *bayanihan* and community resilience may not be effective shock absorbers for disasters. Overall, these findings shed light on the shortcomings of over-relying on community support in the face of disasters.

Conclusion

Comparing the news articles, NGO reports, and government and humanitarian statements with on-the-ground data from both our studies demonstrate that the prevalence and strength of the *bayanihan* spirit after Haiyan was overstated. Our findings show that despite its popular use after disasters by government and NGOs, calling on communities to evoke *bayanihan* is often an inadequate answer to the many problems that commonly exist in post-disaster environments. Indeed, our research found that *bayanihan* is no longer as widely practiced as it was once believed or as reports have stated. In addition, what *bayanihan* looks like in practice is drastically different as traditionally understood. Our findings show that after Haiyan, the most common forms of *bayanihan* observed, such as cash for work, often do not fit the definition of *bayanihan* in the traditional sense. Instead, for Haiyan-affected households, the activities that they considered as *bayanihan* were often organized by government officials such as community clean-ups or paid for by NGOs such as cash for work, as opposed to the organic, voluntary, and altruistic characteristics previously attributed to the spirit of *bayanihan*. Most importantly, our studies uncovered that the strong brand of Filipino resilience and the singular narrative of *bayanihan* often acted to silence criticisms of inadequate risk reduction and preparedness by governments as well as the slow response and recovery by humanitarian organizations that are consistent with other studies and reports (IBON Foundation 2015; Su and Tanyag 2019). Such narratives contribute to shifting the responsibility of relief and recovery from governments and humanitarian organizations with formal duties to the survivors themselves. Our analysis of varying definitions of *bayanihan* and resilience in the literature and the media show both are largely defined as characteristics that are innate, instinctive, and intrinsic to Filipinos. This characterization robs Filipinos of the ability to share their own stories and inhabit a diversity of experiences after disasters. Overall, we caution against the over-reliance of *bayanihan* as a source of post-disaster community resilience and the perpetuation of the myth that *bayanihan* is a safety net for Filipinos after crises.

Note

1 "We go the extra smile" is on the signage that greets foreigners and nationals alike when they arrive at the Ninoy Aquino International Airport.

References

Abenoja, N. and G. Lacamiento (2014) "♯YolandaPH: The Spirit of Bayanihan," *The Philippine Star*, November 5, www.philstar.com:8080/cebu-news/2014/11/05/1388309/yolandaph-spirit-bayanihan [Accessed January 20, 2018].

Adviento, M. L. G. and J. M de Guzman (2010) "Community resilience during typhoon Ondoy: The case of Anteneoville," *Philippine Journal of Psychology* 43(1): 101–113.

Advincula, A. (2013) "In Yolanda's wake, memories of Filipino resiliency," *Inquirer.Net*, November 21, https://globalnation.inquirer.net/91885/in-yolandas-wake-memories-of-filipino-resiliency [Accessed January 20, 2018].

Ang, G. R. (1979.) "The bayanihan spirit: Dead or alive?" *Philippine Quarterly of Culture and Society* 7(1/2): 91–93.

Bankoff, G. (2007) "Dangers of going it alone: Social capital and the origins of community resilience in the Philippines," *Continuity and Change* 22(2): 327–355.

BBC. (2013) "Typhoon Haiyan: Tacloban survivors wait for aid," November 15, https://www.bbc.com/news/world-asia-24960836 [Accessed March 7, 2017].

Christian Science Monitor. 2013. "A Philippines at the ready for typhoons," November 11, https://www.csmonitor.com/Commentary/the-monitors-view/2013/1111/A-Philippines-at-the-ready-for-typhoons [Accessed August 13, 2020].

CMFR (Centre for Media Freedom and Responsibility). (2013) "Reporting 'Yolanda': More needed from the media (1st of 2 parts)," http://www.cmfr-phil.org/2013/12/23/part-1-reporting-yolanda-more-needed-from-the-media [Accessed February 12, 2018].

CNN. (2013a) "Typhoon Haiyan: Six days later," November 13, http://ac360.blogs.cnn.com/2013/11/13/typhoon-haiyan-six-days-later/ [Accessed March 7, 2017].

CNN. (2013b) "In every report we have shown how strong the Filipino people are," November 14, http://ac360.blogs.cnn.com/2013/11/14/in-every-report-we-have-shown-how-strong-the-filipino-people-are/?hpt=ac_bn13 [Accessed January 20, 2018].

Cupin, B. (2013) "Aquino: Yolanda 'serious threat,' don't take chances," *Rappler*, November 7, http://www.rappler.com/nation/43154-aquino-statement-yolanda [Accessed March 7, 2017].

Curato, N. and J. Ong (2015) "Disasters can lift veils: Five issues for sociological disaster studies," *Philippines Sociological Review* 63: 1–25.

Dalisay, S. and M. De Guzman (2016) "Risk and culture: The case of typhoon Haiyan in the Philippines," *Disaster Prevention and Management* 25(5): 701–714.

Davis, I. and D. Alexander (2016) *Recovery from Disaster*, London: Routledge.

Eadie, P. and Y. Su (2018) "Post-disaster social capital: Trust, equity, bayanihan and typhoon Yolanda," *Disaster Prevention and Management: An International Journal* 27(3): 334–345.

Evangelista, P. (2013) "The land of the mourning," *Esquire Magazine*: 123–124.

FAO (Food and Agriculture Organization). (2015) "Typhoon Haiyan – portraits of resilience," http://www.fao.org/emergencies/resources/documents/resources-detail/en/c/357721/ [Accessed August 13, 2020].

Gawad Kalinga. (2013) "Breaking barriers through bayanihan: Typhoon Yolanda (Haiyan) update as of Nov19," http://gk1world.com/breaking-barriers-through-bayanihan [Accessed August 13, 2020].

GMA News Online. (2013) "The bayanihan spirit live on in the Kapuso community despite Yolanda," December 13, www.gmanetwork.com/news/cbb/content/339680/the-bayanihan-spirit-lives-on-in-the-kapuso-community-abroad-despite-yolanda/story/ [Accessed January 20, 2018].

Heydarian, R. (2013) "Philippines' Haiyan tragedy: What went wrong?" *HuffPost*, November 15, https://www.huffpost.com/entry/philippines-haiyan-typhoon-response_b_4283845 [Accessed January 20, 2018].

IBON Foundation. 2015. *Disaster upon Disaster: Lessons beyond Yolanda*, Quezon City, Philippines: IBON. https://www.medico.de/fileadmin/user_upload/media/en/Disaster_Upon_Disaster_Final.pdf [Accessed February 12, 2018].

IOM (International Organization for Migration). (2014) "Typhoon Haiyan: Portraits of recovery," https://publications.iom.int/books/typhoon-haiyan-portraits-recovery [Accessed March 7, 2017].

Lozada, B. (2013) "Preserve the spirit of bayanihan, UN urges Filipinos," *Inquirer.Net*, November 18, https://globalnation.inquirer.net/91517/preserve-the-spirit-of-bayanihan-un-urges-filipinos [Accessed January 5, 2020].

Lukes, S. (2005) "Questions about power: Lessons from the Louisiana hurricane," http://understandingkatrina.ssrc.org/Lukes/ [Accessed January 5, 2020].

Morales, H. R. (2017) "These photos show life for displaced typhoon victims forced into the Trade," *Washington Post*, April 18, https://www.washingtonpost.com/news/in-sight/wp/2017/04/28/these-photos-show-life-for-displaced-typhoon-victims-forced-into-the-sex-trade/ [Accessed February 12, 2018].

NDRRMC (National Disaster Risk Reduction and Management Council). (2014) "Final report re effects of typhoon 'Yolanda' (Haiyan)," http://www.ndrrmc.gov.ph/attachments/article/1329/FINAL_REPORT_re_Effects_of_Typhoon_YOLANDA_(HAIYAN)_06-09NOV2013.pdf [Accessed March 7, 2017].

Nguyen, K. (2014) "One hundred days since typhoon Haiyan," *Al Jazeera*, February 18, https://www.aljazeera.com/indepth/features/2014/02/one-hundred-days-since-typhoon-haiyan-20142188436327673.html [Accessed January 5, 2020].

OCHA (United Nations Office for the Coordination of Humanitarian Affairs). (2013) "Philippines: Typhoon Haiyan (Yolanda) storm surge (based on actual typhoon track) as of 12 Nov 2013," https://reliefweb.int/map/philippines/philippines-typhoon-haiyan-yolanda-storm-surge-basedactual-typhoon-track-12-nov [Accessed March 7, 2017].

Oliver-Smith, A. (1979) "The Yungay avalanche of 1970: Anthropological perspectives on disaster and social change," *Disasters* 3: 95–101.

Ong, J., J. M. Flores, and P. Combinido (2015) "Obligated to be grateful: How local communities experienced humanitarian actors in the Haiyan response," www.alnap.org/system/files/content/resource/files/main/obliged-to-be-grateful-report.pdf [Accessed February 12, 2018].

Quismundo, T. (2013) " 'Bayanihan' impresses UN official," *Philippine Daily Inquirer*, December 2, http://newsinfo.inquirer.net/538505/bayanihan-impresses-un-official [Accessed February 12, 2018].

Salvation Army Vision Network. (2014) "Bayanihan: The spirit of community," https://www.youtube.com/watch?v=ZOMxmh48Q0A [Accessed August 13, 2020].

Sarmenta, L. and S. Hirano (1999) "Bayanihan: Building and studying web-based volunteer systems using Java," *Future Generation Computer Systems* 15: 675–686.

Sevastopulo, D. and S. Kerr (2013) "Filipino diaspora rallies after typhoon Haiyan,". *Financial Times*, November 15, https://www.ft.com/content/aaa1f450-4db9-11e3-b15d-00144feabdc0 [Accessed January 20, 2018].

Su, Y. and L. Le De (2020) "Whose views matter in post-disaster recovery? A case study of "build back better" in Tacloban City after typhoon Haiyan," *International Journal of Disaster Risk Reduction* 51: 1–10.

Su, Y. and L. Mangada (2016) "Bayanihan after typhoon Haiyan: Are we romanticising an Indigenous coping strategy?" *Humanitarian Practice Network*, August 10, http://odihpn.org/resources/bayanihan-after-typhoon-haiyan-are-we-romanticising-an-indigenous-coping-strategy/ [Accessed March 7, 2017].

Su, Y. and L. Mangada (2017) "Surviving survival: The limits of bayanihan and disaster solidarity after typhoon Haiyan in Tacloban City," in A. Opdyke, A. Javernick-Will, and M. Koschmann (eds) *Typhoon Haiyan: Shelter Case Studies*, Boulder, CO: Mortenson Center in Engineering for Developing Communities, University of Colorado, pp 104–107.

Su, Y., L. Mangada, and J. Turalba (2018) "Happy-washing: How a 'happiness campaign' hurts disaster survivors," *New Mandala: New Perspectives on Southeast Asia*, April 27, http://www.newmandala.org/happy-washing-happiness-campaign-hurts-disaster-survivors/ [Accessed January 5, 2020].

Su, Y. and M. Tanyag (2019) "Globalising myths of survival: Post-disaster households after typhoon Haiyan," *Gender, Place & Culture* 20(11): 1513–1535. DOI: 10.1080/0966369X.2019.1635997

Viray, P. (2014) "Roving round: 'Bayanihan' spirit in Yolanda-hit communities," *The Philippine Star*, November 7, www.philstar.com/headlines/2014/11/07/1389140/roving-around-bayanihan-spirit-yolanda-hit-communities [Accessed January 20, 2021].

White House. (2013) "Statement by the president on super typhoon Haiyan/Yolanda," Office of the Press Secretary, November 10, https://obamawhitehouse.archives.gov/the-press-office/2013/11/10/statement-president-super-typhoon-haiyanyolanda [Accessed March 7, 2017].

14

Increasing Resilience in Communities Affected by Typhoon Haiyan: World Renew's Response in the Philippines

Grace Wiebe

Introduction

Typhoon Haiyan (locally known as Yolanda) hit the Visayas regions of the Philippines on November 8, 2013. With wind speeds peaking at 315kph, the storm remains one of the strongest and most destructive typhoons ever to make landfall (Santos 2013). While the international community responded quickly to typhoon Haiyan's devastation, many nongovernmental organizations (NGOs) were present until December 2014, for the emergency response and initial recovery phases. By June 2015, most humanitarian organizations had departed or ended their activities, thus reinforcing a notable gap between short-term humanitarian responses to disasters and the long-term development needs of affected communities. In their 2016 study of resilience and disaster trends in the Philippines, Alcayna et al (2016, 5) found that the Philippines had received "less than half of the $788m needed for recovery" six months after typhoon Haiyan struck the region.

World Renew's (WR) initial funds allowed it to respond in the first two years to assist over 12,000 families, followed by additional funding from Global Affairs Canada (GAC) in 2015 for the 46-month development project. It was a consortium project with Adventist Development and Relief Agency (ADRA) Canada.[1] The final project, named "Restoring, Empowering and Protecting Livelihoods" (REAP), aimed to reduce vulnerability and increase community resilience to mitigate the impacts of future hazards that might be exacerbated by the effects of climate change.[2] ADRA and WR each had their target villages to work in.

This chapter describes how WR Canada's Senior Program Manager and its two Philippines Program Managers worked across various phases of disaster response to build community resilience, moving communities toward the achievement of several Sustainable Development Goals.[3] The behavioral challenges that WR encountered and overcame as communities transitioned through each phase of recovery are described. Examples illustrate sustainable development goals (SDGs) on how gender equality and participation of women promoted women's empowerment (SDG 5); stronger engagement with local stakeholders along with village savings contributed to more sustainable economic growth and empowerment (SDG 8); climate adaptive agricultural activities helped combat the effects of climate change (SDG 13); environmental protection impacted the conservation of oceans and marine resources (SDG 14); and soil biodiversity (SDG 15).

This chapter presents a background discussion about WR and its work in the Philippines as presented in five sections. The first, "Engagement with local stakeholders to build resilience," describes the linkages of various stakeholders at all stages of post-disaster recovery that ensured efficiency and effectiveness, as well as alignment with the Philippine government's approach to disaster recovery and resilience-building. The second section, "Behavioral change across phases of disaster recovery," highlights adopting the core humanitarian standard (CHS)[4] on quality and accountability, which places communities and people affected by crisis at the center of humanitarian action. WR emphasized participation and ownership in the design and implementation of the program, encouraging beneficiaries to become partners in project activities. This necessitated a paradigm shift in the behaviur of affected communities. The third section, "Environmental protection and climate adaptation," presents the environmental protection practices that were undertaken to prevent further environmental degradation and to protect key resources essential to household income and the sustainability of their livelihoods. The fourth section, "Value chain and economic empowerment," focuses on WR's project interventions that replaced productive assets and helped community members move further along the value chain to increase their economic recovery and become more resilient. Meanwhile, the final section, "Gender and women's empowerment to build resilience," shows gender integration was a key factor in promoting long-term resilience through project interventions programming that contributed to additional income for not only men but also for a larger number of women.

Background of WR: Work in the Philippines

WR is an international faith-based organization founded in 1962 by the Christian Reformed Church of North America. It works through local partners, in-country church networks, other international Christian organizations, or through its own country offices to manage and lead humanitarian and community development programming. Its primary offices are located in Burlington, Ontario, Canada, and in Grand Rapids, Michigan, USA. It has also 17 other country offices worldwide. In the 2020 fiscal year, WR reached 1.4 million people in 35 countries.

WR has responded to various disasters in the Philippines since 1964, starting with floods in Pulupandan, Negros Occidental. In 1970, WR sent its first staff who started the operational presence and retains its nongovernmental organization (NGO) registration to this day. Although most of WR's disaster responses have focused on typhoons, including extreme flooding, it has also responded to earthquakes and volcanic eruptions. WR's assistance for typhoon Haiyan was the largest disaster response in the organization's history in the Philippines (deRoo Vanderburg 2009). The response from November 8, 2013, until June 30, 2019, reached over 110,000 individuals in the islands of Leyte, Samar, and Panay, with projects totaling US$8,799,677 or Can$11,428,000 (Wiebe and VanVeen 2018; World Renew Annual Report 2018; Global Affairs Canada 2020).

WR's strategy to disaster response

WR's approach to disaster response is holistic. Rather than programming all its funding in the immediate aftermath of the disaster to reach the greatest number of people possible, it focuses on particular communities to walk with them through various phases of disaster management and to contribute to the resilience building of disaster-affected communities. Figure 14.1 illustrates WR's holistic approach that was applied in the Philippines for the typhoon Haiyan disaster response.

WR provides disaster management assistance through as many phases as possible, depending on available funding. The duration of each phase generally falls into the pattern of months with the actual dates of WR's response to Haiyan listed in Table 14.1.

During the emergency response phase, WR provided US$85,000 to its local partner, the Christian Reformed Church of the Philippines – Disaster Response Team (CRCP-DRT) for the distribution of relief items. WR then established offices in all four municipalities, to continue the post-disaster response directly and hired some of the CRCP-DRT volunteers. Over the next three phases of recovery, rehabilitation, and reconstruction, WR mounted US$6.7 million of projects. In 2015, Global Affairs Canada (2020)

Figure 14.1: The five phases of World Renew's response to an international disaster and community resilience building (Wiebe 2018)

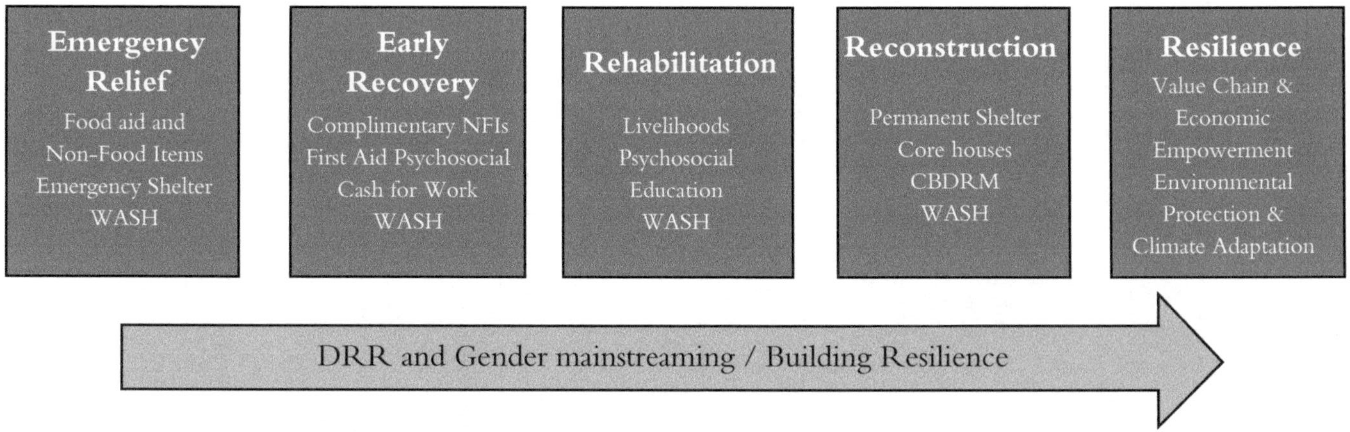

Note: WASH: Water Supply, Sanitation and Hygiene / NFI: Non-Food Items / CBDRM: Community-Based Disaster Risk Management.

Table 14.1: Pattern of months with the actual dates of World Renew's response to Haiyan

Phase	Duration	Dates
Emergency **R**elief	0–3 months	November 13–December 31, 2013
Early **R**ecovery	3–6 months	Starting January 1, 2014
Rehabilitation	6–12 months	April 1, 2014–March 31, 2015
Reconstruction	12–24 months	March 2014–June 2015
Resilience Building	24–72 months	September 2015–June 2019

provided the additional US$1.85 million that allowed WR to continue into the development (resilience building) phase (see Figure 14.2).

Methodology

All WR disaster projects were informed by surveys, which gathered both quantitative and qualitative data during a baseline assessment with results compared at midline and endline. Focus group discussions, key informant interviews, and secondary data analysis helped to triangulate survey findings and ensured a wide range of collected data.

Inclusion

Gender and inclusion were key focal points in WR's project interventions. Each project activity was designed to be sensitive to the needs, circumstances, and unique vulnerabilities of men, women, boys, and girls, the elderly, and people living with disabilities. At all stages of the response, the WR team engaged both men and women. The WR project staff facilitated separate consultations with women and adolescent girls and ensured that their expressed needs, priorities, and concerns were heard. WR also ensured community meetings could facilitate consultation based on the most culturally appropriate way.

Informed consent

WR ensured that permission and informed consent was always obtained when gathering photos, stories from project participants, or quotations from other stakeholders—including local politicians and community representatives.[5] Informed consent included a clear explanation that WR would share their stories and pictures with those who have supported the project.[6]

Figure 14.2: Linking relief, recovery

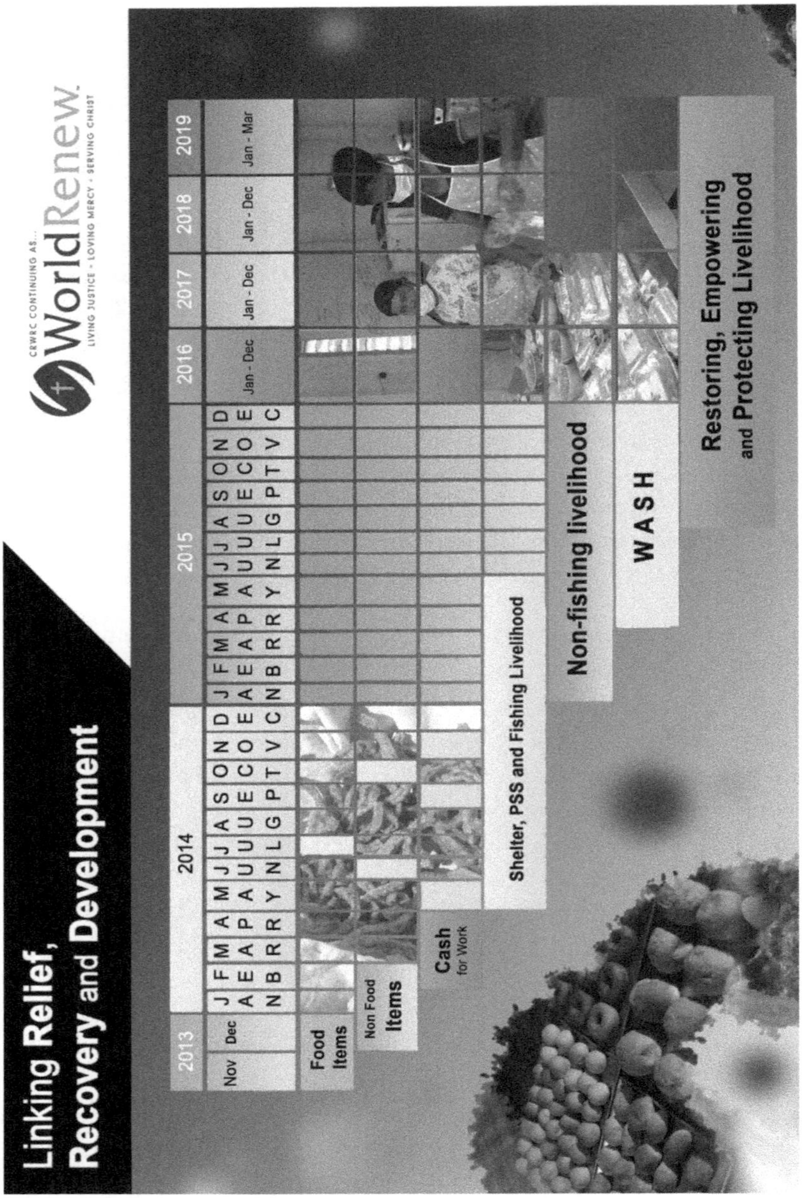

Engagement with local stakeholders to build resilience

Two years before typhoon Haiyan struck the central Philippines, a research team led by Father Norberto Carcellar published a case study on community-level disaster risk reduction (DRR) from Iloilo City, the capital of the province on the island of Panay where WR's Resilience project took place. The study concluded that community-level DRR required "strong social networks, alternative finance facilities and technical professional networks that support community processes and community-managed information systems" (Carcellar et al 2011, quoted in Alcayna et al 2016, 4).

The REAP project was oriented to longer-term development needs and focused on fostering improvements in many of the areas that Father Carcellar's team found to be essential in reducing community vulnerability to the effects of disasters. Stronger linkages to various stakeholders including local governments, academia, and other NGOs at all phases of the Haiyan response helped develop the capacity of community-based organizations (CBOs) and improved community access to technical support, savings, and loans.

WR aligned its plans with the Philippines Disaster Risk Reduction and Management Act of 2010, by promoting "a multi-sectoral and community-based approach [to DRR], and empowering sub-national governments and civil society actors as partners in DRR" (Harkey 2014, 68–69). Its activities supported the Strategic Response Plan of the Republic of the Philippines, which covered the initial 12-month period after typhoon Haiyan made landfall (Philippines Humanitarian Country Team 2013). Aimed to contribute to SDG 8, which is "To promote sustained, inclusive and sustainable economic growth,"[7] the livelihood activities implemented in the final resilience-building phase interfaced with the municipal economic development planning efforts that complemented the five-year National Development Plan (Republic of the Philippines National Economic and Development Authority 2017). The following subsections show the details of WR's engagement with local stakeholders in resilience building in every phase of its disaster response.

Emergency response phase

WR's long-time local partner was the CRCP-DRT, which has six Regional Disaster Response Teams across its over 50 churches and thus had the presence to respond in four affected municipalities of Tacloban (Leyte), Guian (Samar), Concepcion and Estancia (Panay). WR consulted with the local municipal and barangay officials in the planning and design of the disaster response and oriented them on the activities. Its coordination with the United Nations Office for the Coordination of Humanitarian Affairs (UN OCHA) cluster system, where other NGOs are registered, prevented any overlapping of response interventions.

Early recovery phase

The Philippines Children's Ministries Network (PCMN) provided qualified psychotherapists to conduct the psychosocial support sessions in WR target barangays during the early recovery phase of its Haiyan disaster response.[8]

Rehabilitation phase

WR registered its project plans in the Bureau of Farming and Aquatic Resources (BFAR) Office No. VI in Iloilo, Panay, to find out what fishing inputs the government or other NGOs had already provided to avoid duplication. It was found that BFAR had distributed boats but not motors or nets. Thus, WR covered this gap. The BFAR also helped in identifying potential target recipients who were registered as fisherfolks before typhoon Haiyan, as some people might claim to have had a boat and would incorrectly receive the assistance that was not intended for them.

Reconstruction phase

The Department of Social Welfare and Development (DSWD) provided WR with information on the number of houses destroyed and/or damaged per barangay. However, poor government records and the destruction of census documents during the typhoon created confusion. A professor of architecture from Capiz State University (CSU) helped to design four models of typhoon-resistant permanent houses. WR vetted these designs with another stakeholder, the Engineering Ministries International (EMI), who reported their shelter assessment that WR designs would withstand up to 200kph winds. Technical changes were made to improve the structural integrity of the design, which increased the cost of each unit. The CSU professor then trained the WR engineering staff and local carpenters on new typhoon-resistant construction techniques. An architect in Tacloban City designed a similar house model which was also vetted by EMI that could resist higher winds of up to 250kph. Typically, typhoon winds coming across the open Pacific Ocean are the strongest when hitting the Eastern Visayas before dissipating as they travel westward across Western Visayas.

Resilience (building) phase

Collaboration with municipal government

Effective partnerships with municipal offices were critical to the REAP project's success. Strong alignment between the project's goals and those of policy initiatives like 'Zero Poverty 2020' which started in 1999 by the Municipality of Concepcion, Iloilo (Inclusive Cities Observatory 2010) were crucial to securing buy-in from mayors. In turn, their support facilitated WR's enlistment of municipal extension workers for farmers and fisherfolks training, such as the Department of Trade and Industry (DTI), the Provincial Planning and Development Office – Local Economic Development and Investment Promotion Center (PPDO – LEDIP Center), and the Department of Tourism (DOT).

WR had critical success reviving the Municipal Fishery and Aquatic Resource Management Council (MFARMC) and the Municipal Agriculture and Fishery Council (MAFC). It engaged with the University of the Philippines Visayas (UPV) Iloilo Campus School of Technology, to assist in developing or further enhancing the local economic development plans of Municipal Local Government Units (MLGUs)-partners. The UPV shared excellent documentation on the profiles of poor provinces and municipalities that became productive and self-reliant through local enterprise development such as resorts, farm and marine-based products, and culture-based tourism.

WR advocated for the inclusion or incorporation of risk and resiliency measures as part of the MLGUs-partners' ordinances and local economic development plans. It also assisted them in the development of new ordinances that would better protect the livelihoods of vulnerable community members. For example:

- Concepcion Municipal Ordinance No. 1, Series of 2008 on the Sustainable Management, Conservation and Development of Coastal and Fishery Resources of the Municipality of Concepcion, Iloilo.
- Estancia Municipal Ordinance No. 2004–12, Series of 2012 on the Sustainable Management Conservation and Development of Coastal and Fishery Resources of the Municipality of Estancia, of Iloilo.

A participatory planning workshop was conducted in each municipality where CBO leaders at the barangay level were encouraged to draw their visions, identify, and analyze their strengths, weaknesses, opportunities, and threats, and formulate work plans for the next three years to help them move more toward forming into cooperatives.

Linkages with academia

The REAP project also forged valuable connections with professors and consultants from the Central Philippine University (CPU), the UPV Iloilo Campus of Technology, and the Northern Iloilo Polytechnic State College (NIPSC), who provided valuable technical training to villagers on livestock raising, seaweed, and oyster production. The professors also coached members

Figure 14.3: Local stakeholders

Note: First row, from left to right: Hon. Agustin 'Jun Bee' Garilva Jr.—SB Member Concepcion, Rhett Zerrudo—DA Consultant, Miss Fe Seta Encantro—West Visayas State University, Ms. Emeliza Lozada—Dean, School of Technology, University of the Philippines Visayas Middle row: Janvee Amido—Founder Youth First Initiative Philippines, Inc., Miss Nan Cho Cho Thel—Project Manager (ADRA), Miss Luna Bela-ong—Development Management Officer II, PPDO, Atty. Hansel O. Didulo—Assistant Secretary for the Regulations and the Visayas, DA, Mayor Milliard S. Villanueva—Concepcion, Mr. Jerry Requillo—Program Manager (ADRA) Upper row: Engr. Jeff F. Cosico—Program Manager (World Renew), Roy Charles Amacio—Ajuy LGU representative, Engr. Edmundo Q. Medoza Jr. —Municipal Agriculturist—Estancia, Mr. Reynaldo Ambao—Retired Director (BDS Department—TSKI), Dr. Erwin Ilaya—Provincial Fishery Officer—BFAR—Iloilo, Yulee Adan Salvanita—Labor Officer (DOLE). (Photo credit: courtesy of World Renew, Municipal gymnasium of Concepcion, Iloilo, July 2017).

of small and medium-sized businesses on meat and fish product development, labeling, packaging, hygiene and quality assurance in food handling, and market linkages. The Fishery Department of NIPSC also assisted WR in attempting to revive the Municipal Farming Committee of Estancia.

Sustained interaction with many local stakeholders from both the municipal offices and academia (see Figure 14.3) was critical to the success of WR's Resilience Building phase. These stakeholders have committed to continue their linkages with the community members in strengthening their economic development after the exit of WR.

Partnerships with other NGOs

WR partnered with ADRA Canada for further strengthening of the disaster management plans and early warning systems, as well as the implementation of the market information system among WR-supported barangays.

World Vision trained WR in its Community Managed Savings and Credit Association (CoMSCA) method of village savings.[9] WR Philippines became a member of the Philippines CoMSCA Network and attended two national CoMSCA conferences in Cebu. WR consulted with Magsasaka at Siyentipiko para sa Pag-unlad ng Agrikultura (MASIPAG 2013),[10] which is a farmer-led network of people's organizations (POs), NGOs, and scientists researching agricultural methods.

The Southeast Asian Fisheries Development Centre (SEAFDEC n.d.) is an NGO partner with several departments and expertise in marine fisheries and aquaculture that was tapped by WR to train its target fisherfolk.[11] Likewise, WR collaborated with the Panay office of the Zoological Society of London (ZSL),[12] which is devoted to the worldwide conservation of animals and their habitats.

Behavioral change across phases of disaster recovery

In many early disaster response programs, community members are in shock and typically receive assistance with little or no expectation for personal contribution or involvement in the activities. As WR aimed to create long-lasting, sustainable change, a paradigm shift in the behavior of community members was necessary. To create project ownership to sustain WR's post-disaster recovery interventions, it was crucial that community contribution or counterpart and participation were emphasized and integrated into the design. With WR's REAP project, the beneficiaries would no longer receive unconditional inputs but rather, would become partners in the program and take more ownership for efforts to be sustainable. This approach conforms to WR's adoption of the CHS, which places communities and people affected by crisis at the center of humanitarian action.

Emergency response phase (0–3 months post-disaster)

The priority approach in emergencies is to provide food and non-food items unconditionally, with limited or no expectations of affected populations' involvement or participation, to quickly help them meet their basic needs. As such, in the first six weeks following typhoon Haiyan, WR responded with the help of volunteers from its local partner, the CRCP-DRT, to provide two weeks of food assistance and non-food items to 4,000 households, as well as 12 water filtration systems and generators. However, in line with the CHS Commitment 4, which emphasizes the involvement of beneficiaries in the design of humanitarian responses, the CRCP-DRT conducted a rapid assessment 10 days after typhoon Haiyan that involved the affected population in target communities, which resulted in adjusting the traditional food assistance package to include mung (monggo) beans and medicines, and to increase the amount of rice. Listening to beneficiaries' needs showed respect and it laid the foundation for the next phases of working with the affected communities and fostered a sense of trust.

Early recovery phase (3–6 months post-disaster)

As available funding increased and the original church volunteers were not available full-time to lead a longer-term response, WR established four offices in the locations where relief items were distributed. During this period, some international humanitarian actors such as WFP[13] and UNICEF[14] had started giving unconditional cash transfers to families, a modality that has been increasingly recognized as preferable to distributing pre-identified goods, as it puts purchasing power into the hands of beneficiaries themselves, allowing them to decide on goods according to their own needs. However, being unconditional still inculcated a dole-out mentality and risked fostering a behavior of aid dependency, similar to those dependent on the social welfare system from the Department of Labor and Employment (DOLE). Though affected people were very thankful, being mere recipients of relief goods or cash usually decided by NGOs referred to characteristically as the 'good Samaritans,' but such an approach did not respect their dignity if they did not participate in the design of the response.

To address this concern, WR expected more participation of their target partner-beneficiaries. Between January 1 and March 31, 2014, there were 4,155 families involved in cash-for-work activities in projects they identified, such as clearing land to start kitchen gardens to grow vegetables and repairing walkways for safety. In line with WR's commitment to inclusion, all necessary considerations were given to ensure that vulnerable populations, including the elderly and people with disability, were not excluded from assistance because of the cash-for-work expectation.

Over 8,200 families received flashlights, solar lamps, and laundry washbasins, as well as cooking stoves, after being consulted as to which type of stove they were familiar with. Psychosocial support was started with the training of 40 community cadres on psychological first aid to recognize trauma in others and help them to process their thoughts and experiences surrounding the typhoon.

Rehabilitation phase (6–12 months post-disaster)

The goal of this third phase (Rehabilitation) was to reduce vulnerabilities by improving the economic security of fisherfolk and to improve the psychological health of families. With increased participation required of community village council leaders, group leaders, and 508 fisherfolk recipients, starting in April 2014, came increased ownership which makes livelihoods more sustainable. For example, previously existing weak or defunct fishing associations were re-established and strengthened; new oyster farmer groups were formed; existing CBOs were trained to have officers and develop their policies. The Fish Vendors CBO were taught to go to the suppliers as a group to secure better prices for fishing inputs by scale; while for those receiving boats, their contribution or counterpart was to paint the boat. Moreover, those who received piglets constructed their pigpens; and for sari-sari store beneficiaries, they attended basic entrepreneurial skills training before they received recapitalization.

As communities must recover from trauma to move forward in their recovery, qualified psychotherapists from the Philippines Children's Ministries Network were contracted to run group sessions for 3,072 participants, focused on the needs of adults, children, the elderly, and persons with disability (see Figure 14.4).

Subsequently, 182 more community members attended the 'train the trainer' sessions to become psychosocial support (PSS) cadres, including barangay leaders, teachers, police officers, and nursery school teachers. Two daycare staff in Barangay 6 of Guiuan, Samar, trained as PSS cadres, were able to conduct group therapy sessions for children using techniques including comic books and drawing. As WR exited from Leyte in June 2015, a follow-up reinforcement PSS training was conducted for existing PSS cadres along with the village leaders. As a result, one barangay (Camote) had included PSS training and activities in their next annual barangay council budget. This is a clear example of how the behavior change of community members translated into a greater level of ownership and commitment to sustain the PSS efforts to care for the emotional welfare of their community members.

Reconstruction phase

During WR's fourth phase, a total of 1,392 permanent typhoon-resistant shelters were constructed (see Figure 14.5). Surveys identified the number of carpenters in each of the target barangay that could be trained on new typhoon-resistant construction techniques, which improved the carpenter's marketable skills for future employment.

Shelter committees in each barangay comprised of both females and males helped to identify the first recipients as only 40 houses could be built at a time. Shelter beneficiaries contributed their unskilled labor for clearing and excavation of construction sites, transporting materials, hauling water, and mixing concrete. Where possible, they provided coco-lumber, and some painted the corrugated galvanized iron roofing sheets with the paint provided. Counterpart or beneficiary participation encouraged ownership. Thus, they

Figure 14.4: Children participate in child-focused art therapy sessions

Source: Photo courtesy of World Renew, July 2014, Panay.

Figure 14.5: Shelter

Source: Photo courtesy of World Renew, October 2014, Panay.

would not consider dismantling the house and selling the materials provided as they too had contributed to the solution.

Resilience (building) phase (24–72 months post-disaster)

The final phase of WR's typhoon Haiyan post-disaster response required even more participation, which encountered some further resistance. However, WR continued to emphasize that the project belonged to the community and supported them on things that they could not provide themselves.

Overcoming resistance

Initially, there was resistance to the requirement to attend training or exposure visits before livelihood assets were replaced. WR kept reinforcing that free education was a blessing; after attending, many testified how much they learned from such project interventions. There was also resistance to making a contribution or providing a counterpart when this was not required in previous phases. Physical effort or sweat equity was encouraged or the contribution of local materials such as bamboo and wood where possible for the demo farms, pigpens, and so on. WR reminded them that if the project could save money, more families could be helped. The benefit of expecting local indigenous materials as a contribution to the production of organic fertilizers and pesticides created an understanding of local resources available.

CBOs

In the Resilience phase, there were still CBOs such as fisherfolk, *baskal* (boat) operators, seaweed or oyster culture farmers, or women's groups that were defunct, non-functioning, or non-existent that should be strengthened and reorganized first, so that the community spirit would be revived and enhanced. CBO leaders were led to rewriting governance policies and agreements, electing officers, taught how to facilitate meetings, keep good minutes, and basic financial management. CBO leaders assisted in identifying which members would be the first to receive livelihood inputs. Members were expected to participate in different committees like the model farm committee, to ensure that project activities remained on track.

Each CBO became registered with the DOLE, which gave them the legal identity that would link to different government agencies, academics, and local NGOs. These simple steps boosted the morale of the CBOs and helped the members to become active participants. The WR Philippines Program Manager who visited some communities in December 2019, six months after the end of the REAP project, said it was such a blessing to see that what was started by WR has been sustainable, as often interventions do not last. The model farms in Nino Barangay in the Municipality of Concepcion, Iloilo Province, were even expanded and enhanced from before.

Community-based savings

In barangays that did not have access to formal financial services, after explaining the benefits of saving, the community took strong ownership of the CoMSCA. It is a form of a locally owned bank where members can save and earn interest from those who borrow at a very low interest. To reduce their vulnerability to future disasters, CoMSCA also includes a DRR fund that can be drawn when needed. Similarly, some CoMSCA groups have established a sustainability fund as a buffer to cover if some members' payback is in arrears and the operational costs for group members' travel and participation in seminars or other unforeseen expenses.

Becoming CoMSCA agents was another level of ownership that some members took on, which expected them to start up new CoMSCA groups themselves. Thirteen CoMSCA agents (10 women and 3 men), were selected to join the National CoMSCA Summit on May 28–30, 2018, in Tagbilaran City, Bohol. The summit was aimed to strengthen CoMSCA agents, NGOs, and consortiums to be more effective. It provided a venue for participants to learn best practices and lessons learned from other implementers. At present, there are more than 8,500 CoMSCA groups and 11 development organizations that make up the CoMSCA network in the Philippines (CoMSCA 2020).

Environmental protection and climate adaptation

Many environmental protection practices were undertaken in the REAP project to prevent further environmental degradation and to protect key resources that are essential to household income and the resilience and sustainability of their livelihoods. The following discussions indicate the REAP project interventions for marine and agricultural protection that would contribute to environmental protection and climate adaptation.

Marine protection

In line with SDG 14, to "Conserve and sustainably use the oceans, seas and marine resources for sustainable development,"[15] WR addressed the challenges that fisherfolk still faced on the reduced volume of fish after typhoon Haiyan. Fisherfolk identified that they caught PHP400-worth of fish daily versus PHP1,000 before Haiyan. WR coordinated with the Bureau of Fisheries and Aquatic Resources Regional Office No. VI in Iloilo City, Panay, since they are responsible for the development, improvement, management, and conservation of fisheries and aquatic resources. WR also engaged the ZSL that helped the Municipalities of Concepcion and Estancia developed their own Coastal Resource Management Plans, which included the development of Marine Protected Area (MPA). To promote sustainable fishing livelihoods, the project also strengthened the Municipal Fisheries and Aquatic Resources Management Council (MFARMC) in its mapping and management of MPAs and enforcement of the fishery laws to prevent overfishing and destruction of coral reefs.

The ZSL also helped in the development of an ordinance for the MPA in the island barangays of Daculan, Loghingot, and Gogo in Estancia and the island barangays of Plandico, Loong, and Nipa in Concepcion. Then the WR REAP project conducted workshops in each barangay that focused on improving their localized barangay Coastal Habitat Management Plans and training on environmental protection and unfriendly fishing practices. According to a participant:

> "I am a fisherman for years, but it is only now that I understand that corals are not rocks but living things and an important part of fish habitat, that we need to protect them." (Eduardo Datan, 58 years old and the president of a 79-member fisherfolk association)

The community members are hopeful that the number of fish will increase and reach pre-Haiyan proportions as the ecosystem recovers over the next few years.

Agricultural protection

In line with SDG 15,[16] to "Protect, restore and promote sustainable use of terrestrial ecosystems, sustainably manage forests, combat desertification, and halt and reverse land degradation and halt biodiversity loss," several new environment-friendly agricultural practices were introduced such as organic farming, increasing soil biodiversity, and reducing soil erosion.

Organic fertilizer, fodder, and pesticide made from local indigenous materials were introduced to reduce dependence on harmful and expensive chemical fertilizers and pesticides. A workshop participant said:

> "As farmers, we usually expect to harvest more without minding the effect of commercial fertilizers in our land. World Renew taught us natural farming and the importance of being aware of the condition of our environment. We need to consider restoring our land for our children's future, thank you for waking us up through sustainable agriculture training." (Concepcion farmer attending WR training on natural farming, April 26, 2016)

Several men and women working on WR model farms were recognized by the Technical Education and Skills Development Authority (TESDA) for organic agriculture production and they received the National Certificates II (NC II) accreditation. They are being tapped to conduct training in other barangays as support to the Department of Agriculture's extension workers, thus generating new income for them.

To increase project partners' crop yields, vermiculture was introduced in model farms to increase soil biodiversity and the adoption of Sloping Agricultural Land Technology (SALT) to control soil erosion by utilizing nitrogen-fixing trees as soil binders and diversified farming using perennial crops grown in between the hedgerows. Sustainable Corn Production in Sloping Areas (SCoPSA) was also taught.

In line with SDG 13,[17] to "Take urgent action to combat climate change and its impacts," WR partnered with the Magsasaka at Siyentipiko para sa Pag-unlad ng Agrikultura (MASIPAG) to introduce 50 traditional rice varieties (TRV) planted in a demo farm in Concepcion to test their resilience on adverse agroclimatic conditions brought about by climate change. Seven of the varieties were identified as climate-adaptive or tolerant to flooding, 17 were drought-resistant, 19 had high saline tolerance, and 23 were resistant to pests or diseases.

Value chain and economic empowerment

Simply replacing lost inputs after a disaster does not guarantee that beneficiaries have enough household income to support their basic needs. In the Resilience [Building] phase, WR helped community members move further along the value chain, which improved their economic recovery. The scholarly article for *Earthscan*, "Markets and Rural Poverty: Upgrading in Value Chains," aided in our understanding of the rural development challenges in the Philippines (Mitchell and Coles 2011). The REAP project has implemented specific interventions in each of the three major entry points in the value chain such as production, processing, and marketing to strengthen beneficiary entrepreneurs (see Figure 14.6). In some cases, beneficiaries needed help right from the starting point of production, others required assistance in processing, while some only sought marketing support since they could manage production and processing well enough already.

Production

Technical training by the WR agronomists and various academic and government stakeholders provided new skills to increase production in various livelihoods. Model farms showcased how a small parcel of land could be more productive by having different year-round integrated

Figure 14.6: Intervention entry points along the value chain to build the capacity of entrepreneurs

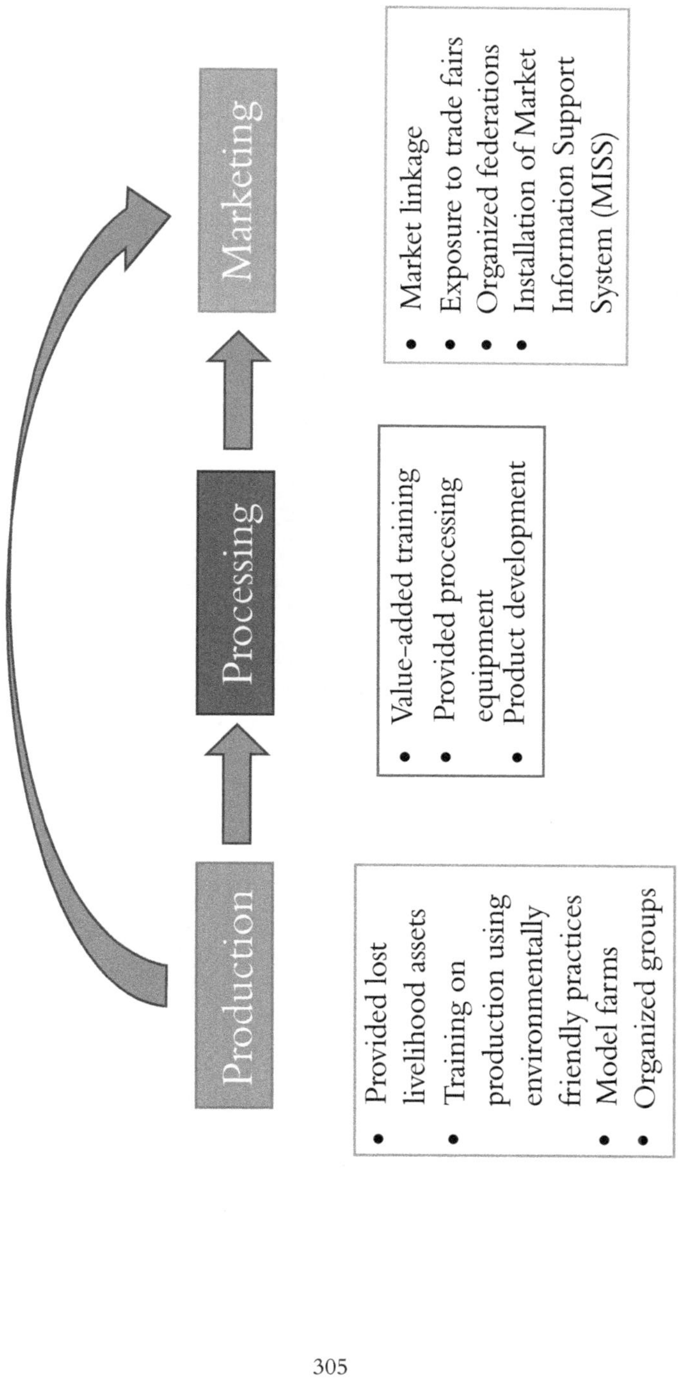

Note: This diagram was created by World Renew Philippines' Program Manager, Jeff Cosico.

farming components. Micro-Small-Medium Enterprises (MSMEs) within the CBOs received basic training in entrepreneurship, accounting, financial literacy, and business proposal writing before productive assets were replaced.

To further increase the household economy, training and productive assets were distributed to complement existing livelihoods or to start alternative livelihoods. For fisherfolk, these included weighing scales and fish processing equipment and tools while there were ovens and dough mixers for bakers. The REAP project also provided more economical access to organic feeds. A mill in the community was established rather than taking grain to a third-party miller. Other productive assets assistance included dryers for seaweed growers and tools for small engine repair shop owners/workers.

Processing

Through partnerships with government agencies and academics, CBOs attended a series of value-chain analysis workshops. Linkage with the Department of Science and Technology, the Department of Trade and Industry, and the experts from the UPV, the WR partner communities were assisted in the process of product development. They were taught that by processing raw products, they could gain higher incomes. For example, a simple banana can be processed into a banana loaf or banana crackers. Rather than dried seaweeds sold to a middleman at a very low price, now fishermen's wives process the seaweeds into pastries and pickles. Hog farmers sold livestock at a very low rate, typically before the yearly opening of classes for tuition fees, when oversupply caused prices to plummet; but now, processed meat products are sold throughout the year, earning higher revenues. In addition to training on product processing, participants also learned about standards on good manufacturing and hygiene practices. A brand name of Nanay's Best or Mother's Best was developed with the tagline "Mothers love cooking" to promote squid, fish, or meat processed products (see Figure 14.7).

Marketing

Not only did WR focus on production and processing, but also assisted several CBOs in marketing their products. They recognized that too many products in the market are often counterproductive. However, market-matching activities with restaurant owners and other buyers identified the gaps in the market for in-demand products. Thus, women and men in sewing businesses are making school uniforms, bags, curtains, bedsheets, pillowcases, and more. Likewise, women started baking *pandesal*, a favorite

Figure 14.7: Nanay's Best

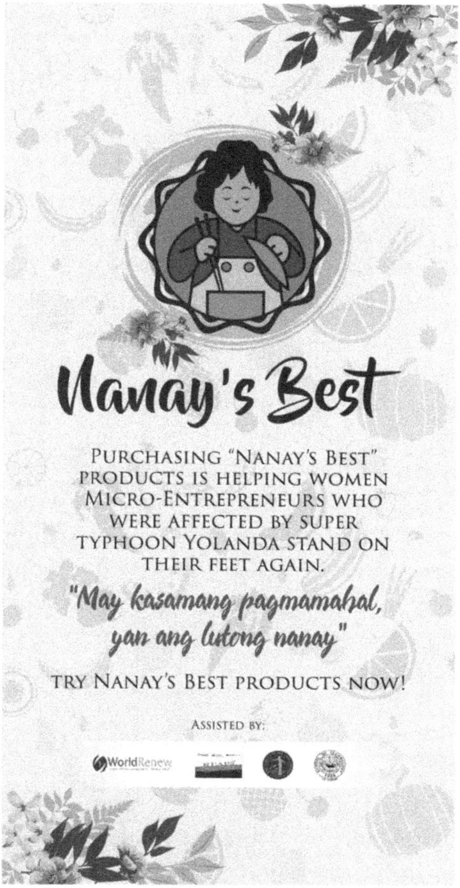

Source: Photo courtesy of World Renew REAP project in the Philippines.

Filipino breakfast bread. Orders for coconut pie have increased outside of their *purok* or village and have extended to the municipal town center. The WR project staff worked with the CBO members and exposed them to trade fairs and different markets in nearby cities to sell their products.

In line with SDG 8, WR also organized municipal federations, where each CBO is a member. The municipal federation of CBOs serves as a middleman, buying and selling members' products, not just in Iloilo province, but also in Negros and Manila. All profits of the federation are divided among its CBO members. To ensure that all assisted CBOs and families are updated on current market trends and prices, WR's consortium partner ADRA established a market information support system (MISS) in the community and municipal markets. The MISS shows what products are available in each barangay, the listing of prices, and who to contact about the market and its products.

Gender and women's empowerment to build resilience

Gender integration was a key factor in promoting the resilience of communities. In line with SDG 5,[18] to "Achieve gender equality and empower women and girls by 2030," the women supported through the REAP project have engaged in livelihood activities that were complementary to their husbands', or in alternative livelihood activities, which increased the household economy. The WR REAP project aligned its livelihood activities with the Philippine Government Women's Economic Development Gender Equality Plan (WEDGE) 2013–2016 (Philippine Commission on Women n.d.).[19] The "Commission on Women" report by the Iloilo Provincial Economic & Investment Promotion Officer helped guide the project with good gender integration. Women's Economic Empowerment (WEE) is described as "Women's access to and control over high-value productive resources and enjoyment of benefits from wealth-creating opportunities equal with men" (Department of Economic and Social Affairs 2009). The following subsections show how the REAP project implemented gender mainstreaming and empowered women toward building resilient communities.

Equal access to education

For women who had been unable to afford a college or university education, they felt proud to have accessed the academic experts in their remote barangays. The men and women of the REAP project participated in oyster production training from NIPSC and the training on seaweed farming from the Iloilo State College of Fisheries.

Equal access to income

Livestock and poultry are major assets for rural households, typically managed by women, and are increasingly regarded as a means of reducing poverty and/or obtaining alternative sources of income. As many rural areas have limited or no access to veterinary services, WR arranged ten local community members (eight females, two males) to be trained as para-veterinarians or Community-Based Animal Health Technicians for Poultry and Livestock by the Iloilo Assistant Provincial Veterinarian. This has generated income and the potential to impact on preventing livestock morbidity and mortality through treatment or prevention of diseases and animal health problems. The Conception Municipal Agriculture Office (MAO) recognized their qualifications and consequently hired them as municipal vaccinators to cover all 25 barangays for 17 days during an anti-rabies campaign and other awareness-raising on health management of livestock and plants. The MAO plans to provide additional training to further enhance its veterinary capabilities. According to a participant:

> "I studied four years in high school but was not able to get a certificate because I got pregnant a few months before our graduation. But, I am so happy that on my training for local veterinarian with World Renew that lasted for a few days, I got a certificate." (Sheila Jean Are)

Complementary livelihoods

The women (wives) were taught by a professor from NIPSC how to knot and sew new fishing nets, repair, mend, and patch torn nets, to support their husbands' fishing livelihoods. Traditionally a male practice, it provided women an alternative income for the household. With experts from UPV, the REAP project assisted two women's groups in Barangay Macalbang, Concepcion, and Barangay Gogo, Estancia, to receive training on meat processing. With their new skills, processing equipment, and tools, they are earning more from their meat-based processed products than just selling their livestock (pigs) to middlemen.

Alternative livelihoods

Before typhoon Haiyan, the Mat Weavers' Association members in Barangay Tagbak, Concepcion, wove mats to augment household income mainly from fishing. However, Haiyan destroyed the boats and fishing gear, houses, and even the pandan trees from which they used the leaves for their weaving. According to Cleofe Tasoy, the president of the Tagbak Village Weavers' Association:

> I am thankful to World Renew that we were allowed to be trained with new baking skills that can provide additional income for our family. They also provided us with productive assets like oven and dough mixer. We were also taught with the good manufacturing practices. (World Renew REAP Magazine June 2019)

At two o'clock in the morning, the new bakery starts its business operations when six mothers begin to prepare their bread and coconut pies (see Figure 14.8). Nowadays, their products reach nearby towns, and orders for their special coconut pies continue to increase.

Figure 14.8: Bakery

Source: Photo courtesy of World Renew REAP project in the Philippines.

The Tagbak Mat Weavers' Association is just one of the many women's groups that WR supported in becoming new entrepreneurs. They do not only rely on their husbands' seasonal income but together can have their independent income.

Feminist International Assistance Policy

Global Affairs Canada (GAC) was the major funder of WR's final REAP project in the resilience (building) phase. Midway through the 46-month project, the announcement of Canada's Feminist International Assistance Policy highlighted the need to further reinforce women's empowerment.[20] Then, WR adjusted project activities and helped women to have greater access and control over productive resources, and to enjoy benefits from income-generating opportunities on an equal footing with men. Gender was a foundational part of the REAP project that WR contributed to changes in collective decision-making of couples concerning their shared livelihoods. Thus, many men are more often found helping the household chores and women are more often engaged in productive activities outside the home.

Eunace Barbo, a project participant in Barangay Nino, Concepcion, Iloilo, is an example of women's empowerment as she is now a certified natural farmer earning an average of PHP1,500 per week from her farm. She is a role model in her community. She trains others on organic farming and advocates for integrated and diversified farming for a more sustainable income versus monocropping. Eunace shared with WR staff that:

> There are lots of changes now in our small village. Women were given individual livelihood projects. A six-woman group ventures into a banana processing as their enterprise. Another group is in ice cream production. We are now using our time in a more productive ways. One of the significant changes that I observed in our barangay is the women's participation in savings. We are thankful with this savings and credit program called CoMSCA. We can now save and borrow money without leaving our community. The best part of it is that all earnings will still be divided among us. (World Renew REAP Magazine June 2019)

Conclusion

Through a holistic approach of working across multiple phases of post-disaster response, WR successfully increased the resilience of its Haiyan-affected communities in the Philippines. It significantly improved the economic well-being of its target beneficiaries, reduced their vulnerability to future disasters, and moved communities toward achieving several SDGs.

One of the most significant aspects of transformation was the paradigm shift in the mindset and behavior as beneficiaries became partners in project implementation to sustain the REAP interventions. Conditional cash transfers helped restore community facilities while putting purchasing power into the hands of 4,155 families. Attending training before receiving inputs increased the knowledge of project beneficiaries. Their participation in shelter committees, use of unskilled labor in the reconstruction phase, and the counterpart contribution as part of their restored livelihoods have enhanced their ownership of the project. The REAP project restored and strengthened the CBOs, built their capacity, put communities in the lead, and linked them to government agencies, academics, and local NGOs.

Gender equality and women's empowerment through the REAP project contribute to SDG 5 as women became more engaged in livelihood activities that were complementary to their husbands' or alternative livelihoods, which increased the total household economy. Women had equal access to training and income-generating activities that are significant changes to the reality that many women faced prior to typhoon Haiyan. Stronger engagement with local stakeholders contributed to more sustainable economic growth that is aligned to SDG 8 and has allowed beneficiaries to move further along the value chain to increase their household economy. A total of 4,779 individuals generated income through the replacement of lost productive assets or the introduction of alternative livelihoods. All 12 model farms have an average monthly net income ranging from PHP500 to 3,500. All MSMEs within the 30 target CBOs developed marketing strategies and business plans, and 1,453 individuals were engaged in primary or secondary livelihood activities. Twenty-six MSMEs run at least one type of value-added agri-marketing activity, and 58% of livelihood participants, of whom more than half are women, have increased their income by at least 10%. Eleven ordinances were enacted by the municipalities to support and protect livelihoods. All CBOs have agreed to set aside at least 5% of the profit from their income for a disaster preparedness fund. The establishment of 72 CoMSCAs promoted economic empowerment which contributes to SDG 8 through access to savings and loans. Its 1,426 members contributed more than PHP14 million (US$280,000) in savings by the end of the REAP project. Proceeds from loans are utilized for business expansion, children's education, health-related needs, and acquisition of new assets, among others.

In line with SDG 13, the REAP project's identification of several traditional rice varieties that were most climate-adaptive or tolerant to flood, drought, saline soils, disease, or pests, has enabled the beneficiaries to prevent or reduce the effects of climate change on their crop production. Environmental protection considerations impacted the conservation of oceans and marine resources that conforms to SDG 14. The municipalities of Concepcion and Estancia developed their Coastal Resource Management Plans, which included the development of a Marine Protected Area. The barangays improved their more localized Coastal Habitat Management Plans.

Several new environment-friendly agricultural and aqua-cultural practices improved soil biodiversity as well as reverse land degradation, which contributes to SDG 15. Furthermore, several men and women received the National Certificate (NC II) in Organic Agricultural Production.

As a result of combining the disaster response humanitarian inputs with development practices in environmental protection and climate adaptation, WR's response has contributed to reduced vulnerabilities and increased protection of key resources that build a more resilient community. The increased income is aligned with the National Philippine Development Plan—2017–2022 goal of economic improvement.

During the REAP End of Project Stakeholders gathering in Iloilo in March 2019, Mayor Milliard Villanueva of Concepcion, affirmed that:

> [...] given the support and capacity building [that his community has received] to preserve the gains five years after Haiyan's devastation, especially on livelihoods programs, [he is] very positive that [they] can continue these programs despite limited resources. (World Renew REAP Magazine June 2019)

Members of the Iloilo Provincial Planning and Development Department who provided enterprise management training as part of the project's value chain work echo Mayor Villanueva's comments in the same meeting, that the REAP project, accordingly:

> [...] surely contributed to the process of local economic development in Northern Iloilo where poverty was an everyday test of strength and spirit. (World Renew REAP Magazine June 2019)

The success that WR had in increasing resilience and decreasing community vulnerability was largely credited to the additional funding provided by GAC through the final REAP project. With this funding, longer-term development-focused livelihood activities were implemented after the initial surge in post-typhoon humanitarian funding ended. The links and networks made with local stakeholders were a critical component for CBOs and communities to sustain the improvements after WR exited in mid-2019. We are hopeful that the experience and lessons learned by WR through the Haiyan program can be applied not just in other parts of the Philippines but also in other high disaster risk-prone countries that WR will respond to in the future.

Notes

1 ADRA is the official humanitarian agency of the Seventh-day Adventist Church. See https://www.adra.ca/.

2 Episcopal Relief and Development implemented relief programs in Tacloban in the Philippines that transitioned to a livelihoods' focus, including the development of community disaster preparedness plans (Delaney 2013). CARE also co-implemented a GAC-funded response to typhoon Haiyan, which transitioned from an initial emergency phase to successive recovery phases that included support for women's enterprises and value chain development (Center of Resources for Innovation and Development – University of San Agustin 2017, 6).

3 "The Sustainable Development Goals are the blueprint to achieve a better and more sustainable future for all. They address the global challenges we face, including poverty, inequality, climate change, environmental degradation, peace and justice" (United Nations 2021).

4 Core Humanitarian Standard on Quality and Accountability (CHS 2014) was officially launched on December 12, 2014, in Copenhagen which sets out common humanitarian standards that organizations and individuals can use to improve the quality and accountability of the assistance they provide. The three founding bodies behind CHS are the CHS Alliance, Group URD, and Sphere.

5 WR follows a strict approach to data protection and privacy, described in more detail in its Data Protection Policy, available at https://worldrenew.net/wp-content/uploads/2019/10/Programme-Data-Policy-June-2019.pdf

6 WR's Director of Disaster Response, Ken Kim, has consented to the submission of this chapter on its typhoon Haiyan disaster response in the Philippines.

7 See https://sdgs.un.org/goals/goal8.

8 PCMN is an agency/commission of the Philippine Council of Evangelical Churches.

9 See http://comsca.org/.

10 See https://masipag.org/about-masipag/.
11 See Southeast Asian Fisheries Development Centre, https://www.seafdec.org.ph/.
12 See https://www.zsl.org/.
13 See World Food Programme (2014).
14 See Reyes, Albert, and Reyes (2018).
15 See https://sdgs.un.org/goals/goal14.
16 See https://sdgs.un.org/goals/goal15.
17 See https://sdgs.un.org/goals/goal13.
18 See https://sdgs.un.org/goals/goal5.
19 WEDGE 2013–2016 reviews persistent and emerging gender issues in the Philippines, and guides agencies in mainstreaming the gender dimension in their regular programs, as well as in implementing gender-focused activities. See Philippine Commission on Women.
20 See Global Affairs Canada (2017).

References

Alcayna, T., V. Bollettino, P. Dy, and P. Vinck (2016) "Resilience and disaster trends in the Philippines: Opportunities for national and local capacity building," PLOS Currents Disasters. https://currents.plos.org/disasters/article/resilience-and-disaster-trends-in-the-philippines-opportunities-for-national-and-local-capacity-building/ [Accessed August 10, 2020].

Carcellar, N., J. Christopher, R. Co, and Z. O. Hipolito (2011) "Addressing disaster risk reduction through community-rooted interventions in the Philippines: Experience of the Homeless People's Federation of the Philippines," *Environment and Urbanization* 23(2): 365–381.

Center of Resources for Innovation and Development – University of San Agustin. (2017) "Typhoon Haiyan Response Program – Summative Evaluation Report". July. Iloilo City, Philippines.

CoMSCA (Community Managed Savings and Credit Association). (2020) "What is CoMSCA," http://comsca.org [Accessed August 10, 2020].

CHS (Core Humanitarian Standards). (2014) "Core Humanitarian Standards on Quality and Accountability," CHS Alliance, Group URD and the Sphere Project. https://corehumanitarianstandard.org/files/files/Core%20Humanitarian%20Standard%20-%20English.pdf [Accessed August 28, 2015].

Delaney, S. (2013) "Even Stronger Next Time – Comprehensive Programming in the Philippines Builds Resilience for the Future". *Episcopal Relief & Development*. https://www.episcopalrelief.org/stories/even-stronger-next-time-comprehensive-programing-in-the-philippines-builds-resilience-for-the-future/ [Accessed 16 December 2019].

Department of Economic and Social Affairs. (2009) *2009 World Survey on the Role of Women in Development: Women's Control over Economic Resources and Access to Financial Resources, including Microfinance*, New York: United Nations, https://www.un.org/womenwatch/daw/public/WorldSurvey2009.pdf [Accessed 10 August 2020).

deRoo Vanderburg, K. (2009) "Church at work: Answers to prayer in the Philippines," *The Banner*, July, https://www.calvin.edu/library/database/crcpi/fulltext/banner/2009-0700-0050.pdf [Accessed December 19, 2020].

Global Affairs Canada. (2017) *Canada's Feminist International Assistance Policy*, Ottawa, ON: Global Affairs Canada, https://www.international.gc.ca/world-monde/assets/pdfs/iap2-eng.pdf?_ga=2.130633487.1885491883.1645760517-906754262.1634508333 [Accessed October 24, 2017].

Global Affairs Canada. (2020) "Project Bowser: search World Renew projects". https://w05.international.gc.ca/projectbrowser-banqueprojets/filter-filtre [Accessed August 10, 2020].

Harkey, J. (2014) *Experiences of National Governments in Expanding Their Role in Humanitarian Preparedness and Response*, Somerville, MA: Feinstein International Center, Tufts University.

Inclusive Cities Observatory. (2010) "Zero poverty 2020: The case of Concepcion, Philippines," United Cities and Local Governments, https://www.uclg-cisdp.org/sites/default/files/Concepcion_2010_en_final.pdf [Accessed January 12, 2016].

MASIPAG. (2013) "Programs," http://masipag.org/programs/

Mitchell, J. and C. Coles (eds) (2011) *Markets and Rural Poverty: Upgrading in Value Chains*, London and Washington, DC: Earthscan.

Philippine Commission on Women. n.d. "Women's Empowerment, Development and Gender Equality Plan 2013–2016," Republic of the Philippines, https://pcw.gov.ph/womens-empowerment-development-and-gender-equality-plan-2013%E2%88%922016/

Philippines Humanitarian Country Team. (2013) "Typhoon Haiyan (Haiyan): Strategic response plan," Humanitarian Response, https://www.humanitarianresponse.info/en/operations/philippines/document/strategic-response-plan-philippines-haiyan-Haiyan-27-dec-2013 [Accessed November 1, 2014].

Republic of the Philippines – National Economic and Development Authority. (2017) *Philippine Development Plan 2017–2022*, Pasig City, Philippines, http://pdp.neda.gov.ph/ [Accessed February 15, 2018].

Santos, R. Jr. (2013) "Haiyan and the World's strongest storms," *Rappler*, November 10, https://www.rappler.com/science-nature/43351-world-strongest-cyclones-history [Accessed March 9, 2014].

SEAFDEC n.d. Southeast Asian Fisheries Development Centre. https://www.seafdec.org.ph [Accessed August 10, 2020].

United Nations. (2021) "The 17 Goals," Sustainable Development, Department of Economic and Social Affairs, https://sdgs.un.org/goals [Accessed December 1, 2015].

Wiebe, G. (2018) "World Renew's holistic approach. Building resilience – the five phases of WR response to an international disaster".

World Food Program (WFP). (2014) "Philippines: Super typhoon Haiyan," Philippines Situation Report #24, February 27, https://documents.wfp.org/stellent/groups/public/documents/ep/wfp263235.pdf [Accessed May 12, 2014].

World Renew. (2018) "World Renew 2018 ministry report," https://worldrenew.net/wp-content/uploads/2019/10/37653-WR-Annual-Report-2018-LO-RES-FINAL-for-web.pdf [Accessed December 19, 2019].

World Renew. (2019) "World Renew 2019 ministry report," https://worldrenew.net/annual-reports/2019-annual-report [Accessed August 10, 2020].

World Renew. June 2019 "REAP Magazine".

15

Religion in the Public Sphere in the Aftermath of Typhoon Haiyan

Maria Cecilia T. Medina

Introduction

One of the major challenges faced by several developing countries such as the Philippines, Bangladesh, and Pacific Island countries like Tonga and Vanuatu during the contemporary period is the perennial onslaught of cyclones, typhoons, or hurricanes which have grown in intensity and frequency of occurrence over the years due to climate change (Alliance Development Works 2016). In 2016, the Philippines ranked third among 145 countries as being most at risk due to climate change (Alliance Development Works 2016). These natural hazards have brought about tremendous destruction to lives and properties. The magnitude of the crisis spawned by natural calamities is heightened by the conditions of poverty and sociocultural and political divisions and issues (Wisner et al 2004; Gaillard 2007; Gaillard et al 2009). The most vulnerable to disasters are communities that are marginalized economically, geographically, and politically, often living in hazardous places and disregarded by those in power (Gaillard et al 2009). In addition, they usually have inadequate social protection and limited social capital (Gaillard et al 2009). Thus, disasters reflect the failure of development (Gaillard et al 2009).

On November 8, 2013, super typhoon Yolanda (Haiyan), with a speed of 315kph and powerful winds of 378kph, made its first landfall in Guian, Eastern Samar. It had a 48km width and brought about sea surges rising to 4m, with 70–80% of homes, buildings, and churches in its path demolished (Caritas-Philippines 2014, 5). It was dubbed the strongest tropical cyclone at landfall in recent history (Caritas-Philippines 2014, 5). It brought devastation to the central portion of the Philippine archipelago, the Visayas. Aside from the province of Samar, Haiyan brought havoc to Leyte province particularly its capital, Tacloban City, as well as the southern towns of Palo, Tanauan, and Ormoc City, and the islands of Cebu, Panay, and Palawan. United Nations officials estimated 11 million people were affected with many left homeless and at least 6,300 deaths as bodies were still being recovered months after the typhoon struck (Caritas-Philippines 2014, 5). Two years later, however, the number of fatalities in the archdiocese of Palo in Leyte province alone rose to 7,500 with many persons still missing (Reyes 2015). Others estimate the fatalities to have reached at least 10,000 persons in Tacloban City alone as its residents bore the brunt of the storm surge (IBON 2015). This was way above the official death count of 6,293 (IBON 2015). In addition, according to Caritas-Philippines, over 1.9 million homes were damaged and over 6 million persons displaced (2014, 5).

In the light of challenges brought by Haiyan in Leyte province, this chapter is based on a study guided by the following questions: What had been the role of the Catholic Church, its religious leaders, and organizations in the public sphere in the aftermath of super typhoon Haiyan? How did the prevailing socio-political context influence the role of religious leaders in the public sphere and their functions as part of civil society in response to the disaster? As a case study, various methods were utilized in collecting data. Primary data was gathered in May 2015, two years after the super typhoon struck. Key informant interviews were conducted with two priests working at the Catholic Bishops' Conference of the Philippines (CBCP) in Manila and the archbishop, four priests, a religious sister, a seminarian, and a lay parish coordinator, four parish lay leaders, and staff of the Relief and Rehabilitation Unit (RRU) from the Archdiocese of Palo in Leyte province. Key informant interviews were also conducted with a government officer in Tacloban City, two municipal mayors, four residents of Barangay Magay, Municipality of Tanauan, and focused group discussions among women in Barangay Liberty, Municipality of Mayorga and Barangay Capiz, Municipality of Alang-Alang all in Leyte province. Other key informants include a staff member of the Catholic Relief Services (CRS) and the editor of the *Samar-Leyte Daily Express*, a local newspaper in Tacloban City.

Secondary data was also gathered from local and national newspapers, websites of the Archdiocese of Palo and the CBCP, and other print and digital media outlets and online sources. In addition, I also witnessed the turnover of housing units in two barangays, communal gardens, and water systems installed in affected communities as I joined the RRU activities and observed a meeting/briefing of the United Nations Development Program (UNDP) and other partner NGOs of the RRU.

The chapter contains a section on the framework of analysis used in the study and background on the Catholic archdiocese of Palo in the province of Leyte. The various initiatives and networks of the religious and lay leaders and organizations of the Archdiocese of Palo during the emergency and rehabilitation phase were analyzed to draw implications for disaster relief, rehabilitation, and management.

Framework of the study

The overall framework of analysis utilized in this study is from Jose Casanova's religion in the public sphere with contributions from the field of religion and development, and concepts of social capital and resiliency. By weaving these various theoretical perspectives, this chapter analyzes the changing role of the Catholic Church leaders and organizations in the Archdiocese of Palo in Leyte province in the aftermath of Haiyan.

Religion in the public sphere

Haustein and Tomalin (2017, 84) mention the influence of Talcott Parsons's thought on the secularization theories that emerged in the 20th century. In particular, they pointed out that while Parsons "ascribed religion with central integration and legitimisation functions, these were impeded by processes of differentiation and pluralisation that necessarily followed with modernization" (Haustein and Tomalin 2017, 84). Parsons "saw religion retreating from the public sphere through privatization, leaving behind a moral community built around value-generalization" (Haustein and Tomalin 2017, 84; see also Parsons 1966). According to Haustein and Tomalin (2017, 84):

> This is at the core of the neo-classical theory of secularization, which was worked out in the following years by sociologists like Peter Berger, Thomas Luckmann, and Niklas Luhmann.

These earlier theories of modernity and secularization, especially in the West which relegated religion to people's private lives as society progressed have increasingly been debunked considering the growth of many religious movements and the public role of religion in contemporary times.

One of the most influential works on religion in the public sphere is Jose Casanova's book, *Public Religions in the Modern World*, wherein he described his central thesis using the term "deprivatization" about religion in the modern world (1994, 5). He explains:

> By deprivatization I mean the fact that religious traditions throughout the world are refusing to accept the marginal and privatized role which theories of modernity as well as theories of secularization had reserved for them. (Casanova 1994, 4)

He adds that it "is the process whereby religion abandons its assigned place in the private sphere and enters the undifferentiated public sphere of civil society to take part in the ongoing process of contestation, discursive legitimation, and redrawing of the boundaries" (Casanova 1994, 65–66). Casanova (1994, 4) cites the appearance of social movements which are religious in nature as evidence of this especially when they challenge "in the name of religion, the legitimacy and autonomy of the primary secular spheres, the state and the market economy." It is also evident when:

> religious institutions and organizations refuse to restrict themselves to the pastoral care of individual souls and continue to raise questions about the interconnections of private and public morality and to challenge the claims of the subsystems, particularly states and markets, to be exempt from normative considerations. (Casanova 1994, 4)

Casanova (1994, 66) notes how religions during the 1980s were at the forefront of "various forms of public collective actions" and became "both the subject and object of contestation and debate." In his view, religious elites also have a role in consolidating democratic regimes and public religion is to "defend the very right of a democratic civil society to exist" (Casanova 1996, 357).

In the case of the Philippines, the Catholic Church played an important role in the return and strengthening of democracy after the Marcos dictatorship (Moreno 2006). Buckley (2014) also points out how a variety of Public Catholicism in the Philippines has centered on the advocacy of combating corruption, monitoring of elections and protection of human dignity (2014, 321). He states that:

> A second approach to public Catholicism, comprehensive mobilization is rooted in a broad Catholic social agenda and pursues a diverse policy agenda coupled with universalistic rhetoric that instructs policymakers on the protection of human dignity. Catholicism here resolves the democratic dilemma by adopting a diverse advocacy agenda through both elite and grassroots organizations. This agenda solidified itself at the Second Vatican Council. (Buckley 2014, 321)

With this advocacy approach Catholic authorities "assume the vacant role of spokes (people) for humanity, for the sacred dignity of the human person, for world peace and for a fairer division of labour and power in the world system" (Buckley 2014, 321). Various local churches in the Philippines have also mobilized their members and the public in defense of the environment, Indigenous peoples' rights, and human rights (Moreno 2006; Claver 2009).

In the face of the massive destruction brought by super typhoon Haiyan to a wide geographical region in the Philippines, understanding the role of religious leaders and influential institutions like the Catholic Church in times of disaster, especially in the public sphere become very important. Religious leaders who survived the tragedy provided disaster relief as first responders in the local communities, opening their churches to serve as refuge for the displaced, and engaging various stakeholders in various phases of the rehabilitation process.

Religion and development

Religion and development have only emerged as a field of study recently, especially in academic circles in the West with the relegation of religion in the private sphere and the popularity of the secularization thesis. Development policies over the years have set aside religion as a factor in development giving more emphasis on the economic, political, or other social and cultural realms. According to Haustein and Tomalin (2017, 84):

> The inability of Western political theory, sociology, and theology to account for religion in the public sphere certainly impacted donors, development theory and programmes, but it should not be confused with the absence of religion in development thought and processes in the global South.

During the colonial period in Africa and South Asia "religions were invoked as an ally or even central ideological justification for the colonial 'civilizing' project" (Haustein and Tomalin 2017, 81). Religious institutions also became "key providers of the welfare services which functioned as crucial indicators of the 'civilizing' project providing health care, education, vocational training, as well as information and advocacy" (Haustein and Tomalin 2017, 81). Religions complemented the failures and needs of the colonial economy and took a similar structural position that NGOs occupy at present (Haustein and Tomalin 2017, 82; see also Manji and O'Coill, 2002).

Many times, other religions were marginalized or displaced by "a narrative of modernity and 'civilizing' religion" unless they managed to adapt their religious outlook to the modernizing project (Haustein and Tomalin 2017, 77). From the division of the world between colonizers and the colonized, a bipolar world emerged after the World War II, with the Cold War between the US and Russia. This now led to a relabeling of the world between the 'developed' and 'underdeveloped' countries with the main measure of inequality being the Gross Domestic Product (Houstein and Tomalin 2017, 82). Accordingly, "[d]ecolonization and the rise of the US as a major global power was marked by the emergence of an optimistic vision of progress for 'underdeveloped' countries imbued by notions of a modernity governed by science and rational macro-economic principles" (Haustein and Tomalin 2017, 77).

However, in the Global South, religion and religious processes have not disappeared and have been present. Haustein and Tomalin (2017, 84) explain that:

> Latin American Liberation Theology (Gutierrez 1974) and its many adaptations in Korean Minjung Theology, Dalit Theology, and South African Liberation Theology became major amplifiers of the anti-modernist Dependency Theory of development (Preston 1996, 179–195) and provided religiously inspired visions of development and social organization that ran counter to the narrative of the privatization of religion.

In the Philippines, as in many countries in Asia, the colonial experience under Spain and later the United States also contributed to defining the role of religion in society. It was during colonial times that Catholic and Protestant educational institutions and hospitals were founded. These institutions have continued to provide social and welfare services to the population even after decolonization and up until the present. Claver (2009) cites how Catholic social action was pioneered in the Philippines and Asia through the efforts of Fr. Walter Hogan, SJ, an American missionary, in founding the Institute of Social Order (ISO) in 1947 after World War II. Aside from promoting the social teachings of the Church, Fr. Hogan even went further in being actively involved in the creation of such groups like the Federation of Free Farmers (FFF) and Federation of Free Workers, in working for these sectors' welfare (Claver 2009, 71). After Vatican II, social action was further enhanced in the Philippine church with the involvement of the religious in rural congresses beginning in 1967 (Claver 2009, 71). At present, all the dioceses of the Philippine church have Social Action commissions and programs which at the national level is supported by the Catholic Bishops Conference of the Philippines' Episcopal Commission on Social Action, Justice and Peace also known as the National Secretariat for Social Action (NASSA). Through the work of these social action commissions, various forms of peace and development work have been organized by the church.

Under the emerging field of religion and development, the contribution of faith-based organizations (FBOs) to development has been given importance in their role in addressing poverty and protecting the natural environment. Narayanan identifies "three ways in which religion may play a role in sustainable development—through the values it offers, through its potential for ecological, social and political activism (based on those values) and through its capacity to enable self-development" (Narayanan 2013, 138).

Narayanan (2013, 131) notes that in the last decade "the importance of religion to development practice and policy is no longer being entirely neglected". She cites the substantial scholarly attention paid to the conceptual and methodological aspects of the religion–development nexus (Rakodi 2012), and the special issues on religion and development in journals and books on the subject (Narayanan 2013, 131). This has led many donors to overcome their latent mistrust of FBOs and even in some cases show a preference for funding the development work of FBOs over secular organizations (Tomalin 2012).

Studies on religion and disaster, on the other hand, have also shown the critical role of church leaders and religious organizations in disaster response. In major disasters caused by natural hazards such as hurricane Katrina in the United States, leaders of various FBOs have responded to provide shelter and food to victims as first responders. Many times, moreover, the assistance of church leaders has been sought in helping people cope with traumatic experiences through the psychological and spiritual services they render (Koenig 2006, 53).

Gaillard (2007) points to how religious groups are usually well integrated into local communities and often can respond to disasters in a short time span. In addition, there is a high level of trust among local communities that makes it possible for religious NGOs to be in a good position to contribute to disaster risk reduction (Gaillard and Textier 2010, 83). Many Christian and Muslim NGOs have led in community-based disaster risk reduction with significant results (Wisner et al 2004).

A systematic review analyzing relevant research on religious organizations' participation in disaster risk management was conducted in 2018 which did a content analysis on 22 articles (Ali Sheikhi et al 2021, 1). It resulted in 11 themes dealing with:

> religious institutions' potential for disaster management, preparedness of religious institutions for disasters, response, recovery, the social capital, partnership with all stakeholders, collaboration and communication between mental health professionals and faith-based organization leaders, unity of information and message, religious institution shelters, dealing with disaster in old and new approaches, barriers and challenges. (Ali Sheikhi et al 2021, 1)

Social capital and community resiliency

There are many definitions of social capital. For Putnam, Blair, and Roberts (2012), "social capital refers to the social networks and norms of reciprocity and trustworthiness that develop alongside these networks or as a result of them." Cultural norms and values bring people together in trust, fostering cooperation is also part of social capital. In the context of disasters, social capital has the potential to bring people together in response to the disaster and increase community resiliency (Rivera and Nickels 2014, 178). Social capital can develop and enhance alliances among members of FBOs to access resources among each other in times of disaster (Rivera 2018, 50). Religious actions such as mass prayers in churches can enhance communication and connect local communities together. In addition, it may increase the sense of belonging in communities and enhance coping (Ali Sheikhi et al 2021, 10).

Family and other relations are helpful to individuals. The knowledge that they have persons to go to when they are in need also brings a sense of stability, knowing also they are not forgotten and belong to a group. They also know they can serve others and are not just victims but part of the solution. Increasing a person's social network can lead to a more effective disaster response (Rose 2016). According to Ali Sheikhi et al (2021, 10):

> Social capital contributes to improving the community's resilience to disasters through various methods, such as facilitating the dissemination of information, promoting social organizations, and motivating the community to contribute to reconstruction efforts.

It is imperative that religious organizations "link themselves with social capital to gain access to resources, mainly information, and funding opportunities that will help them provide social services" Ali Sheikhi et al (2021,11) as evident in the Hurricane Katrina disaster (Rivera and Nickels 2014). Information on vulnerability, response efforts, and evacuation in poor regions are very important, especially where government information is not accessible (Rivera and Nickels 2014). These perspectives on religion in the public sphere, religion in development and disaster as well as social capital will be used to analyze the case of the Catholic Church leaders and organizations in Leyte province in the aftermath of super typhoon Haiyan.

Archdiocese of Palo

The Archdiocese of Palo geographically covers the province of Leyte in the Eastern Visayas region. Field visits in May 2015 to the Archdiocese of Palo allowed me to observe the activities of church leaders and the conditions in communities two years after the disaster struck. Key informant interviews were conducted in Tacloban City, and the Municipalities of Palo, Tanauan, Alang-alang, and Mayorga in Leyte province. By then, the damaged key physical structures had been repaired such as the Metropolitan Cathedral of Our Lord's Transfiguration, Archbishops Residence, the Sacred Heart

Seminary, the St. Joachim Parish Church in San Joaquin, all located in the Municipality of Palo and the Sto. Niño Church in Tacloban City. There were mass graves and markers surrounding the Church in San Joaquin, the Metropolitan Cathedral in Palo, and public spaces in Tanauan, attesting to the many deaths in these communities. There is a yearly commemoration of this tragic event with masses and prayers organized by the Church and the communities. The narratives of survival of the local people in the communities further showed the people's resilience and determination to pick up the pieces after the horror and destruction they had experienced, as well as the loss of their loved ones.

Emergency response: needs of the dead and living

In Tacloban City, the capital of Leyte province, parishioners, and lay leaders of Sto. Niño Parish narrated how they sought refuge first in the parish social hall and later moved to the parish convent as the church structure was destroyed. The whole city was in ruins and many dead bodies were lying everywhere or being brought to open areas such as the plaza. The Tzu Chi Foundation, a Buddhist organization based in Taiwan, had volunteers and was one of the early responders. They had a cash-for-work program in the city that started immediately following the disaster to support family incomes. People were given money every day for picking up the debris and cleaning up. The parish priest of Sto. Niño Church, Msgr. Alex Opiniano, offered the area where the church once stood and the remaining pews for the use of the Tzu Chi volunteers so they had space for a more orderly distribution of money to their cash-for-work program participants every day. In addition, he celebrated mass daily to strengthen the morale of the people as they tried to pick up the pieces of their lives. The Tzu Chi Foundation later offered to help rebuild the Church which the Archdiocese had accepted. During the blessing and inauguration of the church a year after, the mass became an opportunity that strengthened social relations with the presence of Tzu Chi volunteers and acknowledgment of their assistance. According to Msgr. Opiniano:

> Good that Tzu Chi is not proselytizing. In rebuilding Santo Nino church, they gave us 50 million. … It all started with a Yes. There were so many people coming in and out, making inquiries, NGOs ang dami wala naman ginawa (so many but they did not do anything). But at first Tzu Chi was not having an organized engagement. I was asked: Can we use your place for the distribution of financial assistance? The idea was that since people didn't know what to do with their time, it was still unstable, and with many in a state of shock, they could start with cleaning up their own community. Their request was of course

> granted. I know you are Buddhists, its no problem. If you could be there to give a few words of inspiration and give them your blessing. They were in the other side of courtyard along Real Street. Nakapila sila. (They were lining up.) Go inside, it's a natural way of organizing it- by pews- so they started the distribution inside the church. So I talked, gave the blessing and went out. That's the time we had Global Media Arts (GMA) network ready to feed people with their delicious lugaw (porridge). So kain muna sila (So they ate first). They go to work, I give them their blessing then. … One of them suggested it would be better if you say mass first. So I said mass. After 4 or 5 days when they opened Santo Nino was overflowing with people. … Open another station to decongest. They started moving out of Tacloban. … At first they were giving only PHP 500. But when the community needed some skilled workers, Tzu Chi gave a lump sum already- PHP 10,000, PHP 12,000 per family in the second round. ... With the very quick response of Tzu Chi, Tacloban was cleared with all the debris. So mga tao din talaga nag clear (So it was the people really who did the clearing). Tzu Chi hired big dump trucks for the bigger debris. (Opiniano 2015)

Msgr. Opiniano's family in Manila also sent relief items in a truck that arrived in the early days after the disaster. Only after more than a week did relief goods from the government arrive. To facilitate the distribution of relief items, Msgr. Opiniano also coordinated with the Department of Social Welfare and Development (DSWD) so that their assistance could reach the people in the villages especially those in the coastal village of Barangay San Jose with casualties that reached 2,700 persons due to the storm surge. Utilizing the church network, lay leaders in the different areas of the parish volunteered and assisted in the inventory and distribution of food, water, and other relief supplies. According to Msgr. Opiniano, it was also in the convent of Sto. Niño Parish that diocesan priests working in other parts of Tacloban City gathered to take stock of the tragedy that unfolded a few days after the disaster, sharing their challenging experiences to gather courage as well as support from their brother priests and to strategize how to respond to the most immediate needs of the people in the communities they served. Msgr. Opiniano said:

> On the 11th we were able to set up our relief operations, we had 2–3 vehicles. I was waiting for the delivery of goods from my brothers- 10 wheeler trucks. The problem was we had no relief goods to distribute. ... Hindi makapasok (it could not enter). The DSWD had a standing policy that all relief goods had to pass through them first. So I was telling my contacts this was the situation. But we can't

> entrust the relief goods to them. I told the secretary of the DSWD in the area we could come up with an arrangement but she stood firm with their policy. They had a packing area in Maharlika Highway. We decided to just go on with our operation should we receive the relief goods – we had our system. (Opiniano 2015)

Msgr. Opiniano, one of the first responders adds:

> I called for a meeting in Santo Nino Church. My colleague Fr. Danny had a bicycle to move around the city. There were reports that there were two to three priests among the casualties. So we had a headcount. We set up a system. We are about 20 priests in Tacloban alone. There are 13 parishes within the city and two mission stations. Every seven o'clock in the morning we agreed to open our cellphone to be able to communicate ... I said maybe we can make a compromise with the DSWD. I went to the repacking center and was waiting for the first ten-wheeler truck of goods coming from my brother in Manila. I mentioned we are going to receive our goods and give them an inventory of it and tell them where we did the distribution ... I told people donating to make a list since the DSWD secretary was amenable that we do the distribution but give them the information. I asked them to let us know where we can send our information because I have to tell my counterparts in Manila and outside Tacloban to email to their database. Then I made a suggestion for them to make an email which I sent to my partners. I was in the repacking area for four hours. I was observing their operations and I talked to ground commanders. There are three or four deficiencies. No communication. Kulang sila sa personnel. (They lack personnel). Yung repacking, nag re re pack ka pa! (for the repacking of goods, they lack manpower). ... They only had six military trucks. With the six military trucks they could only carry thirty sacks per trip. With the six they could hardly cover the province ... Especially at that time, the distance would lead to delays. (Opinano, 2015)

A staff member of the RRU of the Archdiocese of Palo explained how he assisted in sorting out the dead from the living in the early days of the disaster with the help of a medical doctor in Tacloban City. The experience was difficult, especially when they handled the bodies of children whom they tried to protect from stray dogs.

A few kilometers south of Tacloban City, the parish of San Joaquin in the Municipality of Palo also suffered several casualties. Their parish priest, Fr. Kelvin Apurillo, was lucky to have survived although several of his parishioners perished due to the storm surge. What was most difficult in the immediate aftermath of the typhoon was seeing all the corpses lined up in the street as he spent most of his time praying and blessing over the dead with the help of a seminarian. Even the parish church was full of dead bodies brought by the people so later they opted for a common grave built in the church compound that buried about 300 dead persons. A seminarian, Brother Ryan Salvacion of the Sacred Heart Seminary of the archdiocese narrates:

> By November 9, 2013, I went home to San Joaquin. And that was the time I helped Fr. Kelvin bless bodies which the people laid side by side along the national highway. I was on the left side and he was on the right side blessing dead bodies lined up from the bridge between Cogon and San Joaquin until the barangay after San Joaquin. So that's how many bodies were along the national highway ... I was able to keep my prayer book, just read the rites and we blessed the dead ... later we started to bury them in a mass grave. ... We buried almost 300. (Salvacion, 2015)

The same seminarian was in the Sacred Heart Seminary in Palo during the onslaught of the typhoon. He shared how they were lucky to have survived as the water rose to very high levels. Since the seminary was on higher ground, several evacuees from Barangay Salvacion were also housed in the seminary. After the storm, there were several dead on the streets and people also started bringing the corpses to their chapel to be blessed. Many were laid to rest in the mass grave adjacent to the Cathedral of our Lord's Transfiguration in Palo. He recalls:

> The sky was still very dark a day after the supertyphoon and we could see the debris, the ruins and dead bodies from the seminary hill. We still had a roof where a lot of people sought refuge. Some families were in the dormitory of the seminary to get water. Immediately after the typhoon we gathered the dead bodies and they were blessed by our spiritual directors. We placed them in the altar of our chapel. We piled them. Many of the cadavers we gathered around the community were those of children. (Salvacion, 2015)

The Catholic Church hierarchy especially in the most affected areas of typhoon Haiyan initiated local initiatives to continue feeding the poor. A year after the disaster, Archbishop John Du of the Archdiocese of Palo said that the allotment of PHP1 million helped to jump-start a feeding program for the benefit of 78 parishes in the archdiocese in the opening of the Philippine Church's celebration of the "Year of the Poor." The project was dubbed "La Mesa ni Martha" (Table of Martha). It is hoped that the people would later raise their own funds to keep the feeding program going (CBCP News 2014).

Archdiocese of Palo: Relief and Rehabilitation Unit (RRU)

According to Fr. Alcris Badana, the director of the Relief and Rehabilitation Unit of the Archdiocese of Palo, a few days after Haiyan, the Archbishop called a meeting of priests in Palo to discuss and take stock of the crisis situation and map out strategies to help the various communities cope with the disaster. It was during this meeting that the Emergency Response Team of the Archdiocese was created, headed by the Rector of the Seminary, and composed of priests working in the Seminary and the theologians. They were tasked to distribute relief items donated by different dioceses and donors in November and December 2013. The priests and seminarians served as drivers and helped load and unload relief aid (Badana 2015). Moreover the adjacent Diocese of Maasin in Southern Leyte Province was the first to send relief items to the Archdiocese of Palo. This was due to the difficulty of trucks lined up North of Tacloban City in entering due to debris and traffic of goods. Relief goods from Mindanao Island, south of the province, passed through the national road connecting Surigao province to Southern Leyte. In addition, a week after the supertyphoon struck, relief goods from the Diocese of Tagum in the province of Davao del Norte was also able to enter by way of the southern route. There were other diocesan and archdiocesan teams who came to their aid from Davao City and Cagayan de Oro City who had various interventions, such as providing relief items and psycho-social stress debriefing. Some dioceses brought in carpenters and saws to help the people reconstruct their homes. The priests from the Archdiocese referred them to different barangays and municipalities (Badana 2015).

Some days after, help from CBCP-NASSA arrived, bringing with them their own staff and operating with resources procured through the help of Caritas Internationalis. The network of Social Action Commissions in dioceses outside the Archdiocese were some of the first responders. There were three ports of entry for relief items—Maasin port, Ormoc port, and Allen in Samar by the San Juanico Bridge. Trucks from Manila and other provinces north of Tacloban had to que for a long time as the roads were still full of debris and took time to clear. It was the mayor of the town of Javier in the Southern part of the province of Leyte that helped clear the road from the south leading to Tacloban City with the bulldozers of the municipality (Badana 2015). Residents in the Municipality of Mayorga recounted that the Mayor of Javier was instrumental in facilitating the road clean-up and clearing activities from his municipality in the southern part of the province up to the north in Tacloban City so that more humanitarian aid could enter.

Caritas Philippines, based in Manila, had coordinated with the Social Action Commissions of nearby dioceses to assist and facilitate the procurement of supplies for the Archdiocese of Palo (Gariguez 2015). Other dioceses and archdioceses in Mindanao also sent relief items via the Mindanao–Southern

Leyte route. There was 'relief congestion' or difficulty in sending relief items through the northern part of the province from Manila due to many obstructed roads; it took some time for the operations to clear them. Media reports in the early days of the typhoon Haiyan disaster have mentioned that there were long lines of trucks with relief items in the Bicol region that had difficulty entering the most affected provinces of Samar and Leyte. In the coming weeks and months, more relief items from Caritas Manila and Caritas Philippines had arrived and were distributed to different affected communities in Leyte province.

In January 2014, Archbishop John Du appointed a diocesan priest, Fr. Alcris Badana, to head the RRU of the Archdiocese of Palo. RRU was mandated to assist the communities during the emergency phase of the typhoon Haiyan disaster response and in the transition toward the rehabilitation phase. It was initially composed of seminarians and lay volunteers who coordinated with the priests in the various communities. The RRU chose to bring the relief goods to areas that other humanitarian aid groups had not covered. They coordinated with the network of parish priests in Leyte province. Emergency relief distribution lasted until the end of January of 2014 (Badana 2015).

Church assistance in the rehabilitation phase

The Catholic Church's rehabilitation program for typhoon Haiyan survivors at the national level was called the Recovery Assistance to Vulnerable Communities Affected by Typhoon Haiyan or REACH. The REACH program, according to Fr. Edwin Gariguez, executive secretary of the National Secretariat for Social Action (NASSA) and head of Caritas Philippines, is:

> by far the biggest and most extensive recovery and rehabilitation program led by the Catholic Church in nine provinces severely affected by the typhoon namely: Leyte, Western Samar, Eastern Samar, Palawan, Aklan, Antique, Capiz, Iloilo and Cebu as a commitment to lay faithful and even non-Catholics affected by Haiyan. (Gariguez 2015)

In November 2018, which was the fifth anniversary of super typhoon Haiyan, there were 30,000 homes built in nine provinces (Leyte, Northern Samar, Eastern Samar, Palawan, Cebu, Iloilo, Aklan, Capiz, and Antique) supported by the shelter project of the REACH program. The REACH program has supported different component interventions that included, aside from shelter: livelihood, water, sanitation, and hygiene (WASH), community organizing, community-managed risk reduction, ecosystems recovery, and institutional capacity building. Caritas Internationalis was instrumental in ensuring the provision of funds and experts from its member organizations like the Catholic Relief Services (CRS), Caritas Switzerland, Caritas Italiana, Caritas Belgium, Caritas Germany, Development and Peace

(Caritas Canada), Caritas Austria, and Cordaid. The Catholic Churches' overall cost of assistance through the REACH program amounted to more than PHP3.2 billion (Legarde 2018). Fr. Edwin Gariguez, Caritas Philippines executive director, stated that "In 2013, we at Caritas Philippines didn't even think that we will be capable of implementing what would be the Church's most massive, largest funded and most comprehensive humanitarian response" (Legarde 2018).

According to Fr. Edwin Gariguez, the first phase of the REACH program began in April 2014, and it was centered on building back structures and systems destroyed by typhoon Haiyan. Among its accomplishments in the nine dioceses a year after the program was launched were: building and repair of 2,794 shelter units, 4,749 sanitation facilities, 436 water supply facilities, 179 hygiene promotion activities, and the formation of 81 WASH committees. There were 28,743 units of livelihood interventions, which benefited 51,441 individuals. The REACH program implemented 42 capacity-building and organizational development activities that reached 971 individuals, and Disaster Risk Reduction (DRR) interventions in 167 communities. The second phase was focused on sustaining resilient communities and empowering more the communities by integrating DRR and community organizing that began in April 2015. It targeted to reach 23,941 households in the sectors of Shelter, WASH, livelihood, Ecosystem Services and DRR, Institutional Strengthening, Capacity Building and Community Organizing (Gariguez 2015).

The RRU of the Archdiocese of Palo in 2014 was initially composed of a team of six lay workers with Fr. Alcris Badana as its director. The RRU of the Archdiocese of Palo in 2015 had identified priority areas and communities in Leyte province for typhoon Haiyan disaster assistance from aid agencies channeled through Caritas Philippines. In May 2015, the RRU already had an information officer, program manager, program officers for WASH, DRR, Shelter, Livelihood, area managers, community organizers and warehouse staff or a total of 28 staff (Badana 2015). Following the team in its activities in May 2015, I observed the turnover of 30 new shelters in Barangay Liberty in the Municipality of Mayorga and 15 new shelters in Barangay Capiz in the Municipality of Alang-alang in addition to the 30 shelters built and turned over earlier. In both communities, the RRU of the Archdiocese of Palo also implemented other interventions such as WASH, installation of water systems, livelihood for farmers and women, and health and sanitation programs. As one lady resident (who preferred to remain anonymous) in Barangay Capiz, Municipality of Alang-Alang Leyte explained:

> "*Sa relief, marami binigay ang Caritas nagbigay ng kanin, banig, damit, sabon. Pag tapos ng Yolanda, wala naman ang bahay kaya gumawa muna ng bahay, kasi wala naman malipatan. Tapos nagabot na ang relief-bigas,de*

lata. Kabuhayan ng mga asawa naming ay sa copra noon, ngayon wala na. Na wash out lahat ng coconut trees.

Ang ibang mga bata natatakot. Tulong tulong na lang kami sa pagaalaga ng mga bata. Doon pumunta yung ibang mga tao, sa eskwelahan nagpunta. Dalawang lingo kami doon hanggang maayos ang bahay namin. Gumawa lang ng maliit na tirahan. Dumating ang Department of Social Welfare and Development (DSWD), mga opisyal....may isang nasugatan, pero nagamot naman-napako. Pero nagpagamot sila- anti-tetanus. May mga doctor pumunta dito- nag volunteer, medical mission-taga Department of Health (DOH) yata yun. Matagal din bago naming napagawa ang kapilya. Sa Caritas –pagkain,mga gamit sa bahay, gamit sa kusina, mga seeds, yung hayop-may itik, yung iba lang. gamit sa farm- thresher. Marami silang nabigay na gamit sa farm. Nagtratrabaho (laughs) Nagtatanim ng gulay, gabi. Yung palay kakatapos lang ng ani- dalawang beses isang taon. Naka harvest na. mga kababaihan- nagtatanim tanim sa tabi tabi, nagaalaga ng bata." (Anonymous lady resident of Barangay Capiz, 2015)

["In terms of relief, Caritas gave a lot- they gave rice, mats, clothes, soap. After Yolanda we didn't have a house so we first made our house since we have no place to move. And then they gave relief items- rice and canned goods. Our husbands depended before on Copra, now it is gone. All the coconut trees have been washed out.

Some children are afraid. We help one another in taking care of the children. Others went to the schools. We stayed there for two weeks until our house was built. We made a small house. DSWD came, the officials. One was wounded but was treated-stepped on a nail. But they underwent treatment- anti-tetanus. There were doctors who came, volunteers, medical mission-from DOH I think. It took a long time before we could have the chapel rebuilt. From Caritas-food, materials for the house, for the kitchen, seeds, animals-ducks, farm equipment-thresher. They gave a lot of farm implements. Working. Planting vegetables, tubers. Rice was just harvested-twice a year. They harvested already. Women plant in their surroundings, take care of the children."] (Anonymous lady resident of Barangay Capiz, 2015)

The RRU of the Archdiocese of Palo has also networked with the United Nations Development Program (UNDP). Its staff were active as part of the civil society in networking with local government units (LGUs), especially with the mayors and the barangay or community chairpersons in areas where they implemented their projects. During 2016, the RRU's implementation of the REACH program was reported to have assisted 20 barangays in the nine municipalities of Leyte province. In Barangay Dacay in Dulag town, they installed a Level II water system with 17 tap stands and turned over rice post-harvest facilities such as rice mill, solar dryer, and seed bank valued at

PHP 1.3 million (Garcia 2016a, 2). As a counterpart and to prevent a dole-out mentality, the community provided the land for the post-harvest facilities and labor for the construction. Farmers in nearby Barangay Magsaysay also benefitted from the facilities. The Dacay Barangay Chairperson, Rosalinda Alicando, said:

> This is the first time that our barangay will have easy access to clean and safe water since I was born and we are very thankful to Caritas Palo. This time it would be easy for us to fetch water since faucets are just near our house. (Garcia 2016a, 2)

The barangay was also provided rice threshers, a tractor, sprayers, and seedlings in the past. In addition, drinking faucets and handwashing facilities were also provided for the barangay elementary school (Garcia 2016a, 2). Caritas Palo also provided rice post-harvest facilities to Barangay Santol and a Level II water system to Barangay Tabangohoy, both in Alang-alang town (Garcia 2016a, 2).

On February 11, 2017, Caritas Palo handed over 15 permanent houses to families in Barangay Canmogsay, Tolosa, Leyte. In addition, on March 4, 2017, the church-based humanitarian organization handed over 15 permanent houses for families in Barangay Dacay, Dulag, Leyte, in a ceremony where the program manager of REACH, the barangay chairperson, parish priest of St. Joseph Parish in San Jose, Dulag, and the Mayor of Dulag were in attendance (Puno 2017, 1). In its third year of implementation of REACH, Caritas Palo has built a total of 580 permanent houses in 19 barangays from eight municipalities in the Archdiocese (Puno 2017, 1). The two major partners that provided funds that enabled Caritas Palo to implement the programs were the CBCP-National Secretariat for Social Action/Caritas Philippines and Caritas Internationalis (Puno 2017, 1). In summary, the REACH program provided shelter, livelihood, WASH, ecosystem rehabilitation and community-managed disaster risk reduction (CMDRR) programs (Puno 2017, 1). Through the CMDRR, the communities are empowered to assess their vulnerabilities and analyze the support they need. They chose their livelihood programs, built resilient shelters, and rehabilitated their environment through tree and mangrove planting (Puno 2017, 2). RRU also held a bazaar or multi-sectoral products trade fair last February 25, 26, and 28, and March 1, 2017, where various communities sold their agricultural and processed food products at the Santo Nino Parish grounds in Tacloban City (Puno 2017, 2).

On February 10, 2019, there were 263 houses turned over to families in the Pope Francis Village in Tacloban City. The Pope Francis Village is a PHP187.8 million in-city project targeting to put up a total of 566 housing units, six multipurpose classrooms, a chapel, village market, and wastewater treatment and materials recovery facilities. The project is implemented by a consortium of Development and Peace-Caritas Canada, Caritas Philippines, Archdiocese of Palo, Redemptorist Congregation and Urban Poor Associates expected to be completed in June 2019 (Schiavone 2019).

Religious congregations and organizations assisting local communities

Religious congregations in the communities affected by typhoon Haiyan also mobilized support for relief and rehabilitation efforts. The Claretian Missionaries, for example, were very much involved in aiding victims of the Zamboanga Siege a few months before the typhoon Haiyan disaster quickly mobilized for relief aid for Ormoc City where their novitiate was located. Their network of support included the Immaculate Heart of Mary Parish (IHMP) located in the University of the Philippines (UP) Village, Quezon City, that is administered by the Claretian Missionaries. The IHMP's parish priest and lay workers visited Ormoc City and sent aid twice via C-130 planes for affected residents. Archbishop John Du, in an interview on May 13, 2015, mentioned that the Salesian priests of Don Bosco and Canossian sisters also had housing projects in Palo, Leyte (Du 2015). The Salesians also helped rebuild eight schools in Leyte (Salesian Missions n.d.).

The Municipality of Tanauan reported at least 1,207 persons died due to the storm surge (Meniano 2018). A parish coordinator, Ms. Rowena Papellero in the Assumption of Our Lady Parish, provided the researcher with a list of the different forms of assistance given by various religious congregations and groups to communities within their parish. In Barangay Atipolo, the Priest of Sacred Heart of Jesus provided shelter kits to 79 families with galvanized iron sheets, cocolumber plywood, nails, and paint to rebuild their homes. The same congregation also helped 16 families in Barangay Buntay with housing materials. The Sisters of the Adorers of the Blood of Christ has extended housing assistance to 20 families in Barangay Catmon and Licod and gave roofing materials for the Cabuynan Chapel. The Daughters of St. Mary of Providence religious institute has helped supply roofing materials for the Barangay Sto. Niño Chapel. In Barangay Limbuhan Guti, the Franciscan Missionary of Mary provided concrete housing to 122 families. The Order of San Agustinian of Consolacion has supported 50 families with concrete housing, livelihood, and family values formation in Barangay Magay. In Barangay Santa Cruz, the assistance of the Missionary of Mary was an intervention on values formation to the entire community. The Servants of Charity of Jesus has provided livelihood assistance, family values formation, fed children for three months and provided 52 bancas or small boats for livelihood in Zones 2 and 3 of Barangay San Roque. Caritas also helped build the barangay chapel in Barangay Atipolo. Other church organizations that assisted the Municipality of Tanauan included the Couples for Christ in Cagayan de Oro, Couples for Christ in Manila, CRS for relief and livelihood, Caritas for relief and scholarships, the Archdioceses of Davao and Cagayan de Oro, and Dioceses of Tagum and Maasin. The Archdiocese of Lipa has distributed vestments for the priests since most of them lost all their belongings. The Assumption Academy School located beside the old stone church of Our Lady of Assumption Parish served as an evacuation center of 40 families in three months. The school's library and their Day Care Center were rehabilitated by the Congregation of the Society of Divine Word (SVD), which provided assistance amounting to PHP1.2 million (Papellero 2015). The Redemptorist brothers and priests provided emergency assistance to evacuees who sought shelter in their church in Tacloban City. They also collaborated in support of the Pope Francis Village project.

Other religious congregations not working in the Archdiocese provided assistance by collaborating with religious congregations in the Archdiocese of Palo. The Columban missionaries provided 15 sewing machines for income-generating projects of the Mother of Perpetual Help Parish under the Redemptorist fathers, and it was reported that:

> The Columbans also used some of the funds to help relocate the families in the no-build zone coastal areas in Magallanes, Tacloban, which is under the jurisdiction of the Redemptorist Fathers. Relocation and rehabilitation efforts of the government were very slow. Other projects which the Columbans helped include… the purchase of "pedicabs" (bicycles with side cars) for a community in Tacloban. (Villahermosa n.d.)

They also donated financial help for the construction of 110 concrete houses in a project initiated by the Franciscan Missionary of Mary in Barangay Limbuhan Guti. The community also underwent capacity-building seminars, livelihood and leadership training, and spiritual formation (Villahermosa n.d.).

Support to the Archdiocese of Palo: CBCP and local churches

Several Catholic parishes, dioceses, and archdioceses outside the typhoon-affected areas showed their solidarity to the people in the Archdiocese of Palo. These included the Archdiocese of Manila and dioceses around Manila, and Archdioceses and dioceses in the Visayas and Mindanao.

The NASSA under the CBCP has served as a central coordinator of the Catholic churches in the Philippines resource mobilization for the typhoon Haiyan disaster response. The roles of NASSA are cited as:

> NASSA facilitated coordination with the Archdiocese of Manila and other dioceses and connected the dioceses with international partners. The diocese of Manila and others arranged the reception and handling, at its own cost, of in-kind national and international donations. The non-affected dioceses near the disaster area facilitated transit of relief goods to the final destination and allocated volunteers. (Caritas-Philippines 2014, 11)

Resources were mobilized by NASSA through the Archdiocesan and diocesan social action teams then down to the parish level. The Archdiocese of Palo was one of the nine dioceses that implemented the emergency program due to the severe damages of the typhoon. The initial disaster response worked despite the limited number of responders. At the national level, Caritas Philippines/NASSA in 2013 had only two staff members. The Archdiocese of Palo Social Action Center functioned without a professional staff and an incomplete team prior to the disaster (Caritas-Philippines 2014, 7–8). NASSA has deployed Emergency Preparedness and Response Teams (EPRT) composed of two to four volunteer members from the dioceses that were not affected by the typhoon. They were trained in 2013 prior to the Haiyan disaster to advise the local emergency response teams on the preparation of the three lifesaving sectors of food, non-food items (NFI), hygiene kits, and shelter kits (Caritas-Philippines 2014, 7–8).

In a personal interview with Archbishop John Du of the Archdiocese of Palo on May 13, 2015, he acknowledged that the assistance from the Diocese of Maasin in Southern Leyte and the six Dioceses in Mindanao (Malaybalay, Pagadian, Tagum, Mati, Cagayan de Oro, Iligan), the Archdiocese of Davao and the Archdiocese of Cebu in the Visayas. The Archbishop described how he survived and was advised to seek medical attention for his wound in Cebu, where he also first made an appeal for the typhoon Haiyan disaster response in a press conference a few days after the devastation. The Archbishop's residence was repaired by carpenters sent by the Diocese of Pagadian and Malaybalay (Du 2015).

Catholic church leaders from the national capital region (NCR) have extended varied support to the typhoon-affected churches and communities. A week after the disaster, Cardinal Luis Antonio Tagle of Manila invited Filipino Catholics to a day of prayer and fasting on November 16, 2013, to be in solidarity with people suffering due to disasters. He also asked priests to have a special collection at Masses for the survivors and to review parish and school celebrations and to make necessary adjustments in the light of the enormous suffering and needs of the victims (Viehland 2013).

Caritas-Manila, the social action arm of the Archdiocese of Manila provided relief assistance and, in 2014, helped rebuild six chapels in Leyte (Catholic News Agency 2014). Caritas Manila also provided scholarships to poor but deserving students affected by the typhoon disaster in various affected dioceses. In April 2018, a total of 235 students from the Archdiocese of Palo graduated from college through Caritas-Manila's scholarship program called Youth Servant Leadership Educational Program-Educational Assistance Program (YSLEP-EAP) (Garcia 2018).

Support from outside the Philippines

The appeal of Pope Francis no less in the gathering for the Angelus in Rome a few days after the typhoon was a manifestation of the Catholic Churches' support at the international level. He pleaded for aid for survivors saying "sadly there are many victims, and the damage is huge. Let us try to provide concrete help" (Caritas Internationalis 2013).

The Apostolic nuncio, Giuseppe Pinto, who is the representative of the Pope and highest Vatican official in the Philippines came to Tacloban City and celebrated Christmas Day Mass in solidarity with the typhoon survivors in the damaged Sto. Niño Church. He extended the support of the Holy Father to those affected. He distributed Christmas packs to the typhoon survivors. He concelebrated a Mass with the Palo Archbishop and several priests in the damaged cathedral on Christmas Eve (Sabillo 2013).

On January 28, 2014, Cardinal Robert Sarah, the President of the Pontifical Council Cor Unum for Human and Christian Development, arrived in Leyte to visit the areas devastated by typhoon Haiyan and extend aid to the affected population. Cor Unum is an administrative body of the Pope for the Church's charitable activities. It helped finance the renovation of the Archbishop's residence, a small convent for nuns, a chapel, and a dispensary. Through Cor Unum US$150,000 was channeled through CBCP as one of the first to provide aid from abroad (Archdiocese of Palo 2014a).

In the same year, 2014, Archbishop Joseph Kurtz of Louisville, Kentucky, USA, also visited the Archdiocese of Palo as head of the United States Conference of Catholic Bishops (USCCB) accompanied by CRS Chairman of the Board and Oklahoma Archbishop Paul Coakley, Sr. together with Sr. Carol Keehan, DC, the president of the Catholic Health Association and Carolyn Woo, the president and chief executive officer of CRS. The delegation met with church leaders and relief workers, especially of the CRS to survey the needs as the clergy thanked the churches in the United States for their humanitarian assistance and offer to help in the rebuilding of churches in the Archdiocese and channeled their support through the CBCP (Archdiocese of Palo, 2014b). Moreover, Cardinal Theodore McCarrick, archbishop emeritus of Washington, DC, visited Palo, Leyte (CBCP News 2014).

At the national level, CBCP and NASSA under the auspices of Caritas Internationalis implemented a disaster response called "Emergency Assistance to vulnerable households in nine dioceses affected by typhoon Haiyan in the Philippines" valued at €5,633,523 for a period of three months. This disaster response intervention was funded by pledges from Caritas organizations around the world through an appeal by Caritas Internationalis (Caritas-Philippines 2014, 5–6). Over 275,000 persons were reached through emergency assistance as immediate food and household items were provided to 125,000 people and hygiene kits were given to 275,000 people (Caritas Internationalis 2014). About 13,000 families were assisted with the provision of fishing boats and small-scale farming kits while 100,000 people received shelter through emergency kits including plastic tarpaulins and tools (Caritas Internationalis 2014). Some 20,000 extremely vulnerable families were also given cash grants (Caritas Internationalis 2014). Cardinal Luis Antonio Tagle praised the work of Caritas confederation and said:

> Caritas Internationalis and Caritas Philippines/NASSA, have spearheaded the whole coordinating work. The different local Caritas worked together and made Caritas Philippines the clearing house for all their efforts. The emergency program was funded by over 30 different Caritas organizations around the world. (Caritas Internationalis 2014)

A significant expression of solidarity was evident in the visit of Pope Francis to Tacloban City and Palo, Leyte, on January 17, 2015. Several times before and during his visit, Pope Francis had mentioned the real objective of his visit to the Philippines was to be with the poor and in solidarity with the typhoon-affected population. He presided an outdoor Mass at the grounds of the Tacloban City airport, which they referred to as the 'ground zero' of typhoon Haiyan. During the said Pontifical Mass, about a million Filipinos from Visayas and Mindanao braved the rains and winds of typhoon Amang, the first that entered the Philippine area of responsibility in 2015. They found consolation and strength with the presence of their beloved Pope Francis. Then the Pope proceeded to Palo Leyte on a 12-kilometer motorcade on roads filled with crowds of people with white ribbons, and then had a behind-closed-doors meeting and lunch with 30 survivors of typhoon Haiyan and the Bohol earthquake. He blessed the Pope Francis Center for Orphans and the Elderly and met with priests and religious and lay leaders of the archdiocese in the Cathedral and afterwards blessed the mass grave of Haiyan victims at the adjacent lot in the Cathedral compound.

Prior to the Pope's much-awaited arrival in the Philippines, local politicians on some islands were criticized by the media for their welcome tarpaulins put up in public spaces with their own pictures together with the holy father. Media newscasters also reported that the Pope had told local Church leaders that he did not want his picture displayed—a humble gesture consistent with his desire that Christ and the poor be the focus of his pastoral visit. On the part of the Archdiocese, the RRU director whose team was active in the preparations for the Pope's visit reminded the people to prepare for the event spiritually, reflecting on the Pope's message of "Mercy and Compassion." The Archdiocese gave the Holy Father a replica of Our Lady of Hope of Palo, the original image of which was carved by a local artist in time for the Pope's visit. Every November 8, during the anniversary of super typhoon Haiyan, the archdiocese also celebrates the feast of Our Lady of Hope of Palo, which has helped bring spiritual solace and hope to the people, especially those who have lost their loved ones in the tragic event.

Church as part of civil society in the public sphere

Churches, also known in development circles as FBOs, form part of civil society that also includes the media, educational institutions, business, NGOs, and other associations or non-state actors outside of government. The role of religious institutions and leaders as part of civil society becomes very important in disaster response as they engage and collaborate with the state and other stakeholders in addressing the crisis brought by disasters. Paffenholz and Spark (2006) developed a model comprising the seven basic civil society functions in peacebuilding which may also be applied to disaster situations since these often lead to a disruption of basic social services, exacerbating conditions of poverty and inequality, often triggering conflict and crisis in social relations – manifestations of the absence of peace in its broader sense. The functions include 1) the protection of citizens; 2) monitoring for accountability; 3) advocacy; 4) socialization; 5) building community; 6) intermediation and facilitation between citizens and state, and 7) service delivery (Paffenholz and Spurk 2006, 13).

In the case of the Archdiocese of Palo and its religious and lay leaders, their involvement in relief and rehabilitation as part of disaster response has already demonstrated their role in the protection of citizens, community building and service delivery functions. This was manifested in the capacity of the Archdiocese to mobilize support internally through its network of parishes, volunteers, and religious congregations. It has also used its social capital in networking with Catholic organizations outside the archdiocese such as other dioceses in the country, Caritas Philippines and the international network of Caritas Internationalis, Catholic Relief Services, and the Vatican offices.

Moreover it has also been able to work with non-Catholic organizations like the Tzu Chi Foundation during the relief and rehabilitation phases. By utilizing its social capital in networking and promoting the values of cooperation, unity and solidarity with various stakeholders, it has been able to obtain the much-needed resources for the people. In addition, the Catholic Church and its leaders and members in the Archdiocese of Palo, with its regular commemoration of the super typhoon's anniversary through prayer, Masses, the blessing of mass graves, and so on, emphasized the norms and values of human dignity, respect for life and community as these activities helped in strengthening the relationships and support among its members and with the wider public. Church leaders and religious organizations as part of civil society have networked and engaged different state agencies at the local and national levels to respond to the needs of the people in the affected provinces, municipalities, and villages in Leyte.

At the local level, the RRU according to its director, Fr. Badana, has coordinated with local government officials like mayors and barangay/community chairpersons to facilitate the assistance for the typhoon-affected areas with the local parish priests and lay leaders in the church network. They also networked with other humanitarian aid workers to complement and not duplicate the efforts of many other private organizations and churches. The RRU initiatives foster important social change like empowering communities and reduce a dole-out mentality. Hence, they had their counterpart in any projects in their area. They provide 'sweat capital' or their labor in shelter support and interventions, livelihood initiatives, or building water supply systems. They also helped in maintaining these rehabilitation efforts and have a sense of accomplishment and ownership with their participation in project planning, monitoring, and evaluation processes as part of the REACH program.

At the national level, there had been criticisms that rehabilitation efforts of the government had not moved very fast, even years after the disaster. In the online CBCP News in 2014, Fr. Edwin Gariguez, executive director of NASSA, said he was "shocked over disaster-capitalism" in typhoon-hit areas. He revealed that the "public-private partnership (PPP) framework under the government's rehabilitation program in typhoon-affected areas is being used by big corporations for their own benefit" CBCP News, 2014). Fr. Gariguez said:

> There are reports of counter-reform reactionaries and greedy corporations taking advantage of the helpless and hapless situation of Yolanda farmer and fisher folk victims, who are being eased out of their lands and domiciles to give way to new development projects such as resorts or bigger land consolidations of those predators. This is "disaster capitalism" in its ugliest form. (CBCP News 2014)

This reflects that other than the mobilization of resources and religious institutions involved in the emergency relief operations which had been the focus of the initial months and rehabilitation phase in the succeeding three years, the church especially the hierarchy and officials had the call to be critical and engage the state as an important part of civil society with regards to issues and concerns regarding relief and rehabilitation. The Catholic Church in the Philippines has at many times also advocated for transparency, against corruption or abuses in society, even at times when it includes the state.

In May 2015, a few months after the Pope's visit, no permanent shelters had been built in Tacloban City. The people living by the coastal areas complained they were not being allowed to rebuild their houses in those areas because these were danger zones. Likewise, a number of those who lost their homes in the city complained that the government-built houses were in areas far from their means of livelihood in the northern part of the city, with no water and no electricity. The typhoon-affected people explained the slowness of the government's rehabilitation efforts was due to the long-standing conflicts between the Aquinos and Marcoses with former President Benigno Aquino at the national level and local government headed by Mayor Alfred Romualdez, who is a nephew of the former first lady Imelda Marcos. The changes in functions of coordination, monitoring, and evaluation of all disaster-related programs in April 2015 from the Office of the Presidential Assistant for Rehabilitation and Recovery under Senator Panfilo Lacson, to the National Economic Development Authority (NEDA) indicates the state at the national level had many problems in getting their act together. By the end of the year, NEDA Chief Arsenio Balisacan noted that the resettlement of survivors from the danger zones was the most challenging among recovery efforts:

> To date, only a total of 13,335 housing units have been completed with construction of 79,216 houses ongoing and scheduled for completion in December 2016. Among the issues slowing down the building of resettlement sites are policies on procurement and land acquisition and the many required permits and clearances needed to start certain projects. (NEDA 2015)

In 2016, three years after the typhoon hit, the Archbishop of Palo, in an interview with Radio Veritas (the Catholic radio station aired nationwide), appealed to the National Housing Authority (NHA) to provide the remaining PHP20 billion housing assistance to almost 200,000 survivors. He stated:

> "*dahil sa mabagal na pagtugon ng nakaraang administrasyon ay marami pa rin sa mga kababayan natin sa Central Visayas ang walang masilungan at*

> *ang iba ay nananatili pa rin sa mga temporary shelters na nanganailangan ng disenteng pabahay.*" (Radio Veritas 2016)
>
> ["because of the slow response of the previous administration there are still many of our countrymen in Central Visayas who have no shelter and others are still in temporary shelters who need decent housing."]

According to the DSWD Secretary Judy Taguiwalo, 100,000 people from Leyte and Samar and 81,000 in Panay Island are still without shelter because their homes are near the danger zones and local government units have not funded their relocation in their municipality. Archbishop Du in a radio telecast similarly called on the government to act on the housing problem stating:

> We can only make an appeal because that is really government funding. ... We can just make an appeal "na ibigay nila" (that they give) the allotment for the people who are homeless. I know that there are so many people and families who are still homeless, others staying in temporary shelters and they really need a decent house to live. (Radio Veritas, 2016)

Moreover, according to Veritas Patrol, Senator Loren Legarda mentioned that Congress had allotted PHP25.6 billion for housing assistance in the 2016 budget. However, as of June 2016, it seemed that the fund's utilization rate was slow with the remaining PHP20.7 billion still to be spent, which implies that many were still homeless and living in temporary shelters or evacuation centers. The National Disaster Risk Reduction and Management Council (NDRRMC) was allotted a PHP18.96 billion budget, but as of September 30, 2016, PHP18.43 billion was unutilized. On the other hand, the Catholic Churches through the CBCP/NASSA or Caritas Philippines with Caritas Manila built 74 chapels for selected communities in Central Visayas (Radio Veritas 2016).

During the same year, 2016, which was also the third year anniversary of the typhoon disaster, the Catholic Church, through its social action arm NASSA/Caritas Philippines together with other groups in a coalition of 163 community organizations and ten NGOs, otherwise known as Community of Yolanda Survivors and Partners (CYSP), appealed to the City Government of Tacloban to stop their imminent plans to forcibly evict about 3,500 typhoon-affected families from the city's 15 coastal barangays identified as "danger zones" (Garcia 2016b). The CYSP believed that the hurried relocation was due to the executive order of President Rodrigo Duterte to complete the shelter project for typhoon survivors by the end of December 2016. Most of the survivors resisted the planned transfer to the northern barangays, because of the absence of livelihood opportunities, difficult access to schools, markets, and hospitals, and lack of water and electricity at the relocation sites. Furthermore, in a letter to Cabinet Secretary Leoncio Evasco Jr., CYSP asked that he "convene an inter-agency dialogue to address the situation and clarify key issues on Yolanda reconstruction" (Garcia 2016b).

While working hand in hand with the local government officials in rebuilding the typhoon Haiyan disaster areas through mobilization of resources and provision of relief and rehabilitation services in selected communities, higher church officials, most especially the Archbishop of Palo and the executive secretary of NASSA/Caritas Philippines, advocated for the welfare of homeless survivors of the disaster. They appealed to local and national government officials for shelter assistance in the city for the homeless and decent homes for those still living in temporary shelters. These church leaders came to empathize and prevent the survivors from being evicted. They also demanded an urgent accounting of the billions of pesos donated or allotted for housing and other needs of evacuees. Fr. Edwin Gariguez, Caritas Philippines executive secretary, noted that an investigation into the unused funds during the administration of the Aquino government should be addressed by the Duterte government that came into power in 2016, since many survivors were still waiting for government assistance (UCA News 2016). Their actions were consistent with the social teachings of the church, especially after Vatican II and the Plenary Council of the Philippines II (or PCP 2) on the Philippine Church becoming a 'church of the poor' (CBCP 1992).

The active participation of the religious leaders and the laity in relief and rehabilitation has shown evidence of the Church in action in the public sphere. Church leaders, while performing their functions as priests in the administration of the sacraments, celebrating Mass, and other religious activities, also encouraged the involvement of the laity in rebuilding their communities and providing assistance not limited to fellow Catholics but also reaching out to non-Catholics. Their use of mass media and visits by prominent church leaders, most especially Pope Francis, further demonstrated the commitment of the Catholic Church in the Philippines to the social dimensions of the mission that led them to undertake more services in the public sphere and continue to engage all the stakeholders, especially the state, to provide adequate services to the people affected by the disaster.

References

Ali Sheikhi, R., H. Seyedin, G. Qanizadeh, and K. Jahangiri (2021) "Role of religious institutions in disaster risk management: A systematic review," *Disaster Medicine and Public Health Preparedness* 15(2): 239–254.

Alliance Development Works. (2016) *World Risk Report 2016*. Berlin: Alliance Development Works.

Archdiocese of Palo. (2014a) "Papal representative visits typhoon-stricken Leyte," January. https://archdiocese of palo.org/news/papal-representative-visits-typhoon-stricken-leyte. [Accessed April 2, 2018].

Archdiocese of Palo. (2014b) "Visit of Archbishop Joseph Kurtz," February, https://archdiocese of palo.org/news/visit-Archbishop-Joseph-Kurtz [Accessed April 2, 2018].

Buckley, D. 2014. "Catholicism's democratic dilemma: Varieties of public religion in the Philippines," *Philippine Studies: Historical and Ethnographic Viewpoints* 62(3/4): 313–339.

Caritas Internationalis. (2013) "Philippines storm survivors in great need," November 13. https://www.caritas.org/2013/11/philippines-storm-surviv ors-great-need/ [Accessed December 20, 2013].

Caritas Internationalis. (2014) "Philippines rises up after typhoon Haiyan," April 16, https://www.caritas.org/2014/04/philippines-rises-typhoon-hai yan/ [Accessed June 5, 2014].

Caritas-Philippines. (2014) "Final evaluation report: Typhoon Yolanda (Haiyan) Philippines emergency phase," October, Manila, Philippines.

Casanova, J. (1994) *Public Religions in the Modern World*, Chicago: University of Chicago Press.

Casanova, J. ((1996) "Global Catholicism and the politics of civil society," *Sociological Inquiry* 66(3): 356–373.

Catholic News Agency. (2014) "Filipinos continue to recover from typhoon Haiyan," March 19, https://www.catholicnewsagency.com/news/29262/filipinos-continue-to-recover-from-typhoon-haiyan [Accessed June 8, 2014].

CBCP (Catholic Bishops Conference of the Philippines). (1992) *Acts and Decrees of the Second Plenary Council of the Philippines*, Manila: Pauline Publications House.

CBCP News. (2014) "Palo Archbishop calls on parishes to feed poor," December 2, www.cbcpnews.com/cbcpnews/? P=4670 [Accessed December 5, 2014].

Claver, F., S.J. (2009) *The Making of a Local Church*, Quezon City: Claretian Publications and Jesuit Communications.

Gaillard, J. C. (2007) "De l'origine des catastrophes: Phenomenes extremes ou aprete du quotidien?" *Natures Sciences Sociétés* 15(1): 44–47. doi:10.1051/nss:2007022

Gaillard, J. C., E. Maceda, E. Stasiak, I. Le Berre, and M. V. Espaldon (2009) "Sustainable livelihoods and people's vulnerability in the face of coastal hazards," *Journal of Coastal Conservation* 13(2/3): 119–129.

Gaillard, J. C. and P. Texier (2010) "Religions, natural hazards and disasters: An introduction," *Religion* 40(2): 81–84. https://doi.org/10.1016/j.religion.2009.12.001

Garcia, R. (2016a) "Caritas Palo provides water system, rice post-harvest facilities to Yolanda survivors in Leyte," February 29, https://www.veritas846.ph/caritas-palo-provides-water-system-rice-post-harvest-facilities-to-yolanda-survivors-in-leyte/ [Accessed June 18, 2016].

Garcia, R. (2016b) "Church appeals stop eviction of 3,500 'Yolanda' families in Tacloban," November 16, https://www.veritas846.ph/church-appeals-stop-eviction-3500-yolanda-families-tacloban/ [Accessed December 10, 2017].

Garcia, R. (2018) "425 Yolanda scholars Nagtapos ng Kolehiyo," *Radio Veritas*, April 18, http://wwwthesplendorofthe church.com/2018/04/18/425-yolanda-scholars-nagtapos-ng-kolehiyo [Accessed June 18, 2020].

Gariguez, Fr. E. (2015) Interview with Fr. Edwin Gariguez, Executive Secretary, NASSA/Caritas Philippines, Intramuros, Manila, May 1.

Gutierrez, G. (1974) *A Theology of Liberation: History, Politics and Salvation*, London: SCM Press.

Haustein, J. and E. Tomalin (2017) "Religion and development in Africa and Asia," in P. Carvalho (ed) *Routledge Handbook of Africa-Asia Relations*, London: Routledge, pp 76–93.

IBON. (2015) *Disaster Upon Disaster: Lessons Beyond Yolanda*, Quezon City: IBON.

Koenig, H. (2006) *In the Wake of Disaster: Religious Responses to Terrorism and Catastrophe*, Philadelphia, PA: Templeton Press.

Legarde, R. (2018) "Church builds more than 30,000 houses for Yolanda survivors," November 16, https://www.cbcpnews.net/church-builds-more-than-30,000-houses-for-yolanda-victims/ [Accessed April 4, 2019].

Manji, F. and C. O'Coill (2002) "The missionary position: NGOs and development in Africa," *International Affairs* 78(3): 567–583.

Meniano, S. (2018) "Yolanda deaths still haunt Leyte town survivors," *Philippine News Agency*, November 8, www.pna.gov.ph/articles/ 1053308 [Accessed May 8, 2019].

Moreno, A., S.J. (2006) *Church, State, and Civil Society in Postauthoritarian Philippines: Narratives of Engaged Citizenship*, Quezon City: Ateneo de Manila University Press.

Narayanan, Y. (2013) "Religion and sustainable development: Analysing the connections," *Sustainable Development* 21: 131–139.

National Economic and Development Administration (NEDA). (2015) "As year ends, several Yolanda rehabilitation projects near completion-NEDA," http://www.neda.gov.ph/?s=2015%20Year%20end%20Yolanda%20projects [Accessed December 5, 2016].

Opiniano, Msgr. A. (2015) Interview with Msgr. Alex Opiniano, parish priest of St. Nino Church in Tacloban City, Tolosa, Leyte, May 6.

Paffenholz, T. and C. Spark (2006) "Civil Society, civic engagement and peacebuilding," Social Development Paper No. 36, Working Paper Series of Conflict Prevention and Reconstruction Unit, Social Development Department. Washington DC: The World Bank.

Parsons, T. (1966) "1965 Harlan Paul Douglas lectures: Religion in a modern pluralistic society," *Review of Religious Research* 7(3): 125–146.

Preston, P. (1996) *Development Theory: An Introduction*, Cambridge, MA: Blackwell.

Puno, V. (2017) "Caritas's program helps rebuild lives in Yolanda-affected area," *Business Mirror*, March 12, https://businessmirror.com.ph/2017/03/12/caritass-program-helps-rebuild-lives-in-yolanda-affected-area/ [Accessed September 6, 2017].

Putnam, K., R. Blair, and R. Roberts (2012) "Perspectives of faith-based relief providers on responding to the needs of evacuees following hurricane Katrina," *Traumatology* 18(4): 56–64.

Radio Veritas. (2016) "Maraming Walang Bahay sa Yolanda Victims Dahil sa Nagdaang Administrasyong- Arsobispo," November 9, https://www.veritas846.ph/maraming-walang-bahay-sa-yolanda-victims-dahil-sa-nagdaang-administrasyon-arsobispo/?fbclid=IwAR3VKeKYYUUJGCScNgbdsI4Ps8brxpJCwDQ wqywzDtord5pJpHrcKcjlpv8 [Accessed May 5, 2019].

Rakodi, C. (2012) "A framework for analysing the links between religion and development," *Development in Practice* 22(5/6): 634–650.

Reyes, R. (2015) "Rebuilding Haiyan-affected areas remains 'A Work in Progress' church groups help typhoon victims along the road to recovery," UCA News, December 12, https://www.ucanews.com/news/rebuilding-haiyan-affected-areas-remains-a-work-in-progress/74710# [Accessed May 1, 2018].

Rivera, J. D. (2018) "Reliance on faith-based organizations for tangible assistance in times of disaster: Exploring the influence of bonding social capital," *Sociological Spectrum* 38(1): 39–50.

Rivera, J. D. and A. Nickels (2014) "Social capital, community resilience, and faith-based organizations in disaster recovery: A case study of Mary Queen of Vietnam Catholic Church," *Risk Hazards: Crisis Public Policy* 5(2):178–211.

Rose, B. (2016) *Developing a Model for Church-based Disaster Relief*, Lynchburg, VA: Liberty University.

Sabillo, K. (2013) "Papal Nuncio spends Christmas with Yolanda survivors", https://globalnation.inquirer.net/95013/papal-nuncio-spends-christmas-with-yolanda-survivors?fbclid=IwAR1apw_RN9MThyI-FSa3DPo6eHcv3ytuev-1IDsFhZVg3ykKFAMW8JB97Cw#ixzz2oMgJz1zD [Accessed December 5, 2022].

Salesian Missions. (n.d.) "Featured mission: Philippines". https://salesianmissions.org/salesian_country/philippines/ [Accessed June 15, 2021].

Salvacion, Bro. R. (2015) Interview with Brother Ryan Salvacion, seminarian of the Sacred Heart Seminary at the Archdiocese of Palo. Barangay Capiz, Municipality of Alang-alang, Leyte, May 7.

Schiavone, M. (2019) "263 new houses for Yolanda survivors at Pope Francis Village," https://newsbook.com.mt/en/263-new-houses-for-yolanda-survivors-at-pope-francis-village/#:~:text=Some%20263%20houses%20have%20been,at%20the%20Pope%20Francis%20Village [Accessed December 4, 2022].

Tomalin, E. (2012) "Thinking about faith-based organizations in development: Where have we got to and what next?" *Development in Practice* 22(5/6): 689–703.

UCA News (Union of Catholic Asian News). (2016) "Duterte pressed to examine missing Haiyan funds," May 16, http://www.ucanews.com/news/duterte-pressed-to-examine-missing-haiyan-funds/76044 [Accessed May 20, 2020].

Viehland, N. J. (2013) "Religious orders in Philippines rush to help typhoon victims," *The National Catholic Reporter*, November 15, https://www.ncronline.org/news/religious-orders-philippines-rush-help-typhoon-victims [Accessed December 5, 2013].

Villahermosa, A. (n.d.) "Missionary society of St. Columban: On mission in the Philippines," https://columban.org/magazine/mission-philippines [Accessed June 8, 2021].

Wisner, B., P. Blaikie, I. Davis, and T. Cannon (2004) *At Risk: Natural Hazards, People's Vulnerability, and Disasters*, 2nd ed, London: Routledge.

Conclusion (For Now): Post-Haiyan Philippines and Beyond

Glenda Tibe Bonifacio and Roxanna Balbido Epe

Super typhoon Haiyan or Yolanda left such profound realities of global climate change, destruction, failures, and foresight of a disaster since November 8, 2013. It was first recorded to hit Palau and Micronesia before making landfall as a Category 5 storm in the Philippines (World Vision 2022). Considered one of the worst disasters in recent decades with 313kph winds and 7m waves in the Philippines (BBC Bitesize n.d.), typhoon Haiyan also made landfall in Vietnam on November 11 as a tropical storm with less intensity; other countries affected included China and Cambodia (Al Jazeera 2013; OCHA 2013). In the Philippines, typhoon Haiyan resulted in over 6 million displaced people, 1.9 million homeless, more than 7,000 deaths, and an economic impact of US$5.8 billion (BBC Bitesize n.d.). We may use many superlatives around Haiyan and the fact remains it is only an example of what unfolds in our midst—a disaster.

Disasters intersect with social, political, and other dimensions of power that make vulnerable people and communities at more risk from natural hazards and continuing environmental degradation in the age of the Anthropocene. We acknowledge the effects of human acts that have put the Earth in peril. In this collection, we have focused on typhoon Haiyan, *before and after*, as a sort of midway discourse of the past disasters that occurred in the Philippines that remained unlearned to keep the country disaster resilient as its fate as one of the most at-risk countries in the world for typhoons, floods, earthquakes, and volcanic eruptions. Typhoon Haiyan points to moving ahead with the challenges of island geographies, political will, centralization, bureaucratic (in)competencies, disaster risk reduction and management, humanitarian, and post-disaster development response, amongst others.

All the chapter contributions in this book are concrete illustrations of the economics, sociology, politics, psychology, and governance that shape the development landscape post-disaster of the affected population, communities, institutions, and the international community. The interplay of disaster risk reduction and climate change adaptation is invariably influenced by the capacity in disaster risk proofing development agenda, the cultural production of hazards, the geophysical and environmental characteristics, and timely response. What goes on in the local communities in the Philippines somehow manifests in the global community with foreign disaster aid and support towards them when disasters occur. The types of programs and services provided through foreign development assistance are also manifested in the local communities, and more importantly, whether such support are inclusive and equitable in post-disaster contexts. As the chapters in this collection imply, these are concerns for everybody, whether private or public sectors.

However, disasters impact us differently—for example, by gender, class, origin, and other systems of inequalities, privilege, and oppression. Women, children, and gender-diverse individuals (e.g., LGBTQ+)[1] are more vulnerable compared to men in terms of access, benefits, survival, and protection before, during, and after a disaster (Companion and Chaiken 2017; Alejandria and Smith 2020). The chapters in this collection manifest the urgent need to always integrate inclusion of the marginalized communities in development planning and humanitarian response and ensure equity and participation in all phases of work. Empowerment and engagement of affected communities, especially the vulnerable groups (e.g., people with disabilities, elderly, children, women), in disaster risk reduction and resilience building are deemed crucial pre- and post-disaster planning and activities.

The Philippines' experiences of disasters over time have become more challenging than ever with limited resources and ineffective governance (Gaillard 2015). With its geographical setting in the Pacific Ring of Fire, the archipelago state is in a critical position to face the continuing threats of an increasing number of hazards with more frequent occurrence of super typhoons, earthquakes, and volcanic eruptions. In December 2021, for example, super typhoon Rai killed 375 people and displaced about 400,000 people in the country (BBC News 2021). Building-back-better initiatives have not fully transformed building designs and standard codes to mitigate hazards. The severity and adverse toll of overwhelming costs of disasters seem to exhaust the country's resources for recovery and rebuilding efforts. The much-awaited Department of Disaster Resilience (DDR) remains stalled at the Senate (Punay 2021) as of February 2022, and perhaps could be a priority measure after the May 2022 elections. Sadly, it looks like the Filipino people will have to wait for its establishment by legislative fiat before a coherent and effective rehabilitation response could be further improved.

In the Philippines and elsewhere, resources are always limited regardless of space, economic status, programs, services, and systems to respond to the increasing impact of global climate change on human lives. Disasters are likely to occur, and these often happen expectedly or unexpectedly. Natural hazards tend to be obvious in communities with uneven growth and development yet remain ignored for the lack of political clout of those mainly affected. While these natural hazards could be responded to in due time, if pursued, the histories of systemic exclusion to address vulnerabilities may take longer than a generation. But we can insist to *act now*; to continue pushing for changes in small measures. We know we can do better for the next generation.

Note

1 Lesbian, gay, bisexual, trans, queer+.

References

Alejandria, M. C. P., and W. Smith (eds) (2020) *Disaster Archipelago: Locating Vulnerability and Resilience in the Philippines*, Lanham, MD: Lexington Books.

Al Jazeera. (2013) "Vietnam escapes worst of typhoon Haiyan," November 11, https://www.aljazeera.com/news/2013/11/11/vietnam-escapes-worst-of-typhoon-haiyan [Accessed March 5, 2022].

BBC Bitesize. (n.d.) "Tropical cyclones," https://www.bbc.co.uk/bitesize/guides/z9whg82/revision/4 [Accessed March 5, 2022].

BBC News. (2021) "Philippines super typhoon Rai death toll surges," December 20, https://www.bbc.com/news/world-asia-59714658 [Accessed March 5, 2022].

Companion, M. and M. S. Chaiken (eds) (2017) *Responses to Disasters and Climate Change: Understanding Vulnerability and Fostering Resilience*, Boca Raton, FL: CRC Press.

Gaillard, J. C. (2015) *People's Response to Disasters in the Philippines: Vulnerability, Capacities, and Resilience*, New York: Palgrave Macmillan.

OCHA (United Nations Office for the Coordination of Humanitarian Affairs). (2013) "Typhoon Haiyan – November 2013," https://reliefweb.int/disaster/tc-2013-000139-phl [Accessed March 10, 2022].

Punay, E. (2021) "Senate prodded on bill creating disaster resilience department," *The Philippine Star*, July 10, https://www.philstar.com/headlines/2021/07/10/2111404/senate-prodded-bill-creating-disaster-resilience-department [Accessed March 5, 2022].

World Vision. (2022) "2013 typhoon Haiyan: Facts, FAQs, and how to help," https://www.worldvision.org/disaster-relief-news-stories/2013-typhoon-haiyan-facts [Accessed March 5, 2022].

Index

References to figures appear in *italic* type; those in **bold** type refer to tables and boxes. References to endnotes show both the page number and the note number (107n5).

T